Terrestrial Rare Gases

ADVANCES IN EARTH AND PLANETARY SCIENCES

Advances in Earth and Planetary Sciences 3

Terrestrial Rare Gases

Proceedings of the U.S.-Japan Seminar on Rare Gas Abundance and Isotopic Constraints on the Origin and Evolution of the Earth's Atmosphere

Edited by
E. C. Alexander, Jr.
M. Ozima

Center for Academic Publications Japan
Japan Scientific Societies Press

Published by:
Center for Academic Publications Japan
Japan Scientific Societies Press
2-10, Hongo 6-chome, Bunkyo-ku, Tokyo 113, Japan

Sole distributor for the outside Japan:
Business Center for Academic Societies Japan
20-6 Mukogaoka, 1-chome, Bunkyo-ku, Tokyo 113, Japan

JSSP No. 01382-1104

ISBN 978-94-010-9830-4 ISBN 978-94-010-9828-1 (eBook)
DOI 10.1007/978-94-010-9828-1

Preface

Physical and chemical studies of the earth and planets along with their surroundings are now developing very rapidly. As these studies are of essentially international character, many international conferences, symposia, seminars and workshops are held every year. To publish proceedings of these meetings is of course important for tracing development of various disciplines of earth and planetary sciences though publishing is fast getting to be an expensive business.

It is my pleasure to learn that the Center for Academic Publications Japan and the Japan Scientific Societies Press have agreed to undertake the publication of a series "Advances in Earth and Planetary Sciences" which should certainly become an important medium for conveying achievements of various meetings to the academic as well as non-academic scientific communities. It is planned to publish the series mostly on the basis of proceedings that appear in the Journal of Geomagnetism and Geoelectricity edited by the Society of Terrestrial Magnetism and Electricity of Japan, the Journal of Physics of the Earth by the Seismological Society of Japan and the Volcanological Society of Japan, and the Geochemical Journal by the Geochemical Society of Japan, although occasional volumes of the series will include independent proceedings.

Selection of meetings, of which the proceedings will be included in the series, will be made by the Editorial Committee for which I have the honour to work as the General Editor. I and the members of the Editorial Committee will certainly welcome any suggestions that will promote the series. Whenever the convener of a meeting related to earth and planetary sciences is in a position to have to look for a medium for publishing the proceedings please contact us.

Tsuneji Rikitake
General Editor

Foreword

During a sabbatical visit by Ozima to Minnesota in the spring of 1975, we collaborated on a review of terrestrial rare gas data. As we worked our way through the literature, three observations on the "state of the art" were painfully obvious. The first was the lack of terrestrial data. In stark contrast to extraterrestrial studies where thousands of modern high quality rare gas analyses had been published, only a few modern analyses of terrestrial samples were available—and many of those data were from exotic rocks and/or minerals. Not a single analysis had been published from whole classes of major crustal rock types. The second observation was that even though in its infancy, the field was already beginning to split into an "experimental" camp and a "theoretical models" camp without good communications between the two groups. The third observation was that a significant reservoir of terrestrial rare gas data existed in the Japanese, Russian, and other non-English language literatures and that most of these data were not being incorporated into English language publications.

On the positive side, it was also clear that terrestrial rare gas studies were in the beginning of a renaissance. New experimental results from the study of helium isotopes were being interpreted as evidence that primordial or juvenile rare gases were still emanating detectably from the mantle. This phenomenon has obvious implications for models of the origin of the terrestrial atmosphere (and therefore the origin of the earth itself) and of circulation processes in the mantle. The long standing dichotomy between "continuous" and "catastrophic" degassing models was again being investigated. The ongoing studies of extraterrestrial samples were leading to a better understanding of the various primordial rare gas reservoirs and were therefore refining the boundary conditions on terrestrial models. Finally, it was hoped (in the spring of 1975) that the then imminent landing of the Viking spacecraft on Mars would provide data from the Martian atmosphere which would allow a comparison of the evolution of the Martian and terrestrial atmospheres.

We therefore decided to organize a small, topical international seminar on terrestrial rare gas studies. Our goal was to bring together, in a small meeting format, experimentalists and theoreticians: (1) to compare data, techniques, and results in a more detailed fashion than is possible through formal publications; (2) to gain a better sense of the validity and limitations of the present data; (3) to plan jointly better programs to explore terrestrial rare gas phenomena and in general to encourage such work; and (4) to start bridging the gap between the English and Japanese (and Russian) language literatures. We discovered that the U.S.-Japan Cooperative Science Program, which is administrated jointly by the Japan Society for the Promotion of Science and the National Science Foundation, sponsored such international seminars and we proceeded to organize a joint seminar.

The joint U.S.-Japan Seminar on Rare Gas Abundance and Isotopic Constraints on the Origin and Evolution of the Earth's Atmosphere was held in the Kowakien Hotel, Hakone National Park, Japan, between June 28 and July 1, 1977. There were twenty Japanese participants, eight representatives from the U.S., and a Canadian at the meeting. Nineteen talks were presented and extensively discussed. This volume is based on the talks given at the Hakone Seminar. It is not, however, restricted to the presentations at the conference. The papers were prepared during the fall of 1977, and the authors had the opportunity to modify their contributions to include the comments and material given at Hakone. We have, in addition, included three papers from rare gas workers who were unable to attend the Hakone Seminar.

The papers herein reflect the "state of the art" in the study of terrestrial rare gases as of the summer of 1977 and, we trust some of the excitement we all felt in Hakone. We hope that this volume will encourage (or provoke): (1) many additional analyses of a broad range of terrestrial samples; (2) refinement and sophistication of degassing models; and (3) international cooperation in the study of rare gases. If the volume does so, it will have served its purpose—even if the models and conclusions presented herein fall by the wayside.

We would like to thank the Japan Society for the Promotion of Science and the National Science Foundation for their sponsorship of the Seminar, the Center for Academic Publications Japan and the Japan Scientific Societies Press for agreeing to publish this volume, and the contributors to this volume for their efforts and in particular for their cooperation in meeting a tight publication schedule.

January, 1978

E. C. Alexander, Jr.
M. Ozima

List of Contributing Authors

Astronomy, The University of British Columbia, Vancouver, B. C. V6T 1W5, Canada

Kazuo SAITO 145
Department of Geology and Geophysics, University of Minnesota, Minneapolis, Minnesota 55455, U. S. A.

Koh SAKAMOTO 137
Department of Chemistry, Kanazawa University, Kanazawa-shi 920, Japan

D. W. SCHWARTZMAN 185
Department of Geology and Geography, Howard University, Washington, D. C. 20001, U. S. A.

R. D. SHERRILL 63
Babcock and Wilcox Lynchburg Research Center, Lynchburg, Virginia 24505, U. S. A.

Masaru SUZUKI 17
Department of Chemistry, Tokyo Kyoiku University, Bunkyo-ku, Tokyo 112, Japan

Nobuo TAKAOKA.................... 65, 71
Department of Physics, Osaka University, Toyonaka-shi 560, Japan

I. N. TOLSTIKHIN 33
Institute of Precambrian Geology and Geochronology, Academy of Sciences, nab Makarova 2, Leningrad, U. S. S. R.

CONTENTS

EXPERIMENTAL STUDIES

A Mantle Helium Component in Circum-Pacific Volcanic Gases: Hakone, the Marianas, and Mt. Lassen

H. Craig,* J. E. Lupton,* and Yoshio Horibe**

*Scripps Institution of Oceanography, University of California at San Diego, La Jolla, California 92093, U. S. A.
**Ocean Research Institute, University of Tokyo, Nakano-ku, Tokyo 164, Japan

Measurements of $^3He/^4He$ ratios in deep ocean water, tholeiitic glasses, the Red Sea geothermal brines, and in hydrothermal emissions from the Galapagos Rift zone in the Pacific Ocean, have provided a model of the transport of primordial volatiles from the mantle to the atmosphere via creation of new oceanic crust along spreading centers, and extraction of gases by convective circulation of seawater in basalts on mid-ocean ridges. Another possible source of primordial volatiles is the locus of convergent plate boundaries, characterized by tholeiitic volcanism in island-arcs and andesitic suites in continental-margin orogenic provinces. In these regions, volcanic gases can be used as exploratory probes for evidence of primordial mantle volatiles by measuring the isotopic composition of the helium. We have analyzed gases from Marianas Island chain volcanoes, Hakone volcano in Japan, and Mt. Lassen in California; helium in these gases, and in the gases analyzed by Polak *et al.* from Kamchatka and the Kuriles, show uniformly high $^3He/^4He$ ratios about seven times the atmospheric ratio, clear evidence of a flux of primordial volatiles from the mantle along convergent plate margins. Consideration of the production of radiogenic ^4He in the oceanic lithosphere indicates that the andesite-dacite suites in these areas cannot have formed by remelting of subducted oceanic crust unless the downgoing crust contains an order of magnitude more primordial helium than has been observed so far in tholeiite glasses from ocean ridges. Andesites may thus be derived from primary asthenosphere magmas, but if hydrous melting is required it is difficult to derive water from subducted crust without radiogenic helium. The "andesite problem" is thus linked directly to the problem of initial volatile concentrations in ocean-ridge tholeiites.

1. Introduction

The occasion of the Hakone Seminar on terrestrial rare gases marks an appropriate time for us to summarize the present status of our research on primordial helium emanations from the mantle. It has been just 10 years since the 1967 Scripps Institution of Oceanography NOVA Expedition, when one of us collected deep ocean water samples from the Kermadec Trench for helium isotope analysis. In these samples Clarke *et al.* (1969) found a 10% enrichment in the $^3He/^4He$ ratio relative to the atmospheric ratio (1.4×10^{-6}) at depths of 4,000 to 8,000 meters, and a maximum enrichment of 22% in a mid-depth sample at 1,740 meters. This discovery led to the then completely speculative hypothesis that "primordial" helium, enriched in ^3He relative to terrestrial atmospheric helium, was injected into the oceans at the crests of oceanic rises where new crust is being formed.

The association of a primordial mantle helium component with sea-floor spreading centers in the ocean basins has been established by work on a series of ocean expeditions since NOVA, culminating in the recent GEOSECS (Geochemical Ocean Sections) expeditions in the Atlantic and Pacific. At the GEOSECS-I station off California CLARKE *et al.* (1970) found a well-developed mid-depth maximum in excess ^{3}He, and CRAIG *et al.* (1975) showed that maximum ^{3}He enrichments in Pacific waters were found on and adjacent to the crest of the East Pacific Rise itself, where a ratio anomaly of 32% was observed. In this area the ^{3}He and ^{4}He concentration anomalies (approximately 50% and 12% respectively, relative to solubility equilibrium with atmospheric helium), are large enough to allow reasonable estimates of the isotopic ratio and flux of the added helium component. The ^{3}He/^{4}He ratio of the added helium flux was thus estimated to be $(1.6 \pm 0.3) \times 10^{-5}$, about 11 times the atmospheric ratio. The mean oceanic flux of ^{3}He into the ocean was calculated to be 4 ± 1 atoms/cm^2 sec, approximately equivalent to the estimated escape rate from the atmosphere. LUPTON and CRAIG (1975) then showed that helium trapped in the chilled glassy rims of basalt pillows on the East Pacific Rise has a ^{3}He/^{4}He ratio ten times the atmospheric ratio, in agreement with the isotopic ratio estimated for the added helium flux in ocean water. Samples from the Lau Basin, an area of recent tholeiitic crust formation at the western margin of the Pacific plate, also showed the ten-fold ^{3}He/^{4}He ratio enrichment, indicating that similar mantle-helium isotopic ratios were associated with widely separated sea-floor spreading centers.

In the Atlantic Ocean, the observed ^{3}He anomalies are much smaller than in the Pacific, probably due to lower spreading rates on the mid-ocean rise, coupled with the more rapid flushing of deep Atlantic water by the formation of new bottom water in both north and south latitudes. (In the Pacific, new bottom water enters only from the south, and this bottom water is already enriched in ^{3}He/^{4}He ratio by about 10% relative to atmospheric helium). CRAIG and LUPTON (1976) have shown that basalts on the Mid-Atlantic Ridge are similar to the Pacific basalts in containing trapped helium with 10 times the atmospheric ratio. However, LUPTON (1976) analyzed a section of stations across the Mid-Atlantic Ridge at 30°N, and found only a 2% ratio anomaly in water on the rise crest, indicating that at this latitude the Mid-Atlantic Ridge is a much smaller source of injected ^{3}He than the East Pacific Rise system. JENKINS *et al.* (1972) observed a small (7%) ratio-anomaly maximum at 3,200 meters depth, at the GEOSECS-II station off Bermuda, and JENKINS and CLARKE (1976) have now traced this anomaly from 51°N, along the western boundary of the Atlantic to the equator, in their detailed study of the GEOSECS Atlantic expedition samples. They believe the source of this anomaly to be in the Gibbs fracture zone just south of Iceland, where the newly-formed North Atlantic Deep Water flows westward to join the south-flowing western boundary current.

The most spectacular ^{3}He and ^{4}He enrichments in ocean waters occur in the Red Sea hot brines (LUPTON *et al.*, 1977a). These brines in the axial rift of the Red Sea exhibit supersaturations relative to atmospheric solubility-equilibrium of factors of 3,200 for ^{3}He and 370 for ^{4}He, with a mean ^{3}He/^{4}He ratio of 12.1×10^{-6}, very similar to the ratio in theoleiitic glasses. The salt enrichment in these brines comes from the evaporites encountered by downward percolating seawater (CRAIG, 1969); helium in these evaporites must be of radiogenic origin with a ^{3}He/^{4}He ratio of 10^{-7} or less, and thus the helium in the brines is clearly mantle helium stripped from the basalts by circulating seawater in an oceanic geothermal system.

The discovery of such spectacular enrichments of ^{3}He and total helium in these oceanic brines led immediately to the application of helium isotope measurements to the problem of detecting and establishing the existence of hydrothermal seawater circulation as a normal feature of newly-created lithosphere on mid-ocean rises. Such a circulation regime involving the convection of seawater through fresh basalts had been invoked in order to explain the low conductive heat flow found adjacent to oceanic spreading areas, but observed temperature anomalies in bottom waters on rise crests were so small as to be essentially indistinguishable from hydrographic mixing spikes caused by turbulent mixing of waters with different prior temperature histories. Accordingly, at the Isotope Laboratory of the Scripps Institution, we constructed a deep-water sampling vehicle to measure in-situ temperature and conductivity during towing above the bottom and to recover water samples whenever spikes in the temperature vs. salinity relationship indicated a possible hydrothermal plume emanating from the basalt (WEISS *et al.*, 1977). Helium isotope measurements on recovered samples showed ^{3}He/^{4}He ratio anomalies up to 99%, relative to background water anomalies which average 30% in this area (LUPTON *et al.*, 1977b). The mean ^{3}He/^{4}He ratio in the added helium in the plume samples was found to be 6.7 times the atmospheric ratio, significantly lower than the factor of 10 found in basalt glasses, and thus indicating that perhaps 30% of the helium derived from the basalts was of atmospheric origin, incorporated in the holocrystalline portions of the basalt during the initial penetration by seawater. The observed isotopic enrichments in the plume samples were correlated with temperature spikes which could not be explained by mixing of ambient water masses in the vicinity of the rise crest, and the existence of a significant hydrothermal circulation with emission of plumes of heated water was therefore clearly established.

2. Volatile Fluxes from the Mantle

From these studies described in the previous section a rather consistent model for the flux of volatiles from the mantle to the ocean and atmosphere has emerged. Water and gases are brought up into the oceanic crust in basaltic melts rising along the spreading centers on mid-ocean ridges. As the lavas cool, the newly-created crust is invaded by seawater through cracks and fissures, and a seawater hydrothermal circulation is set up. The gases are stripped from the basalt by the circulating seawater and transferred to the ocean by plumes of hot water issuing from the basalts.

Associated with the primordial ^{3}He component in the gases trapped in ocean-ridge tholeiites are several other components with non-atmospheric isotopic signatures. The D/H ratio of water in basalt glasses from both the Atlantic and Pacific is 77 per mil lower than the D/H ratio of seawater, and the ^{20}Ne/^{22}Ne ratio in the East Pacific Rise basalts appears to be about 2.5 percent greater than in atmospheric neon (CRAIG and LUPTON, 1976). The measurements of DYMOND and HOGAN (1973) and FISHER (1975) on ^{4}He and ^{40}Ar show that the ^{4}He/^{40}Ar ratio increases from about 0.1 to 10 in basalt glasses as the ^{4}He content increases from about 5×10^{-8} to 5×10^{-6} cc STP/g, and then levels off at a value of 10, seemingly independent of He content. At the same time, plots of ^{40}Ar vs. ^{36}Ar show that these two isotopes are totally uncorrelated in the normal range of ^{36}Ar variation from 0.1 to 1×10^{-9} cc/g; for greater ^{36}Ar concentrations (up to 9×10^{-9} cc/g) the ^{40}Ar/^{36}Ar ratio is atmospheric and clearly reflects exchange with argon dissolved in seawater. These relationships are difficult to explain by any present models. The relationship of ^{4}He/^{40}Ar vs. ^{4}He concentration, up to a ratio of 10, could be interpreted as reflecting loss of ^{4}He, although most of the observed He^{4} concentrations in these data (^{4}He$>2\times10^{-7}$ cc/g) lie in a range in which ^{3}He/^{4}He ratios in similar basalts appear to be unaffected by diffusive loss of He (CRAIG and LUPTON, 1976). What is badly needed to understand these data is a set of basalt glasses measured carefully for both He and Ar isotopic ratios and concentrations, and such data are not yet available.

The present value of the ^{3}He flux from the mantle, approximately 4 atoms/cm^{2} s as estimated from deep-ocean concentration data, provides a numerical measure by which fluxes of other volatile components from the mantle can be calculated. Helium provides this unique tracer for mantle fluxes because it is the only inert gas which has detectable isotopic anomalies in the oceans caused by the input of a mantle component; for other gases the possible mantle signature is completely masked by the large atmospheric contribution. If the ^{3}He number is accepted, then assuming the ^{3}He/^{4}He and ^{4}He/^{40}Ar ratios in the mantle flux to be approximately 1.4×10^{-5} and an upper limit of 10 as discussed above, the He and Ar mantle fluxes are:

$$^{3}\text{He}=4 \text{ atoms/cm}^{2}\text{ s}$$

$$^{4}\text{He}=3\times10^{5}$$

$$^{40}\text{Ar}=3\times10^{4}.$$

For ^{4}He and ^{40}Ar the total input fluxes to the atmosphere will of course include radiogenic production in the continental crust with subsequent escape from the crust. This crustal contribution should be much greater than the mantle flux because of the preferential partitioning of U, Th, and K in the crust, and thus ^{3}He does not provide a quantitative tracer for the total atmospheric input of these radiogenic gases. For the non-radiogenic gases, however, the mantle fluxes must be the dominant source and their input rates can be estimated from the ^{3}He flux if reliable ratios to ^{3}He in the mantle flux can be established.

TOLSTIKHIN (1975) has proposed that the ^{3}He flux from the mantle is actually of the order of 100 atoms/cm^{2} s, about 20 times greater than the value given above. This value is based on the assumption that the *entire* helium flux from the earth has the *mantle* ^{3}He/^{4}He ratio, together with very high estimates for both the mantle ratio and the terrestrial helium flux to the atmosphere. For example, if we assume that half the helium flux is from the continental crust because of the much higher U and Th concentrations, and that the ^{3}He/^{4}He ratio of this flux is $\sim10^{-7}$ (ALDRICH and NIER, 1946), while half is from the mantle with a ratio of 10 times atmospheric, then a total He flux of 10^{6} atoms/cm^{2} s corresponds to a ^{3}He flux of only 7 atoms/cm^{2} s, in good agreement with the estimated oceanic flux and atmospheric escape rate.

Perhaps the most curious aspect of the helium isotope data from tholeiite glasses is the striking uniformity of the ^{3}He/^{4}He ratio in glasses in which the absolute He concentration varies by at least a factor of 50 (CRAIG and LUPTON, 1976). Such uniformity in the ratio of a primordial component to a presumably radiogenic component implies a very high degree of homogeneity in ^{3}He, U, and Th, over long times and in large volumes of the mantle. One possible explanation of the uniform $^3He/^4He$ ratio, however, is that most of the mantle ^4He is actually primordial rather than radiogenic. This possibility requires an early and fairly complete removal of U and Th from a major part of the mantle, and causes obvious difficulties with the understanding of K/U and $^{40}Ar/^{36}Ar$ ratios in mantle and crustal material through time. Further discussion must await more detailed studies of the variability of $^3He/^4He$ ratios in many more mantle-derived samples.

In summary, we now know that there really *are* primordial gases in the mantle which continue to be exsolved to the crust, ocean, and atmosphere, and we have at least a working idea of how these gases get from the mantle into the ocean. What is sorely needed is a careful, precise, detailed study of $^3He/^4He$ ratios *with* concentrations of He and other components, in a wide range of basalt glasses. Until such data are available, we can do little but speculate on the relative and absolute fluxes of volatiles from the mantle.

3. Primordial Helium in Volcanic and Orogenic Areas

Another mode by which volatiles are transported from the mantle is by volcanic activity in areas not associated with oceanic crust formation and divergent plate boundaries. MAMYRIN *et al.* (1969) observed high $^3He/^4He$ ratios in Kamchatka gases and later showed that the tritium content of the associated steam was far too low to produce these ^3He enrichments. The clearest example of primordial gases far from any plate boundary is given by the very high $^3He/^4He$ ratio in Kilauea helium, 15 times the atmospheric ratio, associated with a 5 percent $^{20}Ne/^{22}Ne$ anomaly relative to atmospheric neon (CRAIG and LUPTON, 1976). This "hot spot" helium appears to be clearly different from helium on mid-ocean rises. Similar high ratios have been tabulated for Iceland by POLAK *et al.* (in press), and it may be that such high ratios are characteristic of hot spots whenever they occur.

A third type of igneous and volcanic province of great importance is the island arc and continental margin orogenic province, generally characterized by basalt-andesite-dacite or rhyolite sequences, often interspersed to a greater or lesser extent with tholeiitic and olivine basalts. The only example so far studied for $^3He/^4He$ ratios is the study of the Kurile and Kamchatka regions reported by BASKOV *et al.* (1973) who found ratios ranging from 5.4 to 6.3 times atmosphere in five samples of hot spring gases in these areas. (We have used here the ratios listed in their Table 4, which differ from their "delta values" and data in the text, p. 136). Their report lacks details of the hot-spring sampling localities, and we give here brief descriptions of their four areas, taken from the literature, together with their maximum $^3He/^4He$ ratios relative to atmospheric helium (R/R_A).

1. Uzon Volcano, Kamchatka (VLODAVETZ and PIP, 1959). Quaternary volcano with basalt-andesite-rhyolite-dacite sequence and ignimbrites (post-glacial). $R/R_A=5.7$
2. Ebeko Volcano, Paramushir Is., N. Kuriles (GORSHKOV, 1958, p. 90). Andesitic volcano, erupted 1934–35. $R/R_A=6.1$
3. Baransky Volcano, Iturup Is., S. Kuriles (GORSHKOV, 1958, p. 21). Andesite-dacite volcano. $R/R_A=5.8$
4. Mendeleev Volcano, Kunashir Is., S. Kuriles (GORSHKOV, 1958, p. 5). Basalt-andesite-dacite; explosive eruption mid-19th century. $R/R_A=6.3$ (and 5.4 in springs

elsewhere on the island).

The data on helium associated with volcanoes of orogenic areas are important because of current emphasis on the origin of andesitic rocks and the basalt-andesite-rhyolite sequence which characterizes these areas. If andesites result from melting of subducted oceanic crust, as some workers believe, we may expect the helium associated with andesitic eruptions to differ from helium which would accompany andesitic rocks resulting from melting of pristine mantle material (e.g., peridotite). In order to study this question further, we have measured He isotope ratios in volcanic gases from orogenic and island arc areas in three other regions: the Hakone province of Japan, volcanic islands of the Mariana island arc, and Mt. Lassen in the Cascade province of the U.S. in northern California.

4. Sample Localities

4.1 Hakone volcano, Japan

Hakone is a large stratovolcano in the Huzi (Fuji) province, 80 km southwest of Tokyo. The volcanic rocks of the area are characterized by basalt-andesite-dacite eruptive sequences overlying the Miocene basement of submarine basalts and andesites and Miocene-Pliocene volcanics and sediments (KUNO, 1962; KUNO *et al.*, 1970). Hakone itself is a complex double-caldera resulting from collapse of a Pleistocene volcano, which erupted basalts, andesites, and dacitic lavas, and andesitic and dacitic pumices and tuffs, until termination of activity by a steam blast eruption about 5,000 years ago (KUNO *et al.*, 1970). The present activity consists of hot springs and fumaroles concentrated in the Central Cones area within the inner caldera. This inner region is filled with the Central Cone lavas-andesites-erupted in the final stages of activity (Fig. 1).

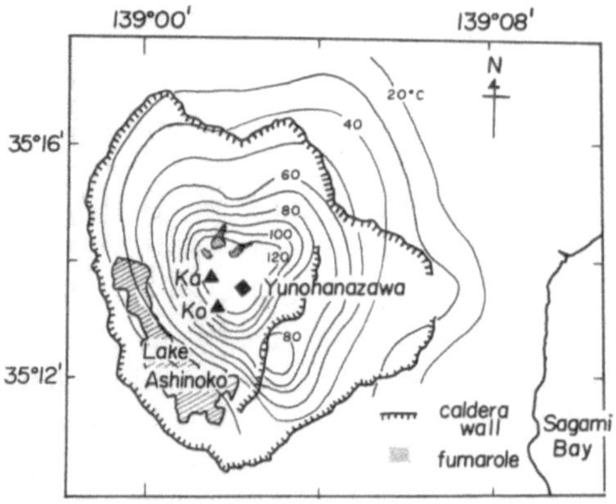

Fig. 1. Map of Hakone Volcano, showing isotherms at sealevel (°C), and the position of Yunohanazawa steam well in the Central Cones region of the caldera. Ka, Mt. Kamiyama; Ko, Mt. Komagatake.

Our sample was collected from Steam Well CW-1 at Yunohanazawa (Flower of Hot Springs), in the highest temperature zone of the Central Cones. The well is 256 meters deep with a bottom temperature of 172°C; steam outflow is at 97.7°C (altitude

968 meters). Samples of total gas and steam were collected in evacuated 1720-glass ampoules (~50 cc volume) closed with a three-way glass stopcock, and carried immediately to Scripps for analysis. (These samples were unfortunately collected in the midst of Typhoon Mary—3 July 1976—so that it was impossible to sample the natural fumaroles of the area). The analytical procedure for helium and neon was as described by CRAIG and LUPTON (1976). The stable isotope geochemistry of steam and water in this area is described by MATSUO *et al.* (in press); chemical characteristics of the geothermal system at Hakone have been studied in detail by OKI and HIRANO (1970).

4.2 Marianas island arc

Samples were collected on two volcanic islands, Agrigan and South Pagan, in the Marianas chain. These islands have been described by TANAKADATE (1940) and KUNO (1962).

Agrihan (18°45′N, 145°40′E) is a stratovolcano made of olivine and olivine-augite basalts, at the northern end of the Marianas chain. The most recent activity was a minor eruption of scoriae in 1917. The sample taken was water from a hot spring at sea level on the east coast, collected in an evacuated 1720 glass flask.

South Pagan (18°6′N, 145°48′E) is the southern island of a double cone in a large caldera. The northern island consists of olivine basalt flows on a platform of augitic andesite; S. Pagan is reported to contain both olivine basalt and andesites. The last eruption recorded was in 1925 on North Pagan. The sample collected was water from a hot spring on the west coast at sea level.

4.3 Mt. Lassen volcano

Mt. Lassen, at the southern edge of the Cascade Range, is a nearly extinct volcano built from a succession of hornblende and pyroxene andesites, rhyolites, dacites, and quartz basalts, terminating in the most recent eruption, 1914–1917, of andesitic and dacitic lava and ash. The general features of the area and a detailed study of the geothermal activity are given in the monograph by DAY and ALLEN (1925). The hot springs are concentrated along two intersecting fissures south of Lassen Peak. There are two principal fumarole and acid hot spring areas: Sulphur Works and Bumpass Hell. These areas were sampled in September, 1974, as part of a larger study which will be reported elsewhere. The samples discussed here were gas samples collected in a small acid spring in each of these two areas. The samples for helium isotope analysis were collected in 1720-glass tubes sealed in the field immediately after collection. Plastic funnels and tubing were used to collect the gas samples.

5. Summary of Measurements

The results of the various measurements on these samples are given in Table 1. Stable isotope measurements (relative to the SMOW isotopic water standard) were made on the water samples from Agrihan and S. Pagan Islands in order to characterize these waters. The water from the Agrihan sample lies exactly on the meteoric water D-^{18}O relationship (CRAIG, 1961) and is an unevaporated, dilute rain water typical of average oceanic rain in this latitude. The Pagan Island sample is enriched in both D and ^{18}O

Table 1. Sample descriptions: He/Ne and $^3He/^4He$ measurements.

				Mt. Lassen	
	Hakone Yunohanazawa	Agrihan Island	S. Pagan Island	Sulphur Works	Bumpass Hell
Type	Steam well: CW-1	Hot spring	Hot spring	Hot spring: L-2	Hot spring: L-8
T (°C)	98	46	52	87	90
pH	—	8.2	7.6	2.4	2.6
Cl (ppm)	—	245	9,000	—	—
δD (‰) (H_2O)	—	−27.1	−17.0	—	—
$\delta^{18}O$ (‰) (H_2O)	—	− 4.92	− 2.47	—	—
Sample:	Gas+Steam	Water	Water	Gas	Gas
Gas/Steam	3.8 cc STP/g	—	—	—	—
He (10^{-6} cc/g)	10.1	0.33	0.046	—	—
He/Ne	40	3.0	0.43	≥900	≥150
(He/Ne)/$(He/Ne)_{ATM}$	140	10.6	1.5	≥3,000	≥520
$(^3He/^4He)\times 10^6$	8.6	9.2	4.4	11.3	10.8
R/R_{ATM}	6.2	6.5	3.1	8.1	7.7
$(R/R_{ATM})_C$	6.2	7.0	5.3	8.1	7.7

relative to the Agrihan water, with a D-^{18}O slope of 4; this is characteristic of evaporation enrichment at the temperature (50°C) of these springs. The Pagan sample may contain some seawater, or the salt content may be residual; the isotopic data do show that this water is not simply a 50/50 mixture of rain water and seawater, as can be seen by plotting the data and the value for mean seawater (not shown). Other tabulated data on the sample collection sites include pH, temperature, chloride content for the Agrihan and Pagan samples, type of sample collected, He concentration in the water samples, the gas/steam ratio for the Hakone sample, and the $^3He/^4He$ and He/Ne ratios.

5.1 Helium isotope ratios

The $^3He/^4He$ ratios are all high compared to the atmospheric ratio of 1.40×10^{-6}; the absolute values range from 4.3×10^{-6} at S. Pagan, to 11.3×10^{-6} at Mt. Lassen, i.e., from 3.1 to 8.1 times the atmospheric ratio. With the exception of the S. Pagan sample, therefore, these samples are all highly enriched in mantle helium.

An estimate of the maximum contribution of ^3He produced by decay of thermonuclear tritium in these samples can be made from the observed 1963 tritium peaks in precipitation in these regions (IAEA, 1967). In the Tokyo area this peak reached 876 TU (tritium units $=10^{-18}$ atoms T/atom H) during February—September, 1963. Assuming the Hakone steam to have been labeled with this tritium concentration in 1963, the decay of 425 TU would add 1×10^{-12} cc STP/g of ^3He to the sample, that is, only 1% of the observed ^3He content. In the oceanic precipitation at the latitude of the Marianas Islands, the maximum tritium peak was about 200 TU in 1963; assuming these waters to have been labeled at that time, the resultant thermonuclear ^3He component is 0.24×10^{-12} cc STP/g, which is only 8% of the ^3He content of the Agrihan sample, but equivalent to *all* the ^3He in the S.

Pagan Island water because of the low He concentration. In the Mt. Lassen samples, the He/Ne enrichments are so high that the tritium contribution to ^{3}He cannot be significant.

5.2 *He and Ne concentrations*

In the Hakone steam, helium is about 200 times supersaturated relative to atmospheric helium solubility at 10°C; neon on the other hand is supersaturated by only about 25%. This neon is thus probably entirely of atmospheric origin, and the Hakone steam probably represents total vaporization of groundwater with little change of the neon to water ratio during transport. According to OKI and HIRANO (1970), the heated groundwater in the area sampled is derived from water from the caldera lake (Ashinoko) infiltrating from the western edge of the caldera, mixed with a small amount of dense steam rising from beneath the Central Cones area.

Neon concentrations in the Marianas hot springs are also close to saturation; about 75% of saturation equilibrium at 50°C in both samples. Helium in Agrihan hot spring is about 8 times supersaturated at the observed temperature, but the Pagan Island water is only 20% supersaturated in helium; some gas loss to the atmosphere probably occurs in both these low-temperature open springs.

5.3 *"Corrected" $^3He/^4He$ ratios*

Assuming that *all* neon in the water samples is of atmospheric origin, the He isotope ratios can be corrected for the corresponding atmospheric helium component to derive the ratio for the added helium component, i.e.,

$$R_C=(RN-R_AN_A)/(N-N_A),$$

where R=^{3}He/^{4}He, N=He/Ne ratio, the subscript "A" denotes the atmospheric ratios, and R_C is the "corrected" ^{3}He/^{4}He ratio. This has been done in the final row of Table 1, using the He/Ne solubility ratios in the water at 10°C for the parameter N_A, assuming initial atmospheric saturation at that temperature. The values are tabulated as R_C/R_{ATM}, i.e., the corrected He^3/He^4 ratio relative to the atmospheric ratio in the final row of the table. For the present samples, the corrected ratios differ significantly from the measured values only for the Pagan Island sample. (It should be noted that the value of R_C/R_{ATM} for Pagan Island becomes 7.1 if the solubility ratios at 52°C rather than at 10°C, are used for the atmospheric component).

5.4 *Gas composition (Hakone, Lassen L-2)*

Table 2 lists the total gas composition of the Hakone and Lassen-Sulphur Works samples. The "non-condensible" gases, which make up 2% and 1% of the total gas, are totaled to 100% in the second part of the table. These analyses were made by manometric measurement of the CO_2+H_2S fraction, gas chromatographic measurement of the H_2S/CO_2 ratio using a thermal conductivity detector, and gas chromatographic analysis of the non-condensible gases using an ultrasonic detector with digital electronic integration.

The Hakone gases are characterized by exceptionally high H_2S and H_2 contents relative to other components. The high H_2 content is noteworthy in view of the high values also observed in volcanoes on Kamchatka (WHITE and WARING, 1963). High CO contents are also reported in many of the Kamchatka analyses; in the Hakone gas sample CO was looked for, but none was observed.

Argon in the Hakone fluid is present in a concentration of 5.5×10^{-4} cc STP/g, about 40% greater than solubility at 10°C, or 10% greater than solubility at 0°C. This gas thus resembles neon in being close to the equilibrium saturation value. Nitrogen, however, has a concentration of 0.044 cc STP/g, and is supersaturated relative to atmospheric solubility at 10°C by a factor of 3. The similarity of the N_2/Ar ratio (=80) to the atmospheric ratio may therefore be coincidental, and N_2 may actually be present in considerable excess relative to Ar, as is clearly the case in the Lassen gas sample (N_2/Ar=256). Further studies, including stable isotope measurements, are in progress on the Hakone gas sample.

Table 2. Gas analyses: Hakone and Lassen L-2.

	Hakone CW-1 Yunohanazawa	Lassen L-2 Sulphur Works
Total gas		
"Non-condensible" (%)	2.0	1.0
CO_2 (%)	58.3	99.0
H_2S (%)	39.7	—
He (ppm)	2.7	7.1
Ne (ppm)	0.067	0.0079
Non-condensible (to 100%)		
N_2	58.48	90.04
O_2	0.28	0.15
Ar	0.73	0.35
H_2	39.10	9.02
CH_4	1.41	0.44
N_2/Ar	80	256
$\delta^{13}C$ (‰) (CO_2)	−0.7	−7.9

$^{13}C/^{12}C$ ratios were also measured on CO_2 in these gases; the data are reported as delta values (per mil) relative to the Chicago PDB standard, in the last row of Table 2. The Lassen measurement is typical of many values measured in this area (CRAIG, 1963), and cannot be attributed to a unique carbon source since it is characteristic of the average type of groundwater carbon derived in roughly equal proportions from carbonate and organic matter. However, the delta value for the Hakone CO_2 is quite significant, since it is completely characteristic of marine carbonate carbon. OKI and HIRANO (1970) have suggested that bicarbonate and CO_2 in the Hakone geothermal system are derived from decomposition of fossil plant material (which should be much lower in ^{13}C), since no limestone is present immediately below the Hakone caldera (KUNO *et al.*, 1970). Nevertheless, the ^{13}C analysis on Hakone carbon dioxide shows clearly that the CO_2 is derived from a limestone source, which must be traversed by groundwater somewhere in its trajectory to the hot spring area.

6. Discussion

The immediate significance of the new ^{3}He data reported here is that, together with the USSR measurements on gases from the Kurile-Kamchatka volcanic chain, they provide a first-look at the isotopic composition of helium along a major section of the convergent boundary of the Pacific plate: the 2,000 mile arc on the west from the Marianas north to Kamchatka, and the southern Cascades at Mt. Lassen on the eastern boundary. All but the Marianas Islands are typical orogenic provinces characterized by andesites and more acid rocks; at Hakone the gases and hot springs are concentrated directly in the immediate area of the Central Cone andesites. In contrast to these areas located relatively close to the continental crust, the Marianas Islands provide a sampling point in an olivine basalt island-arc province in which andesite makes up only a minor fraction of the exposed lavas (KUNO, 1962; T. Dixon and R. Stern, personal communication).

The ^{3}He/^{4}He isotope ratios in all these areas so far sampled are surprisingly high and uniform; the mean ratio is about 7 times atmospheric, ranging from 6 to 8 with the highest value of 8 found in the area most typical of continental-margin volcanism—Mt. Lassen in the Cascade province. The mean ^{3}He/^{4}He ratio, $R=7\ R_{ATM}$, is very similar to the ratio in the helium flux emerging from the Galapagos Rift in the hydrothermal plumes associated with an active spreading center. It is clear, therefore, that the active transport of primordial gases from the mantle to the atmosphere occurs not only in the areas of crustal divergence where new lithosphere is being created, but also in the subduction zones where it is destroyed.

Current arguments on the origin of andesitic and more acid lavas in these orogenic provinces are generally of two classes: those which involve remelting of the subducting oceanic crust (e.g., MARSH, 1976; STERN *et al.*, 1975; GREEN, 1972) or a mixture of the subducted crust and overlying wedge of mantle material (RINGWOOD, 1975) on the one hand, and those on the other (e.g., KUSHIRO, 1973) which are based on melting pristine mantle peridotite above Benioff zones by interaction with water rising from the underlying downgoing slab. We do not wish to discuss andesite models here, but only to point out that, because of ^{4}He production in the crust, either of these models should result in an association of andesitic volcanism with helium of significantly lower ^{3}He/^{4}He ratios than those observed in fresh ocean-ridge tholeiite glasses.

If we assume the age of the subducting plate material to be $\sim 100\times 10^6$ years, a U content of only 0.10 ppm will have produced 1.7×10^{-6} cc/g of radiogenic ^{4}He in the downgoing slab (CRAIG and LUPTON, 1976, p. 376). If we further assume the original He content of the slab to be $\sim 1.5\times 10^{-6}$ cc/g with $R=10\ R_{ATM}$, i.e., values characteristic of fresh East Pacific Rise tholeiite glasses (LUPTON and CRAIG, 1975; CRAIG and LUPTON, 1976), then the ^{3}He/^{4}He ratio in the downgoing oceanic crust will be 4.7 R_{ATM}. However, AUMENTO and HYNDMAN (1971) give U=0.25 ppm for the mean value of fresh ridge basalts, while MACDOUGALL (1977) finds 0.44 ppm as a mean value for older ridge basalts which have undergone alteration by interaction with seawater. In 100 my old ocean crust, these U contents correspond to ^{3}He/^{4}He ratios of $R=2.5\ R_{ATM}$ and 1.7 R_{ATM}, respectively. Although these estimates are subject to considerable uncertainty, it is difficult to avoid the conclusion that with an original basalt helium content of $\sim 1.5\times 10^{-6}$ cc STP/g, the 100 my old lithosphere should have its original ^{3}He/^{4}He ratio diluted by addition of radiogenic ^{4}He down to $R \sim 3\ R_{ATM}$.

The uniform and high ^{3}He/^{4}He ratios, $R \sim 7\ R_{ATM}$, observed in these island arc and orogenic areas thus indicate either that *original* helium contents in new Pacific crust are an order of magnitude greater than we have so far observed, i.e. are $\sim 15 \times 10^{-6}$ cc/g in mean fresh ridge-basalt material, *or* that andesites cannot have formed by remelting oceanic crust. Further, since it is difficult to understand how helium can be separated from water during any transport process from a Benioff zone up into overlying mantle, the hydrous melting of peridotite to produce andesites requires that $\sim 70\%$ of the helium associated with these andesites is original mantle helium from the peridotitic material. Since the He content of ridge-tholeiites presumably represents a significant concentration relative to mantle peridotite, such a model requires a very large dilution of the water and volatiles from the subducting crust.

A more detailed discussion in the face of such an obvious lack of sufficient data is clearly unjustified, although it should be pointed out that if essentially *all* the primordial helium is flushed out of the oceanic crust under the mid-ocean ridges by hydrothermal convection, one then expects to see almost pure radiogenic ^{4}He in gases associated with orogenic andesite volcanism, *if* andesites are remelted crustal material. The best argument one can make to the contrary is that all the data discussed in this paper actually reflect gases baked out of primary mantle *basalts* heated by younger andesites, dacites, etc., which are essentially helium-free. The resolution of this argument requires direct measurements on gases actually held in fresh andesitic and dacitic rocks.

A final point is that the ^{3}He flux from mantle to atmosphere has so far been estimated only as the flux into deep ocean water by submarine volcanism; the total flux will of course include the contributions from subaerial volcanic gas emission in orogenic and island volcano provinces. The approximate agreement of the oceanic flux with the estimated escape rate from the atmosphere (Craig *et al.*, 1975) may indicate that subaerial emission of primordial volatiles is not greater than the flux into the ocean, but both the atmospheric escape rate and oceanic input of ^{3}He are certainly no better than factor-of-two calculations. A considerable refinement of these estimated ^{3}He fluxes will thus be necessary before total fluxes of volatiles from the mantle can be calculated.

We thank S. Matsuo and his colleagues of the Tokyo Institute of Technology for their cheerful assistance to us in sampling the Hakone steam well under intolerable weather conditions, and R. M. Horowitz for his help in the field work at Mt. Lassen. The Marianas Islands samples were collected for us byT. Dixon and R. Stern, both of SIO; we thank them for their interest and for these valuable samples. J. Welhan made the chromatographic analysis of the Hakone gases, and Michiko Hitchcox very capably typed the manuscript. This work was supported by the Oceanography Section of the National Science Foundation under grants OCE 75–04690 and 76-81999 to the Isotope Laboratory, Scripps Institution of Ocean ography.

REFERENCES

Aldrich, L. T. and A.O. Nier, The abundance of ^{3}He in atmospheric and well helium, *Phys. Rev.*, **70**, 983, 1946.

Aumento, F. and R. D. Hyndman, Uranium content of the oceanic upper mantle, *Earth Planet. Sci. Lett.*, **12**, 373–384, 1971.

Baskov, Y., V. Vetshteyn, S. Surikov, I. Tolstikhin, G. Malyuk, and T. Mishina, Isotopic composition of H, O, C, Ar, and He in hot springs and gases in the Kurile-Kamchatka volcanic region as indicators of formation conditions, *Geochem. Int.*, **10**, 130–138, 1973.

CLARKE, W. B., M. A. BEG, and H. CRAIG, Excess ^{3}He in the sea: Evidence for terrestrial primordial helium, *Earth Planet. Sci. Lett.*, **6**, 213–220, 1969.

CLARKE, W. B., M. A. BEG, and H. CRAIG, Excess helium 3 at the North Pacific Geosecs station, *J. Geophys. Res.*, **75**, 7676–7678, 1970.

CRAIG, H., Isotopic variations in meteoric waters, *Science*, **133**, 1702–1703, 1961.

CRAIG, H., The isotopic geochemistry of water and carbon in geothermal areas, in *Nuclear Geology on Geothermal Areas*, edited by E. Tongiorgi, pp. 17–53, V. Lischi and Figli, Pisa, 1963.

CRAIG, H., Geochemistry and origin of the Red Sea Brines, in *Hot Brines and Recent Heavy Metal Deposits in the Red Sea*, edited by E. T. Degens and D. A. Ross, pp. 208–242, Springer-Verlag, New York, 1969.

CRAIG, H. and J. E. LUPTON, Primordial neon, helium, and hydrogen in oceanic basalts, *Earth Planet. Sci. Lett.*, **31**, 369–385, 1976.

CRAIG, H., W. B. CLARKE, and M. A. BEG, Excess ^{3}He in deep water on the East Pacific Rise, *Earth Planet. Sci. Lett.*, **26**, 125–132, 1975.

DAY, A. L. and E. T. ALLEN, The volcanic activity and hot springs of Lassen Peak, Carnegie Inst. Washington Publ. 360, 1925.

DYMOND, J. and L. HOGAN, Noble gas abundance patterns in deep-sea basalts—primordial gases from the mantle, *Earth Planet. Sci. Lett.*, **20**, 131–139, 1973.

FISHIR, D. E., Trapped helium and argon and the formation of the atmosphere by degassing, *Nature*, **256**, 113–114, 1975.

GORSHKOV, G. S., *Catalogue of the Active Volcanoes and Solfatara Fields of the Kurile Islands*, Int. Volcanolog. Assoc., Naples, 1958.

GREEN, D. H., Magmatic activity as the major process in the chemical evolution of the earth's crust and mantle, in *The Upper Mantle*, edited by A. R. Ritsewa, p. 47, Elsevier Pub. Co., Amsterdam, 1972.

IAEA, *Tritium and Other Environmental Isotopes in the Hydrological Cycle*, Int. Atomic Energy Agency Tech. Rept. Ser. No. 73, Vienna, 1967.

JENKINS, W. J. and W. B. CLARKE, The distribution of ^{3}He in the western Atlantic Ocean, *Deep-Sea Res.*, **23**, 481–494, 1976.

JENKINS, W. J., M. A. BEG, W. B. CLARKE, P. J. WANGERSKY, and H. CRAIG, Excess helium in the Atlantic Ocean, *Earth Planet. Sci. Lett.*, **16**, 122–126, 1972.

KUNO, H., *Catalogue of the Active Volcanoes and Solfatara Fields of Japan, Taiwan, and Marianas*, pp. 70–79 Int. Volcanology Assoc., Naples, 1962.

KUNO, H., Y. OKI, K. OGINO, and S. HIROTA, Structure of Hakone Caldera as revealed by drilling, *Bull. Volcanol.*, **34**, 713–725, 1970.

KUSHIRO, I., Origin of some magmas in oceanic and circum-oceanic regions, *Tectonophysics*, **17**, 211–222, 1973.

LUPTON, J. E., The ^{3}He distribution in deep water over the Mid-Atlantic Ridge, *Earth Planet Sci. Lett.*, **32**, 371–374, 1976.

LUPTON, J. E. and H. CRAIG, Excess ^{3}He in oceanic basalts, Evidence for terrestrial primordial helium, *Earth Planet. Sci. Lett.*, **26**, 133–139, 1975.

LUPTON, J. E., R. F. WEISS, and H. CRAIG, Mantle helium in the Red Sea brines, *Nature*, **266**, 244–246, 1977a.

LUPTON, J. E., R. F. WEISS, and H. CRAIG, Mantle helium in hydrothermal plumes in the Galapagos Rift, *Nature*, 603–604, 1977b.

MACDOUGALL, J. D., Uranium in marine basalts: Concentration, distribution, and implications, *Earth Planet. Sci. Lett.*, **35**, 65–70, 1977.

MAMYRIN, B. Z., I. N. TOLSTIKHIN, G. S. ANUFRIYEV, and I. L. KAMENSKII, Isotopic analysis of terrestrial helium on a magnetic resonance mass spectrometer, *Geochem. Internat.*, **6**, 517–524, 1969.

MARSH, B. D., Mechanics of Benioff zone magmatism, in *The Geophysics of the Pacific Ocean Basin and its Margin*, edited by G. Sutton, M. Manghani, and R. Moberly, p. 337, Am. Geophys .Union, Washington, 1976.

MATSUO, S., M. KUSAKABE, M. NIWANO, T. HIRANO, and Y. OKI, Behavior of water in the Hakone geothermal system in view of hydrogen and oxygen isotopes, *Geochim. Cosmochim. Acta* (in press).

Oki, Y. and T. Hirano, The geothermal system of the Hakone volcano, *Geothermics, Special Issue*, **2**, 1157–1166, 1970.

Polak, B. G., V. I. Kononov, I. N. Tolstikhin, B. A. Mamyrin, and L. V. Khabarin, The helium isotopes in thermal fluids, in *Proc. Int. Assoc. Hydrological Science, Grenoble Symposium 1975* (in press).

Ringwood, A. E., *Composition and Petrology of the Earth's Mantle*, McGraw Hill, New York, 1975.

Stern, C. R., W. Huang, and P. J. Wyllie, Basalt-andesite-rhyolite-H_2O: Crystallization intervals, *Earth Planet. Sci. Lett.*, **28**, 189–196, 1975.

Tanakadate, H., Volcanoes in the Mariana Islands in the Japanese mandated South Seas, *Bull. Volcanol.*, **6**, 199–223, 1940.

Tolstikhin, I. N., Helium isotopes in the earth's interior and in the atmosphere: A degassing model of the earth, *Earth Planet. Sci. Lett.*, **26**, 88–96, 1975.

Vlodvavetz, V. I. and B. I. Pip, *Catalogue of the Active Volcanoes and Solfatara Fields of Kamchatka*, Int. Volcanolog. Assoc., Naples, 1959.

Weiss, R. F., P. F. Lonsdale, J. E. Lupton, A. E. Bainbridge, and H. Craig, Hydrothermal plumes in the Galapagos Rift, *Nature*, **267**, 600–603, 1977.

White, D. E. and G. A. Waring, *Volcanic Emanations, Data of Geochemistry*, Chapter K, U. S. Govt. Printing Office, 1963.

Nitrogen to Argon Ratio in Volcanic Gases

Sadao MATSUO,* Masaru SUZUKI,** and Yoshihiko MIZUTANI***

**Department of Chemistry, Tokyo Institute of Technology, Meguro-ku, Tokyo 152, Japan*
***Department of Chemistry, Tokyo Kyoiku University, Bunkyo-ku, Tokyo 112, Japan*
****Department of Earth Sciences, Nagoya University, Nagoya 464, Japan*

N_2/Ar ratios of volcanic gases were examined. The ratio was as high as 2,200 in gases of showa-shinzan, Japan, 10 years after its formation. The ratio decreased yearly, approaching the atmospheric value of 84. This fact was accounted for by yearly increases in the contribution of atmospheric components to the primary volcanic gases with the decline of volcanic activity. Argon was more radiogenic in gases with higher N_2/Ar ratios.

Volcanic gases from island arcs such as Japan and New Zealand have generally higher N_2/Ar ratios than the atmospheric value. The high N_2/Ar ratio is attributed to the addition of N_2 from sedimentary materials supplied by subduction of oceanic sediments to volcanic gases. The N_2/Ar ratios of gases evolved in vacuum on heating of siliceous sediment are as high as 20,000.

Lava lake gases from hot spot volcanoes such as Hawaii exhibited lower N_2/Ar ratios than that of the atmospheric component. On the basis of simple degassing models, the N_2/Ar ratio of the gases retained in the deeper mantle is estimated to be as low as 12. The Hawaiian lava lake gas can be regarded as gas derived from the deeper mantle. Gases from lava lakes of volcanoes along the African rift zone gave N_2/Ar ratios close to the atmospheric value indicating a large atmospheric contribution to the volcanic gases.

1. Introduction

Volcanic gases are one of the most important sources of information on contemporary degassing. The chemical nature of volcanic gases is not only different for each individual volcano as a whole but also different for each fumarole of a single volcano. Several lines of evidence indicate that the chemical composition of volcanic gases from a specific vent of a volcano changes with time (e.g., SAINT-CLAIRE DEVILLE and LEBLANC, 1858; MIZUTANI and MATSUO, 1959).

The complicated nature of the chemical composition of volcanic gases is apparently due to the difference in the mixing ratio of the gases supplied from deeper regions (primary volcanic gases) and the volatile materials furnished by the near-surface part of a volcano.

In very rare cases, such as Showa-shinzan, Japan, which formed suddenly in 1944, volcanic gases can be regarded to be a closed system for some gases (MATSUO, 1961; KUSAKABE, 1969). Atomic ratios between hydrogen, carbon, oxygen, sulfur and chlorine are preserved in gases from fumaroles with different temperatures, twenty years after its formation. Even in this case, however, volcanic gases are not closed to nitrogen and argon, and there is a marked yearly change in N_2/Ar ratios.

In this report, we have summarized the information on N_2/Ar ratios of volcanic gases in order to assess the factors affecting N_2/Ar ratios. Special emphasis was made on the contribution of N_2 derived from sedimentary material to the N_2/Ar ratio of volcanic gases, and on the N_2/Ar ratio in gases derived from the deeper mantle of the earth.

The N_2/Ar ratio of the present atmosphere is 84. The experimental value of the solubility ratio of atmospheric nitrogen and argon depends slightly on the temperature (BENSON and PARKER, 1961). The ratio in pure water is 38 at 0°C and 40 at 30°C. The actual N_2/Ar ratio in ground water ranges from 40 to 50 due to the slight supersaturation of N_2 (SUGISAKI, 1962). If all of the N_2 and Ar in volcanic gases are supplied by the air and the air dissolved in ground water, N_2/Ar ratio should lie in the range from 40 to 84.

2. Experimental

2.1 Analysis of volcanic gases

By means of quartz or glass tubing inserted into a fumarole, volcanic gases were introduced into a water-cooled trap containing carbonate-free 4N sodium hydroxide solution of known volume. The water vapor was condensed and acid-forming gases, such as CO_2, SO_2, H_2S, HCl, etc., were completely absorbed into the solution. The volume of non-absorbed gases, mainly H_2, N_2, CO, and CH_4, was measured by use of a syringe attached to the trap. The condensate solution was analyzed for the acid-forming gases. Combining the incremental volume of the aqueous phase, volume of non-absorbed gases and the result of analyses on acid-forming gases, the gas-water ratio could be calculated.

For the analysis of the non-absorbable gases, the gas samples were collected in the following manner (MIZUTANI, 1962). A 100 to 200 ml flask with inlet and outlet rubber tubings at both ends was connected to the tubing inserted into the fumarole. After flushing the flask for a sufficient time to expel air completely, alkaline solution was introduced into the flask via syringe allowing the water vapor and acid-forming gases to be absorbed with the solution. When the flask was filled with the solution to about half of its volume, the outlet was closed. The flask was tilted so as to allow the volcanic gases to pass through the solution. When the non-absorbed gases were accumulated in the flask to a sufficient volume, both ends of the flask were sealed off with a torch. The non-absorbed gases were analyzed for H_2, N_2, CH_4, CO, and Ar with a mass spectrometer. After 1961, however, the analyses were made by the gas chromatographic method described by SUWA (1965).

In order to measure the $^{40}Ar/^{36}Ar$ ratio, the argon in the non-absorbed gases was purified by the following manner. The gas samples were first circulated through hot CuO, where hydrogen and methane were completely burned to water and carbon dioxide. The remaining gases were then reacted with hot titanium to remove nitrogen. The purified argon was absorbed on active charcoal at 77°K, and the charcoal trap was sealed off. The rare gas portion was then subjected to mass spectrometric measurement of the $^{40}Ar/^{36}Ar$ ratio. The mass spectrometer was operated under non-static conditions. The $^{40}Ar/^{36}Ar$ ratio was determined by at least 5 sweeps over the mass range from 36 to 40. Since the ion beam intensity varied systematically with time, a calibration with time was made.

2.2 Extraction and purification of gases in sediments

Square pillar-shaped samples of siliceous sediments were loaded in a pre-degassed molybdenum crucible and placed in a vacuum system. The samples were heated by an RF-induction furnace to the melting point of the sample. A spike of ^{38}Ar was mixed with the crude gas and the mixture was passed over Cu-CuO (700°C) and a cold trap system to remove active gases. The purified mixture of N_2 and rare gases was adsorbed on active charcoal at 77°K and sealed off from the system.

The desorbed gas sample was introduced into a mass spectrometer through an orifice leak. Results of ion current ratios were calibrated by the artificial mixture of N_2 and Ar to obtain the molar ratio of N_2/Ar in the samples.

3. Results and Discussion

3.1 Yearly changes in the N_2/Ar ratio in fumarolic gases of Showa-shinzan

The results of measurements on the yearly changes in the N_2/Ar ratio in fumarolic gases of Showa-shinzan, Hokkaido, Japan are given in Table 1. A-1 is the highest-temperature fumarole. Its temperature was more than 700°C as late as 1961.

Table 1. Yearly changes in the N_2/Ar ratio of fumarolic gases from Showa-shinzan.

Year	A-1 (600 ~ 760°C)	C-2 (500 ~ 600°C)	C-3 (~ 200°C)
1954	2,200	—	—
1957	405	286	265
1958	334	367	271
1959	302	328	353
1960	151	400	286
1961	449	482	
1962	112	326	
1963	110		
1964	—		
1965	98		

The N_2/Ar ratio of gases from A-1 was 2,200 in 1954, ten years after Showa-shinzan was formed. The ratio decreases yearly as shown in Fig. 1; the yearly change in temperature is shown in Fig. 2. The ratios in gases from C-2 and C-3 seem to be constant or to slightly increase with time, but there are fewer measurements for these fumaroles.

Most of the values of the N_2/Ar ratio of A-1 are the result of single analyses. In 1959, however, we made a continuous collection of gases from A-1 at one hour intervals. The range of the results for continuous measurement is also represented in Fig. 1 by a vertical series of cross marks. It indicates that the N_2/Ar ratio of a single fumarole changes in a rather short period of time. In this connection, the irregularity of the N_2/Ar ratio found around 1960–1961 in Fig. 1 could be partly due to the sample heterogeneity. There seems to be a correlation between the N_2/Ar and CO_2/N_2 ratios of A-1 as shown in Fig. 3. The solid curve in Fig. 3 is a theoretical curve of mixing of the 1954 gas with air. It appears from this result that yearly changes in the N_2/Ar ratio of A-1 fumarole may be the result of an increase in the contribution of atmospheric components to the original gas.

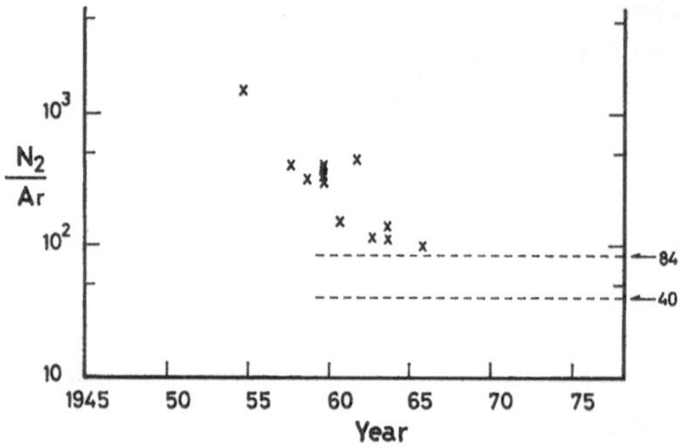

Fig. 1. Yearly changes in the N_2/Ar ratio of gases from the A-1 fumarole of Showa-shinzan. Air value (84) and dissolved air value (40) are indicated in the figure.

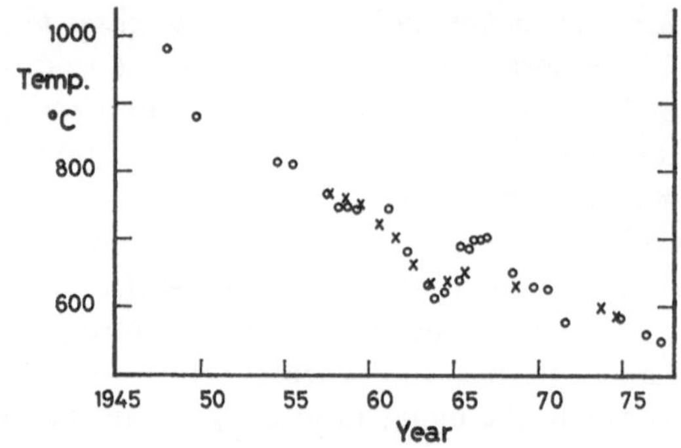

Fig. 2. Yearly changes in outlet temperature of the A-1 fumarole of Showa-shinzan. ○, measured by Geophys. Inst. of Hokkaido Univ.; ×, measured by Dept. of Earth Sciences of Nagoya Univ.

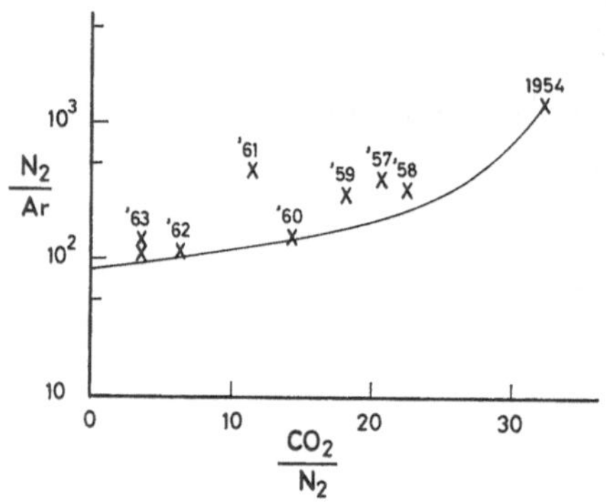

Fig. 3. Relationship between the N_2/Ar and CO_2/N_2 ratios of the A-1 fumarole of Showa-shinzan. The curve represents the theoretical trend of mixing air with the composition of 1954.

There is a correlation between N_2/Ar and $^{40}Ar/^{36}Ar$ ratios, as seen in Table 2. Assuming that the N_2 and Ar actually present in the fumarolic gases can be treated as simple mixtures of primary volcanic gases and an Atmospheric component, for a single fumarole the following two equations can be written:

$$(^{40}Ar/^{36}Ar)_{obs}=(^{40}Ar/^{36}Ar)_{prim}\,X_i+(^{40}Ar/^{36}Ar)_{atm}(1-X_i),$$

$$(N_2/Ar)_{obs}=(N_2/Ar)_{prim}\,Y_i+(N_2/Ar)_{atm}(1-Y_i),$$

where X_i is the fraction of primary ^{36}Ar in the total ^{36}Ar, and Y_i is the fraction of primary argon in the total Ar. Since the atmospheric components are dual, i.e., air itself

Table 2. Relationship between N_2/Ar and $^{40}Ar/^{36}Ar$ ratios in fumarolic gases of Showa-shinzan (1960).

Fumarole	Temperature (°C)	N_2/Ar	$[^{40}Ar/^{36}Ar]_{gas}/[^{40}Ar/^{36}Ar]_{air}$
C-2	605	400	1.08 ± 0.01
B-5	446	400	1.09 ± 0.02
C-3	187	285	1.05 ± 0.01
B-1b (1961)	227	312	1.04 ± 0.01
A-1 (1963)	635	110	1.01 ± 0.01

and the air dissolved in ground water, $(N_2/Ar)_{atm}$ should range from 40 to 84, but $(^{40}Ar/^{36}Ar)_{atm}$ can be safely assumed to be the same as that of the air; $(^{40}Ar/^{36}Ar)_{prim}$ should be higher than the maximum observed value of 1.09 $(^{40}Ar/^{36}Ar)_{atm}$ given in Table 2.

The relationship between X_i and Y_i is as follows:

$$X_i/Y_i \simeq (^{36}Ar/^{40}Ar)_{prim}/(^{36}Ar/^{40}Ar)_{obs}.$$

For the first step, let us assume $X_i = Y_i$. Then, X_i and Y_i can be eliminated from the above two equations, and we have three unknowns, i.e., $(N_2/Ar)_{prim}$, $(^{40}Ar/^{36}Ar)_{prim}$ and $(N_2/Ar)_{atm}$. We have five sets of equations using data given in Table 2, and can estimate these three unknowns. However, $(^{40}Ar/^{36}Ar)_{prim}$ was estimated to be $1.5 \times (^{40}Ar/^{36}Ar)_{atm}$, which requires further iteration, because $X_i \neq Y_i$. The result of iteration gives the most probable set of solutions in and around 1961 to be $(N_2/Ar)_{prim} \gtrsim 1{,}500$, $(^{40}Ar/^{36}Ar)_{prim} \sim 1.5 \cdot (^{40}Ar/^{36}Ar)_{atm}$, and $(N_2/Ar)_{atm} \sim 84$. In other words, $(N_2/Ar)_{prim}$ is quite high and this indicates the presence of excess N_2, probably derived from sedimentary material in and near the magma. $(^{40}Ar/^{36}Ar)_{prim}$ is slightly more radiogenic than the atmospheric argon. $(N_2/Ar)_{atm}$ is just the same as the air value, indicating no contribution of atmospheric component through ground water.

Table 3. N_2/Ar ratio and pertinent information on some of sediments.

Locality of sample	Sample	Age (B.P.)[1] (yr)	N_2/Ar	^{40}Ar (10^{-5} ccSTP/g)	K (ppm)	Radiogenic ^{40}Ar estimated (10^{-5} ccSTP/g)
Kaminiskura (Canada)	Chert I	2.5×10⁹	150	2.0	2,400[2]	4.0
	II		750	3.2		
Gunflint (Canada)	Chert I	1.9×10⁹	2,000	0.7	50	0.05
	II		2,000	0.08		
	III		4,000			
Fukaura (Aomori Pref., Japan)	Cherty shale	2 ×10⁷	21,000	0.7	3,000	0.002
Sagami Bay (Japan)	Sponge spicule <1,000°C	Recent	290	0.8	—	0
	1,000–1,500°C		290			

1) Estimated.
2) The major source of K would be iron oxide phases present in the sample.

As given in Table 3, the N_2/Ar ratio of siliceous sediments is much higher than that of the atmosphere. Siliceous sediments were chosen with the anticipation that the content of organic matter is relatively low among the various types of sediments. As will be mentioned in the later section, the N_2/Ar ratio of the atmosphere decreases with time. Accordingly, the N_2/Ar ratio in sea water also decreases with time. However, the results given in Table 3 indicate that the N_2/Ar ratio in siliceous sediments has no correlation with geologic time. This implies that the N_2/Ar ratio of sediments is controlled by the amount of organic matter contained in the sediment. Contemporary sponge spicules consisting of pure quartz have an N_2/Ar ratio of 290 instead of 40. As exemplified in this case, biogenic siliceous matter contains a significant amount of organic matter which gives off N_2 on heating in a vacuum. The highest value of 21,000 has been found in cherty shale.

When sedimentary material is transferred into the lower crust or upper mantle through tectonic activity, it becomes one of the sources of excess N_2. An active zone, such as an island arc is just the place where excess N_2 tends to be abundant in primary volcanic gases. Relatively high N_2/Ar ratios in the volcanic gases of Japan and New Zealand can be accounted for by the N_2 contribution of organic matter to the primary volcanic gases.

3.2 N_2/Ar ratios of volcanic gases from volcanoes other than Showa-shinzan

There are few reliable measurements on N_2/Ar ratios of volcanic gases. Available data are presented in Table 4. When possible, all of the data were corrected for any oxygen detected. The first five data are for Japanese fumarolic gases. Except for the low value of Chausudake Japanese fumarolic gases are featured by high N_2/Ar values, usually more than 150.

Table 4. N_2/Ar ratios of volcanic gases.

Locality	N_2/Ar	
Showa-shinzan	100~2,200	Present study
Chausudake	36~ 100	Present study
Issaikyō	200~ 350	Present study
Kujū	130~ 170	Present study
Yakedake	~ 170	Present study
Kamchatka volcanoes	41~ 112	KAMENSKY *et al.* (1976)
Kilauea lava lake	5~ 39	SHEPHERD (1938)
Kilauea Iki	54~ 85	HEALD *et al.* (1963)
Katmai	64~ 104	ALLEN and ZIES (1923)
White Island	120, 2,500~2,600	HULSTON and MCCABE (1962)
Surtsey	47~ 94	SIGVALDASON and ELISSON (1968)
Larderello	270	MAZZONI (1951)
Erta'Ale	90	TAZIEFF *et al.* (1972)
Nyiragongo	85~ 109	CHAIGNEAU *et al.* (1960)

In Hawaii, SHEPHERD (1925, 1938) analyzed Halemaumau lava lake gases from the Kilauea volcano, and obtained N_2/Ar values as low as 5. The explanation of the low values will be given separately in the later section. In contrast to the lava lake gases, the ratio in Hawaiian fumarolic gases is also characterized by lower values, but these values fall in the atmospheric range, i.e., 40–84; this result is in agreement with the conclusion that about half of the water vapor of fumarolic gases is meteoric in origin (HEALD *et al.*, 1963).

A joint party of Belgian and French researchers collected lava lake gas from Nyiragongo, Zaire, and obtained the result that $N_2/Ar=85$–109 (CHAIGNEAU *et al.*, 1960), the lower limit coinciding with the air value. Erta'Ale, Ethiopia, is also located along the rift zone that traverses the Afar triangle. The lava lake gas of Erta'Ale was also analyzed by TAZIEFF *et al.* (1972), and almost the same ratio as that of air was obtained.

Japanese and New Zealand fumarolic gases may be regarded as representative of island arc volcanoes, Hawaiian lava lake gases as representative of hot spot volcanoes and Africa gases (Nyiragongo and Erta'Ale) as representative of rift zone volcanoes.

Volcanoes which are located exclusively along the east coast of the Kamchatka Peninsula are inferred to be island arc-type volcanoes. However, most of the N_2/Ar ratios in fumarolic gases from the Kamchatka volcanoes are in the range of the atmospheric component and seldom exceed 84, as seen in Table 4. This may be due to the smaller scale of subduction which results in a poor supply of sedimentary material to the primary volcanic gases. It is interesting to note that the contribution of mantle gas is concluded to be significant in the fumarolic gases of Kamchatka, on the basis of $^3He/^4He$ ratios (KAMENSKY *et al.*, 1976). There is an indication of a mantle gas contribution to the low-temperature fumarolic gases of Hakone, Japan also based on $^3He/^4He$ ratio (CRAIG *et al.*, 1978). If this is the case in fumarolic gases from active zones, N_2/Ar ratios of the "primary" volcanic gases should be determined by the mixing ratio of the mantle gas with a low N_2/Ar ratio (refer to 3.3) and the gases derived from sedimentary material with a very high N_2/Ar ratio.

We can not provide a reasonable explanation for why the lava lake gases from the rift zone give N_2/Ar ratios almost the same as that of air.

3.3 Hot spot: the case of Kilauea lava lake

In connection with the development of the terrestrial atmosphere, a variety of degassing models have been proposed. Early catastrophic degassing was first advocated by DAMON and KULP (1958), and a continuous degassing model was first employed in a quantitative way by TUREKIAN (1959) who used the $^{40}Ar/^{36}Ar$ ratio of the atmosphere and a chondritic earth model. OZIMA and KUDO (1972) improved Turekian's model, using the $^{40}Ar/^{36}Ar$ ratio of the present atmosphere and mantle, 295.5 and 2,000 respectively to estimate a potassium content of the earth of about 100 ppm. Later OZIMA (1975) proposed an early catastrophic-later continuous degassing model.

In this section, the change in N_2/Ar ratios of the atmosphere with time is discussed briefly in relation to some of the degassing models. The result is given in Table 5 and graphically presented in Fig. 4.

Table 5. N_2/Ar ratios of the atmosphere and the solid earth based on two degassing models.

Case I (no initial atmosphere)			Case II (half of N_2 and ^{36}Ar were present in the initial atmosphere)		
Time B.P. (10^9y)	$(N_2/Ar)_A$	$(N_2/Ar)_E$	Time B.P. (10^9y)	$(N_2/Ar)_A$	$(N_2/Ar)_E$
3.0	180	71	3.0	230	38
2.0	120	33	2.0	137	23
1.0	93	18.5	1.0	100	16
0 (present)	84	11.9	0 (present)	84	12.5

Case I is the simplest continuous degassing model in which it is assumed that no atmosphere was present at the time of the earth's formation. In this case, a degassing constant of 8.82×10^{-10} y^{-1} and a K content of the earth of 96 ppm were used (OZIMA and KUDO, 1972). The same degassing constant was applied to N_2 and Ar.

Case II is a modified Ozima-Kudo model (closer to the OZIMA (1975) model) where half of the total N_2 and ^{36}Ar were already present in the atmosphere of the initial earth. In this case the degassing constant turns out to be 5.56×10^{-10} y^{-1}. In both cases, the N_2/ Ar ratio of 84 of the present atmosphere is taken as a boundary condition, and the

$^{40}Ar/^{36}Ar$ ratio of the initial earth is assumed to be 10^{-4} (MATSUO and ONUMA, 1967).

As seen from the results given in Table 5, simple calculations show that the N_2/Ar ratio in the deeper part of mantle where there is virtually no supply of sedimentary materials is much lower than that of the present atmosphere. The estimated ratio of N_2/Ar in the present mantle is about 12 which seems to be insensitive to the change in the initial conditions.

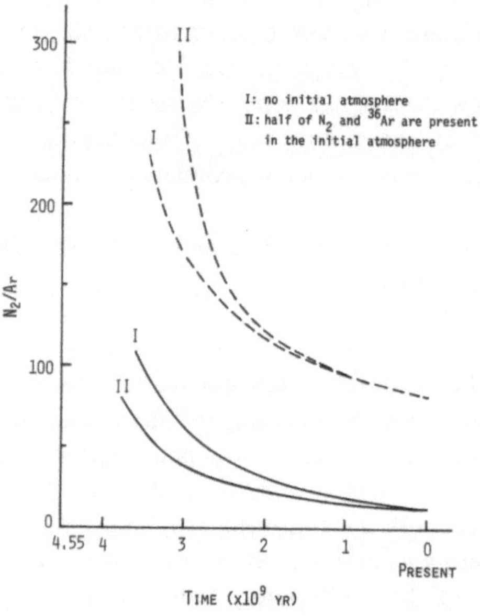

Fig. 4. Changes in the N_2/Ar ratio with time based on the Ozima-Kudo model. Solid lines represent the flux ratio (N_2/Ar ratio in the solid earth). Broken lines represent the cumulative ratio (N_2/Ar ratio in the atmosphere). I, no atmosphere at 4.55×10^9y. B. P.; II, $(^{36}Ar)_{air}/(^{36}Ar)_{earth}$ and $(N_2)_{air}/(N_2)_{earth}$ at 4.55×10^9y. B.P. are taken to be 1, respectively.

In the actual volcanic gases, the N_2/Ar ratio is never as low as 12. However, if there is an example of volcanic gases which is derived from the deeper mantle and subject to little contamination by the present atmospheric component and/or sedimentary material, the N_2/Ar ratio of such gases should be as low as 12. The N_2/Ar ratio ($5 \sim 39$) of Kilauea lava lake gas indicates this gas is of deeper origin (refer to Table 4). The high quality of sampling and analytical results by SHEPHERD (1938) have been confirmed by MATSUO (1962) on the basis that chemical equilibrium had been established among the various gas species at the observed temperature. This result is in accord with the assumption that the island of Hawaii is over a hot spot where material from deeper mantle has been continuously transported to the surface.

We are indebted to K. Kanehira of Chiba University, K. Motojima and F. Uemura of the Geological Survey, Japan, and M. Inoue of the Ocean Research Institute of Tokyo University for providing us with samples of sediments. Our thanks are extended to I. Kaneoka of Tokyo University who measured isotope ratios of Ar extracted from sediments for us. We are grateful to E. C. Alexander and M. Ozima who have read the manuscript and given us many valuable comments. Thanks are also due to the staff members and students of Nagoya University and Tokyo Kyoiku University who joined and helped us during the field work.

REFERENCES

ALLEN, E. T. and E. G. ZIES, *Contrib. Tech. Papers*, Katmai Ser., **1**, 75, Natl. Geogr. Soc., Washington, 1923: Cited in *Data of Geochemistry* (U. S. Geol. Surv. Bull., 770), F. W. Clarke, p. 270, 1924.

BENSON, B. B. and D. M. PARKER, Nitrogen/argon and nitrogen isotope ratios in aerobic sea water, *Deep Sea Res.*, **7**, 237–253, 1961.

CHAIGNEAU, M., H. TAZIEFF, and R. FABRE, Composition des gaz volcaniques du lac de lave permanent du Nyiragongo (Congo belge), *Compt. Rend., Paris*, **250**, 2482–2485, 1960.

CRAIG, H., J. E. LUPTON, and Y. HORIBE, A mantle helium component in circum-Pacific volcanic gases: Hakone, the Marianas, and Mt. Lassen, in *Terrestrial Rare Gases*, edited by E. C. Alexander, Jr. and M. Ozima, Cent. Acad. Publ. Japan, Tokyo, 1978.

DAMON, P. E. and J. L. KULP, Inert gases and the evolution of the atmosphere, *Geochim. Cosmochim. Acta*, **13**, 280–292, 1958.

HEALD, E. F., J. J. NAUGHTON, and K. L. BARNES, Jr., The chemistry of volcanic gases. 2. Use of equilibrium calculations in interpretation of volcanic gas samples, *J. Geophys. Res.*, **68**, 545–557, 1963.

HULSTON, J. R. and W. J. MCCABE, Mass spectrometer measurements in tne thermal areas of New Zealand, Part 1. Carbon dioxide and residual gas analyses, *Geochim. Cosmochim. Acta*, **26**, 383–397, 1962.

KAMENSKY, I. L., V. I. LOBKOV, E. M. PRASOLOV, N. S. BESKROVNY, E. I. KUDRYAVTSEVA, G. S. ANUFRIEV, and V. P. PAVLOV, Components of the upper mantle of the earth in gases of Kamchatka, *Geochimiya*, No. 5, 682–695, 1976 (in Russian).

KUSAKABE, M., Atomic composition and chemical equilibrium of volcanic gases, *Geochem. J.*, **7**, 141–151, 1969.

MATSUO, S., On the chemical nature of fumarolic gases of volcano Showa-shinzan, Hokkaido, Japan, *J. Earth Sci., Nagoya Univ.*, **9**, 80–100, 1961.

MATSUO, S., Establishment of chemical equilibrium in volcanic gases obtained from the lava-lake of Kilauea, *Bull. Volcanol.*, **24**, 59–71, 1962.

MATSUO, S. and N. ONUMA, The source material of the earth, *Kagaku*, **37**, 554–560, 1967 (in Japanese).

MAZZONI, A., *Bologna, Anonima Arti Grafiche*, p. 161, 1951.

MIZUTANI, Y., Chemical analysis of volcanic gases, *J. Earth Sci., Nagoya Univ.*, **10**, 125–134, 1962.

MIZUTANI, Y. and S. MATSUO, Successive observations of chemical components in the condensed water from a fumarole of Volcano Showa-shinzan, *Kazan (Bull. Volcanol. Soc. Japan), 2nd Ser.*, **3**, 119–127, 1959 (in Japanese).

OZIMA, M., Ar isotopes and earth-atmosphere evolution models, *Geochim. Cosmochim. Acta*, **39**, 1127–1134, 1975.

OZIMA, M. and K. KUDO, Excess argon in submarine basalts and earth-atmosphere evolution model, *Nature Phys. Sci.*, **239**, 23–24, 1972.

SAINT-CLAIRE DEVILLE, C. and F. LEBLANC, *Ann. Chim. Phys., 3rd Ser.*, **52**, 5, 1858.

SHEPHERD, E. S., The analysis of gases obtained from volcanoes and rocks, *J. Geol.*, **33**, 289–370, 1925.

SHEPHERD, E. S., The gases in rocks and some related problems, *Am. J. Sci., 5th Ser.*, **35A**, 311–351, 1938.

SIGVALDASON, G. E. and G. ELISSON, Collection and analysis of volcanic gases at Surtsey, Iceland, *Geochim. Cosmochim. Acta*, **32**, 797–805, 1968.

SUGISAKI, R., Geochemical study of ground water, *J. Earth Sci., Nagoya Univ.*, **10**, 1–33, 1962.

SUWA, Y., Chemical analysis of volcanic gases by gas chromatography, *J. Earth Sci., Nagoya Univ.*, **13**, 12–22, 1965.

TAZIEFF, H., F. LE GUERN, J. CARBONELLE, and P. ZETTWOOG, Etude chimique des fluctuations des gaz éruptifs du volcan Erta'Ale (Afar, Ethiopie), *Compt. Rend., Paris*, **274**, 1003–1006, 1972.

TUREKIAN, K. K., Thc terrestrial economy of helium and argon, *Geochim. Cosmochim. Acta*, **17**, 37–43, 1959.

Rare Gas Abundance Pattern of Fumarolic Gases in Japanese Volcanic Areas

Osamu MATSUBAYASHI,[†] Sadao MATSUO,[††] Ichiro KANEOKA, and Minoru OZIMA

Geophysical Institute, University of Tokyo, Tokyo 113, Japan

In order to examine the rare gas composition from the earth's interior, high temperature fumarolic emanations of volcanoes (Tokachidake, Showashinzan, Nasudake, and Satsuma-Iwojima, Japan) were analyzed. All the results show the rare gas abundance pattern of Type 1 defined by OZIMA and ALEXANDER (1976), i.e., progressive enrichment of heavier rare gases relative to atmospheric rare gases. Most of these rare gases are likely to be derived from air saturated meteoric water. We conclude that these emanations, having temperatures from 200°C to 600°C, do not contain significant amounts of rare gases whose abundance pattern is different from that of Type 1.

1. Introduction

The elemental composition as well as the isotopic ratios of rare gases in the mantle appear to yield crucial information for understanding the evolution of the terrestrial atmosphere (BOULOS and MANUEL, 1971; OZIMA and ALEXANDER, 1976; CRAIG and LUPTON, 1976). The mantle-derived rare gases may be sampled from ultramafic rocks which have their sources in the upper mantle. However, because of the extremely small amounts of rare gas trapped in these rocks, atmospheric contamination often masks the trapped rare gases of non-atmospheric origin. Most of the rare gas data on geothermal emanation have been limited so far to hot spring gases with temperatures not higher than 200°C. This low temperature, however, may be generally indicative of contamination extensive of the high temperature magmatic water, which has been derived from the interior of the earth, with low temperature meteoric ground water. The rare gases dissolved in ground water may have masked the rare gases from the interior of the earth. High temperature fumarolic emanations might not be contaminated with low temperature ground water and may contain a more faithful record of the magmatic gases. The recent report of a high $^3He/^4He$ ratio in hot springs and other natural gases in Kamchatka (KAMENSKIY *et al.*, 1976) suggests the possibility of identifying mantle-derived volatiles even in low-temperature emanations. We hoped that high-temperature fumarolic gases might represent a less-contaminated sample of volatiles from the deeper interior of the earth. However, few analytical results for rare gases in high-temperature fumarolic gases, for example those of White Island, New Zealand (WASSERBURG *et al.*, 1963), have been reported. Hence, we analyzed rare gases in fumarolic gases having temperatures higher than 200°C from Japanese volcanoes, in an attempt to understand the rare gas composition in the earth's interior.

Present address: [†] Geological Survey of Japan, Takatsu-ku, Kawasaki-shi 213, Japan and [††] Department of Chemistry, Tokyo Institute of Technology, Meguro-ku, Tokyo 152, Japan.

2. Samples

The sampling localities of fumarolic gases are shown in Fig. 1 and Table 1. Chemical compositions of gases from the same fumaroles are also cited with their data sources in Table 1. Samples for this work were collected by using a standard technique which collects the fraction of the fumarolic gas which is not absorbed by an alkali solution (for detailed procedure see MIZUTANI, 1962). Gases from a fumarole were first introduced into

Fig. 1. Sampling localities of fumarolic gases in this work. 1. Tokachidake, 2. Showashinzan, 3. Nasudake, and 4. Satsuma-Iwojima.

Table 1. Sampling localities of fumarolic gases.

Name of volcano	Tokachidake	Showashinzan		Nasudake	Satsuma-Iwojima
Altitude (m)	2,077	405		1,917	704
Latitude	43°25′N	42°30′		36°07′	30°47′
Longitude	142°41′E	140°53′		139°57′	130°18′
Locality of fumarole	Ansei fumarole	A-1	C-2	M-1	Kuromoe-2
Sample No.	TK-3	SS-3, 6, 16	8051-1	NS-1	KM-1
Chemical composition (ml/l)					
H_2	—	0.83	1.21	0.17	0.72
CO_2	—	0.32	3.66	1.12	4.7
N_2	—	}	0.21	0.19	0.04
CH_4	—	}0.079	8×10^{-3}	32×10^{-3}	8×10^{-6}
SO_2	—	0.080	0.091	0.26	8.2
H_2S	—	0.008	0.030	1.22	0.5
O_2	—	0.018	—	—	—
HCL	—	0.035	—	0.25	4.9
HF	—	0.14	—	—	0.40
H_2O	—	998.2	994.1	996.8	981
Reference	—	unpublished data by Mizutani	MATSUO (1961)	KUSAKABE (1969)	MATSUO *et al.* (1974)

a condenser through glass tubing followed by Tygon tubing. Second, most of the acid components in the uncondensed gas phase, such as CO_2, SO_2, H_2S, and HCl were absorbed by KOH solution in a glass reservoir. The gas phase unabsorbed by the KOH solution, which is mainly composed of H_2, CH_4, CO, N_2, and the rare gases, was collected in a Pyrex container with a volume of about 5 cm^3. The container whose inside pressure is about 700 Torr was sealed off at the site of collection and analysis was made later in the laboratory. Because of the high diffusion rate through Pyrex glass, the gas in the sample containers may have lost He. Hence, the He concentration obtained in this work should be regarded as the lower limit in quantitative discussions.

3. Experimental

Rare gases were analyzed by using a quadrupole mass spectrometer with a secondary electron multiplier. Resolution $M/\Delta M$ is about 150, so that the peak of ^{84}Kr, for example, can be resolved from its neighboring peaks. The lowest detectable amount of rare gas is about 10^{-12} cm^3STP for Xe, Kr, and Ne. The backgrounds of ^{40}Ar and ^4He were of the order of 10^{-8} and 10^{-9} cm^3STP respectively. To remove the active gases from the samples we used Getterloy (Zr-Ti alloy of 36% Ti) at 800°C. The purified heavy rare gases (Ar, Kr, and Xe) were adsorbed on charcoal at liquid nitrogen temperature and the unadsorbed fraction (He and Ne) was introduced into the analyzer tube. The ^{22}Ne peak height was used for the determination of the total amount of Ne since doubly charged ^{40}Ar interferes with ^{20}Ne. The total amount of He was determined using the peak height of ^4He. After analysis of the He-Ne fraction, the Ar-Kr-Xe fraction was released from the charcoal trap and admitted into the analyzer tube to measure together. The ratios of Kr/Ar and Xe/Ar were determined by extrapolation to the time when the sample gas was introduced into the analyzer tube. Analytical precision for each of the rare gases in an atmospheric standard sample in this procedure is better than $\pm 5\%$ for ^4He, ^{22}Ne, ^{36}Ar, ^{40}Ar, and ^{84}Kr. The ratio of $^{40}Ar/^{36}Ar$ was also determined with a precision of $\pm 2\%$. We have not attempted to measure accurately isotopic ratios other than $^{40}Ar/^{36}Ar$ in this work.

The sensitivity of the mass spectrometer for rare gases has been regularly calibrated by analyzing the same amount of atmospheric rare gases. The sensitivity of the mass spectrometer is within $\pm 15\%$ of a linear function of gas amount over the range of gas amounts analyzed in this work, and it is reproducible to about $\pm 10\%$ for ^4He, ^{22}Ne, ^{36}Ar, ^{40}Ar, and ^{84}Kr. We are reporting the content of Xe only when the reproducibility in replicated measurements was better than $\pm 20\%$, since the Xe sensitivity is not as reproducible as those of the other rare gases.

Standard atmospheric samples and fumarolic gas samples were analyzed alternately in the same procedure to exclude errors which might arise from drift of the sensitivity of the mass spectrometer. All the measurements were made under static conditions.

4. Results and Discussion

The data for the elemental abundances of Ne, Ar, Kr, and Xe are presented together with the rare gas fractionation factor F^m defined by HENNECKE and MANUEL (1975):

$$F^m = (^m X/^{36}Ar)_{sample}/(^m X/^{36}Ar)_{air}.$$

Table 2. Analytical data of rare gases.

Amount, ratio (error in duplicated analyses)	8051-1	TK-3	NS-1	KM-1	Average of SS-3, SS-6, and SS-16	Air	Water saturated with air at 15°C
^{4}He[1)]	3,500	1,470	1,680	576	437		
^{22}Ne[1)]	3.57	4.99	9.50	11.0	15.0		
^{36}Ar[1)]	201	193	247	250	328		
^{84}Kr[1)]	5.08	5.27	4.94	5.46	6.15		
^{132}Xe[1)]	—	0.25	—	—	—		
F^{22} (±10%)	0.35	0.51	0.76	0.87	0.90	1.0	0.28
F^{84} (±18%)	1.39	1.50	1.10	1.20	1.03	1.0	1.70
F^{132} (±20%)	—	1.90	—	—	—	1.0	2.98
$\frac{(He/Ne)_s}{(He/Ne)_{air}}$	300	90	54	16	8.9	1.0	0.8
$^{40}Ar/^{36}Ar$ (±3%)	318	311	296	292	296	295.5	295.5
log $(O_2/^{40}Ar)$[2)]	−1.0	−0.5	0	0.9	1.3	1.4	1.4
L (±0.1)[3)]	1.5	2.0	2.4	3.4	3.7	3.8	3.8
Outlet temperature (°C)	645	200	330	596	588		

1) In units of 10^{-11} cm^3/cm^3 total gas including water vapor.

2) Determined before removal of active gas components using a double focusing mass spectrometer under dynamic conditions.

3) By definition $L = \log(O_2/^{40}Ar) + \log(^{40}Ar/^{36}Ar)$.

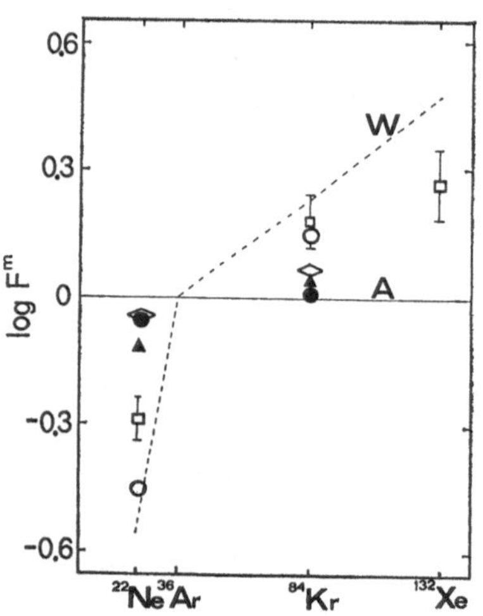

Fig. 2. Rare gas fractionation pattern for fumarolic gases. The dashed line W represents water saturated with air (A). The square symbols represent sample TK-3 from Tokachidake; the solid circles are the average of SS-3, SS-6, and SS-16 from Showashinzan; the open circles are 8051-1 from Showashinzan; the rhombi represent KM-1 from Satsuma-Iwojima. Error bars attached to square symbols indicate typical errors for duplicated analyses in this work.

Table 2 shows the results for samples from Tokachidake, Showashinzan, Nasudake and Satsuma-Iwojima. Mass discrimination in the $^{40}Ar/^{36}Ar$ ratios was corrected by assuming atmospheric $^{40}Ar/^{36}Ar$ to be 295.5. All of the isotopic ratios of Ne, Kr, Xe, and $^{36}Ar/^{38}Ar$ were identical to those of the atmosphere within experimental errors. Figure 2 shows the rare gas fractionation patterns of these samples, which we will use in the rest of our discussion.

OZIMA and ALEXANDER (1976) compiled the published data of elemental abundances of non-radiogenic rare gases in various terrestrial samples and proposed that observed rare gas relative concentrations can be classified into three types of "rare gas fractionation patterns." The rare gas compositions obtained in the present work are generally very similar to Ozima and Alexander's Type 1 pattern which they argue represents the abundance pattern of atmospheric rare gases fractionated by low temperature differential dissolution into water. The simplest explanation is, therefore, that the observed rare gases from these fumaroles are, in general, a mixture of atmospheric rare gases with the fractionated atmospheric rare gases present in local, low-temperature ground water. This conclusion is essentially the same as those obtained by other investigators from the study of natural gases and hot waters (MAZOR and WASSERBURG, 1965; BENNETT and MANUEL, 1970; MAZOR, 1972; MAZOR and FOURNIER, 1973), although the temperature of the fumarolic emanations in the present work is considerably higher than those studied by the previous workers.

If we make a closer inspection of the rare gas fractionation factors F^m, however, we see differences from sample to sample. Since the hot gases ascending from a magma reservoir are reducing, they quickly consume the O_2 present in the ground water via reactions such as $2H_2+O_2 \longrightarrow 2H_2O$, $2H_2S+O_2 \longrightarrow 2S+2H_2O$ etc. The logarithm of the $O_2/^{36}Ar$ ratio (denoted by L) of each of the samples should therefore be a measure of the degree of atmospheric contamination. Values of L less than that of air (L=3.8) clearly indicate various mixing ratios of air with the original magmatic gas. We can also use ^4He as an indicator of the mixing of atmospheric rare gases. Although the ratio of $(^4He/Ne)/(^4He/Ne)_{air}$ for samples from the A-1 fumarole of Showashinzan is as high as 8.9 (average of the three samples), the rare gas fractionation factors for Ne and Kr indicate that almost all of the gases, except for He, are from the atmosphere. It may be reasonable to suppose that magmatic gases with He/Ne ratios of several hundred or more (CRAIG and LUPTON, 1976) are diluted by near surface air contamination to yield gases having low He/Ne ratios as observed in this work. If the $^{40}Ar/^{36}Ar$ ratio of the original gas is higher than 295.5, the fact that the $^{40}Ar/^{36}Ar$ ratios we obtained have a correlation with He/Ne ratios is consistent with this interpretation. According to WASSERBURG *et al.* (1963) the ratio of $^{40}Ar/^{36}Ar$ of fumarolic gases in White Island, New Zealand increases with increasing temperature for three fumaroles. Our results from Japanese fumarolic gases indicate, however, that the temperature of the fumarolic outlet may not be the main controlling factor of either the rare gas abundance pattern or the radiogenic rare gas content, as shown in Table 2.

Since a great variability is known for chemically reactive components from fumarole to fumarole even in a single volcanic area (MATSUO, 1961), it is not obvious that our data are even typical of fumarolic gases in the Japanese Islands.

5. Conclusion

The rare gas abundance pattern of fumarolic gases in Japanese volcanic areas is consistent with a model of a mixture of atmospheric rare gases dissolved in water (local meteoric ground water) and those of the atmosphere itself, even though we chose fumaroles which have high outlet temperatures from 200°C to 600°C. Isotope ratios of rare gases in fumarolic gases, however, may be informative on the problem of how much rare gas is being degassed from the earth's deep interior. It will be especially important to collect $^3He/^4He$ ratios of fumarolic gases with high and low temperature from island arc volcanic zones, including the Japanese Islands, since the He isotopic ratio is not as easily affected by atmospheric and ground water contamination as is the rare gas abundance pattern.

The authors are grateful to Dr. M. Kusakabe for providing them with samples from Satsuma-Iwojima and Nasudake and for helpful assistance during the collecting of samples at other fumaroles.

REFERENCES

BENNETT, G. A. and O. K. MANUEL, Xenon in natural gases, *Geochim. Cosmochim. Acta*, **34**, 593–610, 1970.

BOULOS, M. S. and O. K. MANUEL, The xenon record of extinct radioactivities in the earth, *Science*, **174**, 1334–1336, 1971.

CRAIG, H. and J. E. LUPTON, Primordial neon, helium and hydrogen in oceanic basalts, *Earth Planet. Sci. Lett.*, **31**, 369–385, 1976.

HENNECKE, E. W. and O. K. MANUEL, Noble gases in an Hawaiian xenolith, *Nature*, **257**, 778–780, 1975.

KAMENSKIY, I. L., V. I. LOBKOV, E. M. PRASOLOV, N. S. BESKROVNIY, E. I. KUDRYAVTSEVA, G. S. ANUFRIEV, and V. P. PAVLOV, Components of upper mantle of the earth in gases of Kamchatka, *Geokhimiya*, No. 5, 682–695, 1976.

KUSAKABE, M., Atomic composition and chemical equilibrium of volcanic gases, *Geochem. J.*, **3**, 141–151, 1969.

MATSUO, S., On the chemical nature of fumarolic gases volcano Showashinzan, Hokkaido, Japan, *J. Earth Sci., Nagoya Univ.*, **9**, 80–100, 1961.

MATSUO, S., T. SUZUOKI, M. KUSAKABE, H. WADA, and M. SUZUKI, Isotopic and chemical compositions of volcanic gases from Satsuma-Iwojima, Japan, *Geochem. J.*, **8**, 165–173, 1974.

MAZOR, E., Paleotemperatures and other hydrological parameters deduced from noble gases dissolved in ground water: Jordan Valley, Israel, *Geochim. Cosmochim. Acta*, **36**, 1321–1336, 1972.

MAZOR, E. and R. O. FOURNIER, More on noble gases in Yellowstone National Park hot waters, *Geochim. Cosmochim. Acta*, **37**, 515–525, 1973.

MAZOR, E. and G. J. WASSERBURG, Helium, neon, argon, krypton and xenon in gas emanations from Yellowstone and Lassen volcanic National Parks, *Geochim. Cosmochim. Acta*, **29**, 443–454, 1965.

MIZUTANI, Y., Chemical analysis of volcanic gases, *J. Earth Sci., Nagoya Univ.*, **10**, 125–134, 1962.

OZIMA, M. and E. C. ALEXANDER, Jr., Rare gas fractionation patterns in terrestrial samples and the earth-atmosphere evolution model, *Rev. Geophys. Space Phys.*, **14**, 385–390, 1976.

WASSERBURG, G. J., E. MAZOR, and R. E. ZARTMAN, Isotopic and chemical composition of some terrestrial natural gases, in *Earth Science and Meteoritics*, edited by J. Geiss and E. D. Goldberg, pp. 219–240, North-Holland, Amsterdam, 1963.

A Review†: Some Recent Advances in Isotope Geochemistry of Light Rare Gases

I. N. TOLSTIKHIN

Institute of Precambrian Geology and Geochronology, Academy of Sciences, nab Makarova 2, Leningrad, U. S. S. R.

During the last decades our ideas on the origin of terrestrial helium have changed radically. Systematic studies of helium isotopic distribution in nature resulted in two important conclusions. (i) The isotope ratio of $^{3}He/^{4}He$ in the deep earth's interior is very high and may reach 5×10^{-5} (compared with a ratio of about 10^{-8} in radiogenic helium). This is probably due to the preservation of primordial helium trapped by the earth in the accretion period. (ii) The isotopic composition of most naturally occurring helium represents a mixture of primordial and radiogenic helium. The proportion of each type in these mixtures depends on such natural processes as degassing, concentration of radioactive elements, metamorphism, etc. Young eruptive rocks (of possible mantle origin) and thermal fluids contain helium which consists of $^{4}He_{rad}$ ($\sim 90\%$), $^{4}He_{prim}$ ($\sim 10\%$) and $^{3}He_{prim}$ ($\sim 100\%$). Ancient acid rocks and gases released from such rocks contain pure radiogenic helium.

The isotope ratio of $^{3}He/^{4}He$ seems to be an excellent means to extend our knowledge of the sources and migration paths of other volatiles. On the basis of this ratio the mantle value of $^{40}Ar/^{36}Ar$ is deduced to lie between 500 and 1,000.

The distribution of light rare gas isotopes in gas-liquid micro-inclusions in minerals indicates that the main sources of trapped volatiles are rocks of the earth's crust (releasing volatiles under metamorphism) and atmospheric waters and gases.

A study of the isotope composition of neon in various natural materials (rocks, minerals, gases) suggests a radiogenic origin for excess $^{21,22}Ne$ for those cases where the measured $^{21,22}Ne/^{20}Ne$ ratios are above the atmospheric values. The origin of $^{22}Ne/^{20}Ne$ ratios in cases where they are lower than atmospheric value is not clear yet.

The isotopic ratio of $^{3}He/^{4}He$ in natural gases and waters seems to be a geological-historical parameter reflecting the interaction between volatiles of the mantle and the upper crust. These ratios depend on tectonic activity and are more or less constant within the region of a single tectonic structure. A direct correlation between $^{3}He/^{4}He$ ratios (in gases and waters) and terrestrial heat flow has been established and indicates a large contribution of convective flow to terrestrial heat flow.

Correlations of primordial and radiogenic light rare gas isotopes allow an estimate of the upper limit of helium isotopic abundance in an "undissipated" atmosphere. This estimate results in some useful conclusions: one being a new way to determine the U content in the earth.

† Editorial note: This manuscript was written in Russian and then translated in the USSR into English. It was not possible for the author to supply a copy of the original Russian version of the text so that the translation could be checked in those cases where the English version was not clear. I have made numerous editorial corrections to the English version of the manuscript. These range from simple tense corrections and format changes to the complete rewriting of a few sentences. I have tried to restate the author's meaning, *as closely as I could decipher that meaning*, in passable English sentences. I apologize to the author and to the reader for any inaccuracies I have thereby introduced into the text. ECAJr.

1. Introduction

1978 (when this volume will find its way to the reader) is undoubtedly a jubilee year for helium isotopes. 110 years ago P. J. C. Janssen and J. N. Lockyer independently and simultaneously discovered ^{4}He on the Sun. 40 years ago L. W. Alvarez and R. Cornog first detected the light stable helium isotope ^{3}He (the results were published in 1939). 30 years ago the pioneer studies by L. T. Aldrich and A. O. Nier (1948), and V. G. Khlopin and E. K. Gerling (1948) were published and helium isotope geochemistry was born. 10 years ago extra-terrestrial, solar helium was observed in the earth's interior. Last year the first seminar devoted to rare gases geochemistry took place in Hakone (Japan) and problems of helium isotopic distribution were widely discussed. It was not only my wish to mark this jubilee, however, that made me devote this article mainly to helium isotope geochemistry.

It is also my strong belief that the ^{3}He/^{4}He ratio is a unique key to the study of the earth's volatiles, in particular this ratio is and always will be the single clear isotopic tracer of juvenile fluids. This belief is based on two well established facts. 1) Both helium isotopes escape from the earth's atmosphere into space and the helium content in the atmosphere is extremely low. The isotope ratios of ^{3}He/^{4}He in rocks and gases therefore are not distorted by contamination of atmospheric helium. 2) The isotopic ratio of ^{3}He/^{4}He in primordial helium is about 10,000 times greater than that in radiogenic helium. The ratios in natural helium therefore vary widely—making reliable interpretations possible.

Because of all this I thought it reasonable to give priority here to works which deal with helium isotopic distribution. Some new results in neon and argon isotope investigations are also included, however.

This review is not a systematic analysis of recent advances in the isotope geochemistry of light inert gases. I did not aim at a complete description of these advances. I chose rather to review in this article a few problems that I consider most important—even though they may seem to be unconnected. I hope, however, that a careful reader will see a definite general pattern behind them.

2. Results and Discussion

2.1 A comparison of calculated and measured ^{3}He/^{4}He ratios in rocks

The most effective approach to the problem of the origin of helium isotopes in rocks (which are the main source of helium in nature) seems to be a comparison of measured ^{3}He/^{4}He ratios with calculated ones. Assuming a radiogenic origin for both helium isotopes, Morrison and Pine (1955) discussed the correlation of the chemical composition of a rock with the ratio of ^{3}He/^{4}He. These authors compared the results of a calculation of the ^{3}He/^{4}He ratio in a common, Clark's granite composition with measurements of this ratio in natural gases (Aldrich and Nier, 1948). Later Gorshkov *et al.* (1966) improved the method for calculating the ^{3}He/^{4}He ratio in rocks and made some of the attendant parameters more exact. The first direct comparison of measured and calculated ratios was carried out by Gerling *et al.* (1971, 1972).

Figure 1 (Tolstikhin and Drubetskoi, 1975c, 1976c) is a comparison of measured

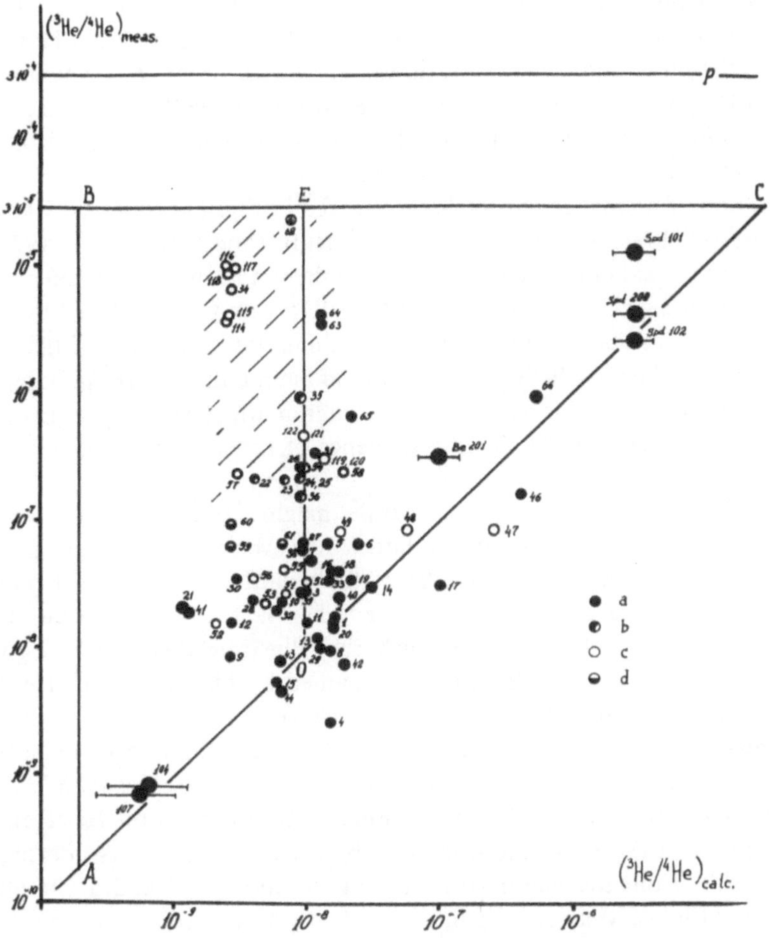

Fig. 1. The calculated and measured ratios of ^{3}He/^{4}He in different rocks and minerals (after TOLSTIKHIN and DRUBETSKOI, 1975c). The isotopic ratio of ^{3}He/^{4}He in primordial helium is shown by (P) and the value in the earth's mantle by (BEC). AOC is the line of concordant ratios of $(^3He/^4He)_{calc}$ equal to (^{3}He/^{4}He) meas. OE is the mixing line between radiogenic (O) and mantle (E) helium in common rocks. Types of rocks: a-acid; b-intermediate; c-basic; d-alkaline. Age and location: 1–18, 54–61-Precambrian, the Ukraine; 19-Paleozoic, near the Baikal Lake; 20–25, 46–53, 119–122-Proterozoic and Archean, the Kola Peninsula and Karelian; 26–33-Neogene-Jurassic, the Caucasus; 34–36, 38–44-Paleozoic and Precambrian, the Saiany; 62–65-Quaternary, the Kamchatka; 66, 101, 102, 200, 201-rocks and minerals rich in lithium (after ALDRICH and NIER, 1948); 104, 107-uranium minerals (after SHUKOLYUKOV, 1970); 114–118-xenoliths in basalts (after TOLSTIKHIN *et al.*, 1974a). Some data have been omitted from Fig. 1 to prevent crowding. The shaded region shows the area occupied by samples of young erupted rocks described by KRYLOV *et al.*, 1974; MAMYRIN *et al.*, 1974; LUPTON and CRAIG, 1975.

and calculated ^{3}He/^{4}He ratios from terrestrial samples. The data approach lines AC and OE. AC is the line where $(^3He/^4He)_m = (^3He/^4He)_c$. The data from: (i) radioactive minerals, (ii) old acid rocks, (iii) minerals and rocks rich in lithium scatter around AC.

The points on Fig. 1 which approach the line OE represent common Clark's rocks of different ages, locations and compositions. The low ratios of $(^3He/^4He)_m$ are close to 10^{-8}, i.e., to the pure radiogenic ratios (in the vicinity of point O). The high ratios of $^3He/^4He \sim 3 \times 10^{-5}$ (in the vicinity of point E) are typical of the "mantle mixture" of primordial and radiogenic helium. The problem of the origin of helium isotope in the mantle was discussed earlier (TOLSTIKHIN *et al.*, 1974a).

A distribution of points similar to that presented in Fig. 1 is observed when ratios of $(^3He/^4He)_m$ are plotted versus lithium content (TOLSTIKHIN and DRUBETSKOI, 1975c). Such a presentation of the data eliminates inaccuracies inherent in the calculation of $(^3He/^4He)_c$ but since in some specific cases high ratios of $(^3He/^4He)_m$ may be produced by radioactive decay and nuclear reactions in minerals poor in lithium (see section 2.6); the points corresponding to pure radiogenic helium may be found among those corresponding to a mixture of primordial and radiogenic helium—in the diagram with $(^3He/^4He)_m$—Li coordinates.

It is possible to suggest the following sequence of processes to explain the observed distribution of points on Fig. 1.

(i) At the moment of eruption a mantle magma releases most of its juvenile volatiles. Some traces of the juvenile volatiles however are preserved in solidifying magma and can be a significant component of the volatiles in very young rocks. (This situation is well known from K/Ar dating.) The data obtained from a ugandite (west rift zone of Africa) and a dacite (Kamchatka) illustrate this point. Low contents of $^4He \sim 10^{-8}$ cm^3/g, in these samples are accompanied by high ratios of $^3He/^4He$, $\gtrsim 2 \times 10^{-5}$ (MAMYRIN *et al.*, 1974; TOLSTIKHIN *et al.*, 1976d).

(ii) Known crustal processes (such as radiogenic helium accumulation and/or metamorphism of rocks) result as a rule in uncompensated losses of mantle ^3He and/or increases in the contribution of radiogenic helium (with low ratio of $^3He/^4He$). Initially high $^3He/^4He$ ratios in the rocks are decreased by these processes. Accordingly, points on Fig. 1 will shift from the mantle region (point E) along EO and accumulate in the radiogenic region (in the vicinity of point O). The rate of decrease depends (provided other things are equal) upon the ratio of $^3He/(U+Th)$ in rocks. For example the ratio of $^3He/^4He = 6.7 \times 10^{-6}$ is observed in olivine gabbro-norite (Table 3) characterized by K/Ar whole-rock age of 750 m. yr. and $^3He/(U+Th) = 3 \times 10^{-5}$ cm^3/g (TOLSTIKHIN *et al.*, 1977a). Olivinite (Monche-Tundra) yields ratios of $^3He/^4He = 0.6 \times 10^{-6}$ and $^3He/(U+Th) = 6 \times 10^{-4}$ in spite of the very old age of the rock, 2,800 m.yr.

In summary I conclude that mixing of mantle (partly primordial) helium and radiogenic helium is the main process which produces the range in the helium isotope composition observed in rocks of the earth's crust.

2.2 *The distribution of ^3He and ^4He in mineral fractions*

The observed $^3He/^4He$ ratios can differ from calculated ones even in the absence of mantle (primordial) helium. Such an effect may be due to differences in the sites of the helium isotopes in rock-forming and accessory minerals. Since ^3He atoms are localized in rock-forming minerals containing Li and ^4He atoms accumulate within accessory minerals rich in U and Th, the two isotopes may be released differentially. Such a difference in the migration of ^3He and ^4He could be studied by analyzing separated mineral fractions. The distribution of rare gas isotopes in the fractions would reflect the results of accumulation and migration of the isotopes into and out of different minerals under the same P-T conditions. Such an investigation of three samples was carried out by GERLING *et al.* (1976).

The data obtained for Rapakiwi granite [a case where $(^3He/^4He)_c=(^3He/^4He)_m$] are listed in Table 1. The contents of those elements (Li, K) and isotopes (^{40}Ar, ^3He) which are concentrated in rock-forming minerals are in accordance with their contents in the whole rock. There is, however, no balance between the radioactive elements (U, Th) and radiogenic ^4He, all of which are concentrated in accessory minerals, and may be lost in process of mineral separation.

Table 1. Rare gases isotopes, radioactive elements and lithium in mineral fractions of Rapakiwi-granite (datum No. 20 in Fig. 1).

Fraction	Content in mineral[1)]							Mineral content in rock (%)
	^3He	^4He	^{40}Ar	K	Li	U	Th	
Orthoclase	0.8	2	960	10.9	6.0	2.5	2.7	56
Biotite	2.3	18	622	6.4	620	7.2	4.8	3
Plagioclase	1.35	27	17.2	0.155	10	5.9	5.3	10
Quartz	0.3	2	11.4	0.186	12	5.1	3.0	30
Amphibole	16.6	368	135	1.31	37	3.5	12.7	1
Zircon	50	25×10^3	6.4	—	—	990	150	0.003
Bulk sample:								
Calculated by mineral fractions	0.91	9.4	565	6.43	27.4	3.74	3.18	
Measured	1.23	85	515	5.06	36	6.7	35	

1) Dimensions: $^3He=10^{-12}$; ^4He, $^{40}Ar=10^{-6}$ (cm^3STP/g). $K=10^{-2}$; Li, U, $Th=10^{-6}$ (g per g).

In spite of this discrepancy the following important fact was established. The ratios of $^4He/(U+0.24\ Th)$ in rock-forming minerals vary widely—unlike the nearly constant ratios of $^{40}Ar/^{40}K$. Biotite and amphibole are antipodes—both minerals contain approximately the same U and Th but differ greatly in ^4He content:

$$\frac{[^4He/(U+0.24\ Th)]^{am}}{[^4He/(U+0.24\ Th)]^{bi}}=25.8.$$

Moreover the ratio of $^3He/Li$ in amphibole is about a hundred times greater than that in biotite:

$$\frac{(^3He/Li)^{am}}{(^3He/Li)^{bi}}=120.$$

Biotite contains almost all of the lithium in the rock but contains a minimal ^3He concentration. It is noteworthy that both minerals show the similar Ar/K ratio of 9.7 and 10.3 ($\times10^{-3}$ cm^3/g) for biotite and amphibole respectively.

These results may be explained as follows: (i) $^3H(^3He)$ is produced in the reaction $^6Li(n, \alpha)^3H+Q$; (ii) the energy of the reaction (Q=4.8 MeV) is split between the kinetic energies of triton ($Q_{^3H}$=2.75 MeV) and α particle (Q_α=2.05 MeV, ZVEREV *et al.*, 1972); $Q_{^3H}$ implies a range of ^3H tracks of $\sim10^{-4}$ cm which is much greater than the thickness of monolayer of biotite structure and atoms of $^3H(^3He)$ can be released from the mineral along the track and then between the layers of crystallic structure.

The analogous explanation of the ^3He distribution appears to be reasonable assuming that: (i) part of the atoms of U and Th are dispersed in the mica and/or are located in small (diameter less than the α range) accessory minerals; (ii) while other atoms are concentrated in comparatively large accessory minerals. In case (i) some atoms of ^4He ought to have been released by biotite but preserved in amphibole (complete analogy with ^3He); while in case (ii) other atoms (located in accessory minerals) ought to have been released by biotite and amphibole in the same proportion. Such an interpretation is compatible with the above mentioned ratios of daughter and parent atoms.

The analysis of monomineralic fractions of another ancient rock, a diorite-gneiss, shows a similar distribution of the helium isotopes (Table 2). Biotite contains a small fraction of ^3He (about 10%) and nearly all of Li (as much as 80%). Almost all of the helium (both isotopes) is concentrated in the amphibole, in spite of the comparatively low contribution to the Li, U and Th content of the rock by this mineral.

Table 2. Rare gases isotopes, radioactive elements and lithium in mineral fractions of diorite-gneiss (datum No. 25 in Fig. 1).

Fraction	Content in mineral[1]							Mineral content in rock (%)
	^3He	^4He	^{40}Ar	K	Li	U	Th	
Biotite	18	83	1,100	7.4	64	0.8	4.0	20
Plagioclase	4.6	19	33	0.132	0.6	1.0	3.0	60
Quartz	5.3	29	64	0.14	0.1	0.8	2.7	10
Amphibole	250	1,250	111	0.6	17	1.0	9.0	9
Chlorite	9.4	39	25	0.2	64	0.6	5.6	1
Bulk sample:								
Calculated by mineral fractions	29.6	144	257	1.64	15	0.93	3.7	
Measured	32	145	350	1.6	15	1.3	6.0	

1) See footnote to Table 1.

It is notable that the measured ratio of $^3He/^4He$ in the rock (0.20×10^{-6}) is about 40 times greater than the calculated one (0.005×10^{-6}). If we assume a radiogenic in situ origin of ^3He and ^4He, it is difficult to explain the high content of the isotopes. The $(U+Th)/^4He$ age of the amphibole is somewhat greater than that obtained by the K/Ar method. The observed ^3He content in this mineral is 30 times greater than the calculated one. These results indicate that juvenile volatiles (excess ^3He among them) took part in the metamorphism of the rock.

The olivine gabbro-norite (Table 3) is an excellent example of comparatively old ($\sim$300–700 m.y.) rock rich in mantle helium. Again nearly all of the helium is concentrated in the dark-colored fraction rich in amphibole. The separation of pure dark-colored minerals from this fraction was not carried out because of the complex structure of the sample. The rock is characterized by high $^3He/^4He$ ratio equal to 6.7×10^{-6} (compared with the calculated ratio 0.003×10^{-6}) and high content of ^4He, 7×10^{-6} cm^3/g. The following model of the genesis of the gabbro-norite was put forward on the basis of these results (TOLSTIKHIN *et al.*, 1977a). The matter of the upper mantle (or the lower crust)

Table 3. Rare gases isotopes, lithium and potassium in mineral fractions of olivine-gabbro-norite (datum No. 34 in Fig. 1).

Fraction	Content in mineral[1)]					Mineral content in rock (%)
	^{3}He	^{4}He	^{40}Ar	K	Li	
Plagioclase	1.39	0.19	7.3	0.26	3	65
Olivine	10.4	1.4	0.84	0.071	5.5	15
Clinopyroxene— Hypersthene	107	16	0.86	0.061	4.5	15
Hypersthene+ Amphibole+ Clinopyroxene	280	42	9.1	0.43	—	5
Bulk sample:						
Calculated by mineral fractions	32	4.8	5.5	0.21	3.5	
Measured	47	7	6.0	0.17	5	

1) See footnote to Table 1.

was squeezed up into the upper crust comparatively slowly and the decreasing of pressure was accompanied by a proportional decreasing of temperature. The conditions for degassing were therefore unfavorable and a large part of the volatiles was preserved.

An alternative explanation is, however, also possible. The rock was sampled near the outcrop of a deep fractured zone and it may have trapped volatiles when this zone was active. It is noteworthy that another (typically metamorphic) rock collected in the same region is very rich in helium ($^{4}He=1.5\times10^{-3}$; $^{3}He=2.7\times10^{-9}$ cm^{3}/g) with a comparatively high ratio of $^{3}He/^{4}He=1.8\times10^{-6}$.

Thus, some difference between calculated and measured ratios of ^{3}He/^{4}He in radiogenic helium may be due to the variations of mineral composition, the peculiarities of Li, the U and Th distribution in a rock, the intensity of helium migration, etc. But these discrepancies by no means disprove the theory of MORRISON and PINE (1955) which, as a whole, correctly describes the origin of radiogenic helium isotopes. Amphibole (and most likely some other dark-colored minerals) appears to be very suitable for the study of the origin and sources of volatiles which are well trapped and preserved by this mineral. The result of SAITO *et al.* (1978, Fig. 3) appears to confirm this. Future investigations in this field when coupled with geochronology will probably make it possible to reconstruct the mode and time history of the earth's degassing.

2.3 Helium isotopes in erupted oceanic rocks

According to modern petrological, tectonic and geochemical notions, oceanic basalts are believed to be rocks of mantle origin. Many of the submarine basalts contain quenched, glassy rims which are rich in volatiles. The volatiles contain juvenile components including ^{3}He. Initially high ratios of ^{3}He/^{4}He are preserved for long times in the quenched rims due to a low content of radioactive elements. High ^{3}He/^{4}He ratios seem to be characteristic in oceanic rocks.

FISHER (1970) attempted to detect ^{3}He in oceanic basalts but he only succeeded in estimating the upper limit of $^{3}He<10^{-10}$ cm^{3}/g.

The first measurements of ^{3}He/^{4}He ratios in oceanic rocks (12 erupted basalts

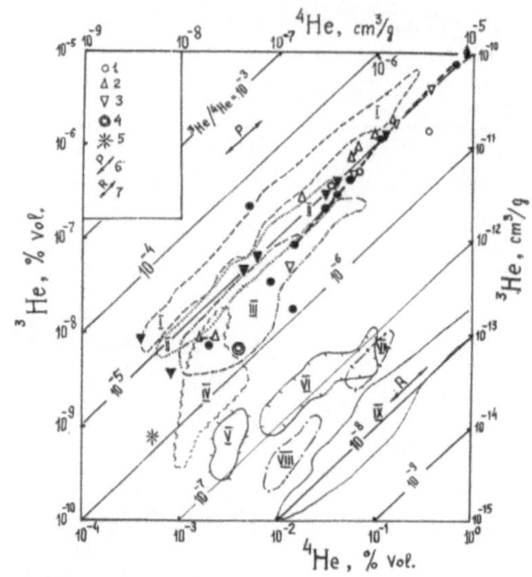

Fig. 2. The concentration of helium isotopes in units of (cm^3/g) in erupted submarine rocks from the Indian (solid symbols), Atlantic (shaded symbols), and Pacific (open symbols) oceans. Sources of data: submarine rocks 1-KRYLOV *et al.*, 1974; 2-LUPTON and CRAIG, 1975; 3-TOLSTIKHIN *et al.*, 1978; 4-oceanic water, LUPTON and CRAIG, 1975; 5-earth's atmosphere, MAMYRIN *et al.*, 1970; isotopic ratios of $^3He/^4He$ in (6) primordial and (7) radiogenic helium. Dotted lines are drawn around the points corresponding to different fractions of one sample (see text). For comparison the zones of helium isotope contents in terrestrial gases (% Vol) are shown for some regions (see also Fig. 7): I-Iceland, thermal fluids, II-the Kurils-Kamchatka, thermal fluids; III-the Caucasus, thermal and mineral springs; VI-the Sakhalin, springs and bed gases, Epihercynian platforms, bed waters and gases; V-West Siberia; VI-Turanian; VII-Ferganian; VIII-Turkmenian, Precambrian plates, bed waters and gases; IX-East-European and East Siberian.

sampled mainly near rift zones in the Atlantic, Indian and Pacific Oceans) were carried out by KRYLOV *et al.* (1974). The $^3He/^4He$ ratios in the basalts were found to be about 10^{-5} in full agreement with the idea of a ^3He excess in the deep interior of the earth (Fig. 2). Some of the samples contained high contents of helium isotopes. The authors pointed out an interesting tendency in the helium isotope distribution in a spherical tholeiitic basalt (Fig. 2). Both the helium content and the $^3He/^4He$ ratio decrease from the glassy edge towards crystalline center. Later a similar observation was reported by LUPTON and CRAIG (1975).

The results of KRYLOV *et al.* (1974) were important experimental evidence bearing on the origin of the excess ^3He in seawater discovered by CLARKE *et al.* (1969). Looking forward to future degassing model investigations, we considered that it would be useful to compare the distribution of helium isotopes in rocks, gases and oceanic water. On the basis of the publications then available (CLARKE *et al.*, 1969; CRAIG and CLARKE, 1970) we estimated a ratio of $^3He/^4He \sim (5 \text{ to } 6) \times 10^{-6}$ in the excess helium introduced into seawater from a hypothetical earth interior (DEVIRTZ *et al.*, 1971; TOLSTIKHIN *et al.*, 1972a). This value was in agreement with those available at the time for hot-springs (MAMYRIN *et al.*, 1969a, b; TOLSTIKHIN *et al.*, 1972b). KRYLOV *et al.* (1974) showed that the

$^3He/^4He$ ratios in basalts and seawater are also similar and that oceanic basalts are a probable source for excess helium in the ocean. Later CRAIG *et al.* (1975) carried out new measurements of ^3He and ^4He contents in sea-water and gave a more precise ratio of $^3He/^4He = 1.6 \times 10^{-5}$ in excess helium. This value is approximately equal to those in oceanic basalts (KRYLOV *et al.*, 1974; LUPTON and CRAIG, 1975).

The analysis of the helium isotopic distribution in submarine rocks (summarized in Fig. 2) has led to the following conclusions:

(i) Concentrations of the isotopes have a wide range. Both high and low contents are found in rocks of the Pacific, Atlantic and Indian Oceans and this implies the absence of any difference in the degassing conditions of various regions of oceanic crust (see section 2.8).

(ii) The isotopic ratios of $^3He/^4He$ remain more or less constant. Nearly all points lie very close to the line of $^3He/^4He = 10^{-5}$. Only one or two points plot somewhat higher. The range of $^3He/^4He$ ratios increases with decreasing helium contents in rocks. This is probably due to contamination of the samples (mainly those poor in juvenile volatiles) by radiogenic helium.

(iii) The isotope ratios of $^3He/^4He$ in oceanic basalts and thermal fluids of Kamchatka, Kurils (KAMENSKII *et al.*, 1976; TOLSTIKHIN *et al.*, 1972a) and Japan (HORIBE and CRAIG, 1977) are similar. Possibly partial melting and degassing of submarine basalts in the Benioff zone provide some volatiles in the thermal fluids (including helium) of the circum-Pacific belt.

(iv) The isotopic ratios of $^3He/^4He$ in Icelandic (KONONOV *et al.*, 1974; POLAK *et al.*, 1976) and Kilauea (CRAIG and LUPTON, 1976) thermal fluids are about a factor of two more than the average ratio for basalts. This indicates that submarine basalts may not be the most suitable material for estimating the mantle helium isotope composition.

2.4 Helium isotopes in some continental rocks

It seems likely that magmas and young erupted rocks of continental (and island) rift and volcanic zones should be more efficiently degassed than those of the oceanic crust. Intense degassing occurs at the moment of magma eruption and during the comparatively slow solidification. Additional degassing of such rocks may take place due to active metamorphic processes caused by the high thermal activity of the regions (see section 8). Continental rocks are more likely to be contaminated by "crustal" radiogenic helium (trapped or accumulated "in situ") because of the high concentration of radioactive elements in the crust. All these considerations indicate that young continental erupted rocks should have comparatively low helium contents and a wide range of $^3He/^4He$ ratios.

As can be seen in Fig. 3 continental rocks do display a wide range of helium contents and isotopic compositions. In examining Fig. 3 attention should be paid to the fact that the ratios of $^3He/^4He$ in recently erupted continental rocks do vary over a wide range from 5×10^{-5} to 5×10^{-8} and that the low ratios are found in rocks of probable mantle origin.

For example, the ratios of $^3He/^4He$ in Icelandic rocks vary from 10^{-7} to 10^{-5} (MAMYRIN *et al.*, 1974) while nearly constant ratios of $^3He/^4He = (1 \text{ to } 3) \times 10^{-5}$ are found

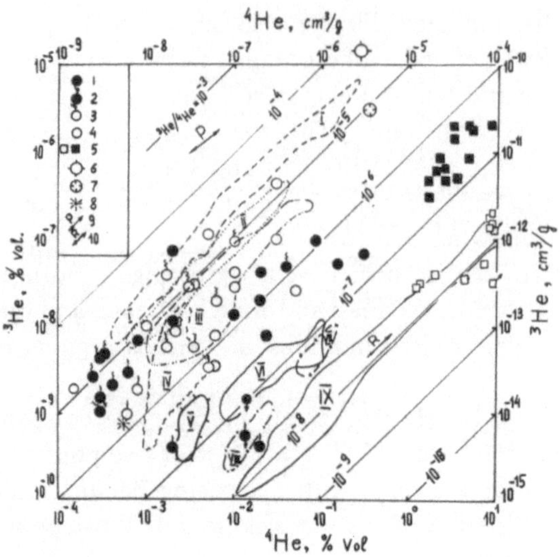

Fig. 3. The distribution of the helium isotopes (in units of cm^3/g) in continental and Iceland's rocks. Young erupted rocks: 1-Africa; 2-Iceland (Mamyrin *et al.*, 1974); 3-Kamchatka; 4-some other regions (Tolstikhin *et al.*, 1972b, 1974a, 1976d); 5-open =East-Europian Plates, ancient granitic rocks, solid=ancient ultrabasic rocks of the Kola Peninsula (Gerling *et al.*, 1971, 1972; Tolstikhin *et al.*, 1975c, 1976c); 6-amphibole, New Zealand (Saito *et al.*, 1978); 7-diamond, South Africa (Takaoka and Ozima, 1978). The rest of the symbols are the same as in Fig. 2.

in hot springs (Kononov *et al.*, 1974; Polak *et al.*, 1976). Since the relatively young rocks are poor in volatile components they cannot be considered as the source of the helium in the hydrothermal fluids of Iceland. Magmas, very young intrusions and deep mantle fluids probably serve as such a source.

It is important to note that some samples of alkaline and acid rocks show very high values of $^3He/^4He$ (Fig. 3) and thus contain evidence for a deep mantle source of their trapped volatiles. An extremely high ratio of $^3He/^4He=4\times10^{-5}$ (which reached 5.6×10^{-5} under step-wise heating experiments) was observed in ugandite, Ruanda (Mamyrin *et al.*, 1974). A similar ratio was obtained for amphibole from alkali basalts, Kakanui, New Zealand (Saito *et al.*, 1978). The ratio of $^3He/^4He=2.2\times10^{-5}$ observed in a dacite (sampled from modern lava flow of Karymskii volcano, the Kamchatka) was twice as high as the maximum ratios observed in Kamchatka hot springs (Tolstikhin *et al.*, 1976d; Kamenskii *et al.*, 1976) and inclusions in basalts (Tolstikhin *et al.*, 1974a).

These results contradict the idea that acid magmas generated from crustal matter. On the contrary, the sources of the volatile components of acid and alkaline magmas appear to be deep layers of the mantle (deeper than those represented by ultra-basic inclusions).

It is important to continue investigations of both helium and strontium isotope ratios in the same rocks. The ratio of $^{87}Sr/^{86}Sr$ in submarine basalts and many other erupted rocks of possible mantle origin range, as a rule, from 0.7020 to 0.7060 and the ratio of 0.7037 is often adopted as an maximum value for present mantle (Faure and Powell, 1972). In dacite the ratio of $^{87}Sr/^{86}Sr$ has the same value and in the amphibole it is even lower. In interpreting the ratios of $^3He/^4He$ and $^{87}Sr/^{86}Sr$ in rocks one should keep in mind that the ratio of $^3He/^4He$ bears evidence on the origin of the volatiles whereas the $^{87}Sr/^{86}Sr$ ratio suggests the sources of lithophylic elements. Thus, the ratios characterize different aspects of rock genesis and their comparison may lead to important new conclusions in future studies.

Some ancient acid rocks sampled from old platform regions (the Ukraine, the Kola Peninsula, etc.) are also presented in Fig. 3. The agreement of the data obtained from rocks and gases of such regions as well as low values of ratios of $^3He/^4He \sim 10^{-8}$ are remarkable and constitute valid evidence that in this case gas-well helium is really accumulated "... neigher from radioactive minerals as such, nor from the atmosphere, nor from preplanetary materials, but from a large mass of ordinary granite rock ..." (MORRISON and PINE, 1955).

In addition, Figure 3 shows the helium isotope distribution in some ancient ultrabasic rocks of Monche-Tundra (the Kola Peninsula). GERLING *et al.* (1962, 1968) reported a high content of radiogenic argon in these rocks as shown by unrealistic values of K/Ar ages up to 10 b.yr. Later careful measurements of argon isotopic compositions were carried out for these rocks (SHUKOLYUKOV and TOLSTIKHIN, 1965; KANEOKA, 1974) and ratios of $^{36}Ar/^{40}Ar$ (0.3 to 1.0)$\times 10^{-4}$ were measured. Interpreting these data KANEOKA (1974) concluded that "most of ^{36}Ar was already degassed from the earth's interior at least 2 or 3 b.yr. ago". The $^3He/^4He$ ratios in these rocks exceed radiogenic values by one order of magnitude but they are still about one-hundredth of the modern mantle ratio (Fig. 3) and are only one-five hundredth of that in the ancient mantle (TOLSTIKHIN *et al.*, 1975b). Assuming (i) that variations of $^{36}Ar/^{40}Ar$ as well as $^3He/^4He$ ratios in rocks occur due to comparative enrichment of gases by radiogenic isotopes and (ii) that $(^4He/^{40}Ar)_{rad}$ is approximately constant (see sections 2.5 and 2.8) one may use the foregoing coefficients to obtain a $^{40}Ar/^{36}Ar$ value for the ancient mantle which is lower than that in present atmosphere. Notably, this rough estimate is in agreement with calculations for the continuous degassing model (TOLSTIKHIN *et al.*, 1975b, 1976b), and it in no way requires a catastrophic outgassing of the earth.

2.5 *The $^3He/^4He$ ratio—a key to isotopic composition of the mantle argon*

Numerous attempts have been made to determine the mantle $^{40}Ar/^{36}Ar$ ratio. The literature shows the problem still to be of interest (SHUKOLYUKOV and TOLSTIKHIN, 1965; CHERDYNTSEV *et al.*, 1967a; CHERDYNTSEV and SHITOB, 1967b; SCHWARTZMAN, 1973a, b; BROWN *et al.*, 1974, 1976; ALEXANDER and SCHWARTZMAN, 1976; OZIMA, 1973, 1975; OZIMA and KUDO, 1972; KANEOKA, 1974; KAMENSKII *et al.*, 1974; SMELOV *et al.*, 1975; MANUEL, 1977; SAITO *et al.*, 1978; and others). A review of these works is far beyond the scope of this section but OZIMA (1975) had summarized them best of all: "... our present knowledge of the ($^{40}Ar/^{36}Ar$) ratio in the mantle is still far from conclusive. This is cheifly due to atmospheric Ar contamination ... and to the difficulty in correction for radiogenic ^{40}Ar. ..."

If we have succeeded in convincing the reader of the existence of a mantle helium (rich in ^3He) then we can use this evidence to estimate the $^{40}Ar/^{36}Ar$ ratio of the mantle (TOLSTIKHIN *et al.*, 1978).

We will start with a plot of ratios (in young rocks of probable mantle origin) of

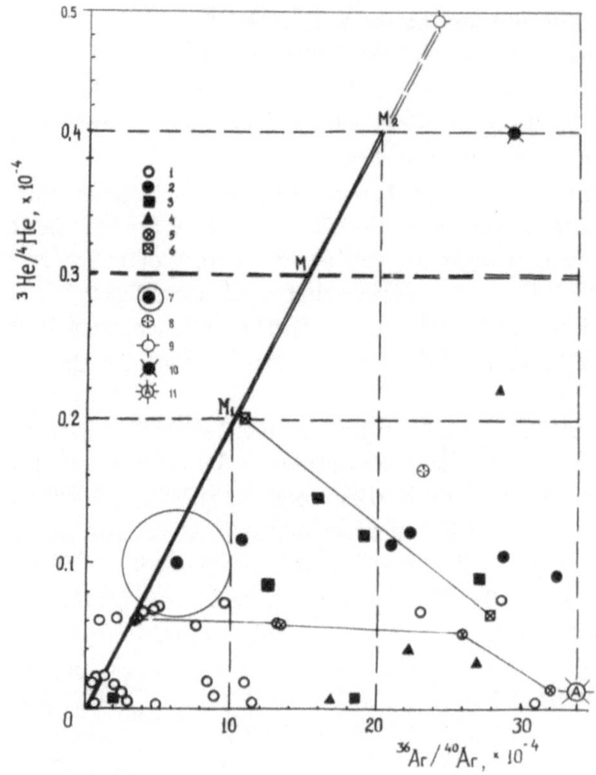

Fig. 4. The determination of $^{40}Ar/^{36}Ar$ in the earth's mantle. Line OM_1MM_2 is the theoretical limit of the area rich in experimental points. The data must lie to the right of the line. Samples: 1-metamorphic rocks; 2-submarine basalts; 3-ultrabasic inclusions; 4-acid erupted rocks of Kamchatka. Results of stepwise heating experiments: 5-sample of gabbro-norite; 6-ultrabasic inclusion; 7-an average value for submarine basalts; 8-diamond; 9-amphibole; 10-josephinite; 11-the earth's atmosphere (the sources of data are listed in the text).

$^3He/^4He$ and $^{36}Ar/^{40}Ar$ (Fig. 4). These data have been corrected only for instrumental background.

Two assumptions appear to be useful at the beginning: (i) helium and argon isotopes are released from the mantle in constant proportion equal to the ratio of their contents in the mantle; and (ii) the samples analyzed are not contaminated by atmospheric argon (atmospheric helium contamination seems to be negligible).

Keeping these assumptions in mind one may find the "mixing" line which will approximate samples with various ratios of trapped mantle and radiogenic crustal rare gases. Samples rich in radiogenic isotopes will plot near the origin because of the low radiogenic ratios of $^3He/^4He \sim 10^{-8}$ and $^{36}Ar/^{40}Ar \sim 10^{-7}$ and the "mixing" line will therefore cross the point (0.0). The angle between the line and the ($^{36}Ar/^{40}Ar$) axis may be calculated from the formula

$$\tan \alpha = \frac{^3He/^4He}{^{36}Ar/^{40}Ar} = \left(\frac{^3He}{^{36}Ar}\right)^{*} \times \left(\frac{^4He}{^{40}Ar}\right)^{**\,-1}.$$

The first ratio (*) may be estimated from well known data: $(^3He/^4He)_{prim}{}^*=3\times10^{-4}$; $(^4He/^{20}Ne)_{prim}=500$ (MAZOR *et al.*, 1970; TOLSTIKHIN and KHABARIN, 1976a); $(^{20}Ne/^{36}Ar)_{atm}=0.525$ (COOK, 1961).

The second term (**) is also well known from numerous publications (ZARTMAN *et al.*, 1961; WASSERBURG *et al.*, 1963; GERLING *et al.*, 1967; VORONOV *et al.*, 1974b; TOLSTIKHIN and DRUBETSKOI, 1975c; FISHER, 1975; etc.). In trapped and well gases the ratios of $(^4He/^{40}Ar)_{rad}$ normally approach 10. This ratio in rocks (free from trapped gases) and in thermal fluids is about one or lower. The theoretical value for common (Clark's) rocks (with U/K ratio $\sim10^{-4}$) is roughly 3–7 and remains nearly constant with time (see also section 2.8). In the discussion below a value of $(^4He/^{40}Ar)_{rad}=4$ will be used.

The line OM_1MM_2 (Fig. 4) is drawn in accordance with the foregoing parameters; M_1 and M_2 are the points of intersection of this line and the limits of the $^3He/^4He=(2$ to $4)\times10^{-5}$ zone adopted as the interval of mantle values; M is the point of intersection with the line of the most reliable mantle ratio $(^3He/^4He)_{mant}=3\times10^{-5}$ (see section 2.8). If the above assumptions were correct, the samples of mantle rocks would have shifted in time along the "mixing" line (from the mantle segment M_2M_1 toward the "radiogenic" segment in the vicinity of point 0) with the accumulation and/or trapping of radiogenic isotopes.

If the reader feels that the initial assumptions are improbable, we will give them up. We can still conclude the following:

(i) Our knowledge of processes of rare gas migration enables us to state: either helium and argon are released by rocks and minerals in the constant proportion or helium is released preferentially and the ratio of $^iHe/^jAr$ decreases with time in solid matter. For our purpose, this statement means that either our estimate of $(^3He/^{36}Ar)_{prim}$ ratio is correct for the recent mantle (and points will approach the line OM_1MM_2) or this ratio in the mantle has decreased (then points will always lie to the right of the line).

(ii) The contamination of the samples by atmospheric argon will deflect the points, as in (i), to the right of the line OM_1MM_2.

Thus, the theoretical "mixing" line OM_1MM_2 becomes the left limit of the area containing the experimental points; but the above determination of the line is not yet completed. Exceptions (especially for metamorphic rocks contaminated by radiogenic argon and helium in unusual proportion, i.e., $^4He/^{40}Ar_{rad}\ll4$) are possible. So the line OM_1MM_2 acquires the meaning of a boundary between areas of few (to the left) and numerous (to the right) experimental points.

The experimental data in Fig. 4 are in good agreement with the foregoing considerations. In addition to the results by Tolstikhin *et al.* (1–6, Fig. 4) the following observations are shown: 7) the probable ratios for basalts—the ratio of $^{40}Ar/^{36}Ar\sim1{,}000$ to 5,000 were compiled from OZIMA (1975) and FISHER (1975), and the ratios of $^3He/^4He\sim(0.6$ to $1.4)\times10^{-5}$ were taken from Fig. 2; 8) the data for diamond by TAKAOKA and OZIMA (1978); 9) the results for amphibole (SAITO *et al.*, 1978)—the "record-holder" as on the previous Fig. 3, is beyond the limits of the figure; 10) DOWNING *et al.*'s (1978) meas-

* ANDERS *et al.* (1970) suggest that the ratio of $(^3He/^4He)_{prim}$ in "planetary" helium is equal to 1.4×10^{-4} which is now accepted by some investigators. We have reliable evidence that the value above is not decisive and give preference to $^3He/^4He$ ratio measured in those carbonaceous chondrites richest in gas after the correction for cosmogenic ^3He (TOLSTIKHIN and KHABARIN, 1976a).

urements of rare gases isotope composition in josephinite; 11) atmospheric ratios.

Examining Fig. 4, we conclude that the theoretical "mixing" line OM_1MM_2 is close to that which would be drawn solely on the basis of experimental data.** We can then estimate the range of mantle ratios of $^{40}Ar/^{36}Ar=500$ to 1,000 (with an average value of 680) as the projection of the segment M_1MM_2 on the axis of $^{36}Ar/^{40}Ar$ ratios. It is encouraging that the value obtained here and that predicted by our model calculations (TOLSTIKHIN, 1975a; TOLSTIKHIN *et al.*, 1975b; 1976b) are the same.

2.6 The distribution and origin of the helium and argon isotopes in gas-liquid microinclusions in minerals

Having been released by minerals, helium, argon and other volatiles occur as components of terrestrial gases and may be subdivided into two main categories: (i) gases in closed inclusions (porosity) of minerals and rock (more than 90% of gases) and (ii) free and oil-water-dissolved gases (about 10%) (SOKOLOV, 1966).

In this section we will discuss the first category of natural gases which have been the subject of many investigations (LIPPOLT and GENTNER, 1963; LEUTWEIN and KAPLAN, 1963; LIPPOLT, 1966; RAMA *et al.*, 1965; NAUGHTON *et al.*, 1966; FUNKHOUSER and NAUGHTON, 1968; KRUMMENACHER, 1970; HARPER and SCHAMEL, 1971; NAIDENNOV *et al.*, 1972; PRASOLOV and TOLSTIKHIN, 1969, 1970, 1972; PRASOLOV, 1976; TOLSTIKHIN and PRASOLOV, 1971; TOLSTIKHIN *et al.*, 1974b; and others). The ^4He contents in microinclusions vary over a wide range from 10^{-8} to 10^{-3} cm^3/g of mineral. The average value is $\sim 10^{-5}$ cm^3/g. The corresponding values for ^{40}Ar contents are 10^{-8} to 10^{-4} and $\sim 10^{-6}$ cm^3/g. The isotopic composition of helium and argon indicates that they have mainly radiogenic and/or atmospheric origins. These data confirm the modern notions that the main sources of volatiles in the earth's crust are (i) rocks (releasing volatiles under metamorphism) and (ii) atmospheric (oceanic) waters and gases. The occurrence of radiogenic isotopes in microinclusions may be a major source of error in the dating of young rocks and minerals as well as those poor in potassium. The study of rare gas isotopes in microinclusions is useful for dating purposes (usually by the isochron variant of the K/Ar method) and to increase our knowledge of the fluid regimes of the earth's crust.

We have carefully investigated a collection of minerals from a chambered pegmatite in the Ukraine. Some nontrivial results were obtained from this study. Figure 5 shows that the ratios of $^3He/^4He$, $(0.20$ to $0.24)\times 10^{-6}$ are typical of the majority of the samples but that some ratios are higher (up to 7×10^{-6}, i.e., very close to those in young erupted rocks and thermal fluids). The question on the origin of such high ratios is of obvious interest. The pegmatites are located among ancient granites (the age of granites and pegmatites is about 1.8×10^9 year). The distance between the pegmatites and the nearest region of recent volcanic and tectonic activity is more than 300 kilometers. The $^{40}Ar/^{36}Ar$ ratios in the included argon are about 10,000 and even higher (compared with these ratios in volcanic rocks and gases). Very high ratios of $^3He/^4He$ are found only in some samples of honeycomb quartz and topaz. All of these facts are in disagreement with the hypothesis that mantle helium is the source of the high $^3He/^4He$ ratios in the

** This conclusion indicates that the ratio of $^3He/^{36}Ar$ in the mantle was always approximately constant and close to the atmospheric value—a conclusion very important to the study of mantle outgassing process.

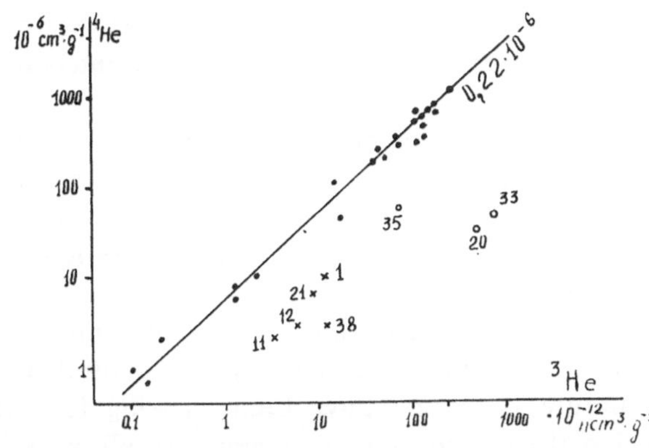

Fig. 5. The helium isotopes in gas-liquid microinclusions in quartz and topaz sampled in chambered pegmatites (after TOLSTIKHIN *et al.*, 1974b). Numbers near points on Fig. 5 and in Table 4 are the same.

Table 4. ^{4}He contents and ^{3}He/^{4}He, Li/^{4}He ratios in quartz and topaz from chambered pegmatites.[1)]

Mineral, No.		^{4}He ($\times 10^{-6}$ cm^3/g)	^{3}He/^{4}He ($\times 10^{-6}$)	Li/^{4}He (g/cm^3)
Quartz	15	210	0.19	0.0071
	2	1,090	0.25	0.037
	36	1,100	0.26	0.039
	28[1)]	640	0.23	0.067
	27[1)]	300	0.40	0.21
	33	75	7.0	0.67
	20	46	6.5	1.31
Topaz	1	15	1.5	1.6
	11	2.2	1.5	—
	12	3.2	1.8	—
	21	10	1.1	—
	38	5.7	4.3	—

1) See Fig. 5.

pegmatite.

For an explanation of high ^{3}He/^{4}He ratios we proposed the following model based on a heterogeneous distribution of radioactive elements between the crystals ($U<10^{-8}$ g per g) and the enclosing rocks ($U>10^{-6}$, $Th>10^{-5}$ g per g). (i) Since the range of α particles produced by radioactive decay in the enclosing rocks is about 10^{-2} cm, the α particles do not penetrate deeply into crystals whose sizes are approximately 10 cm. α particles are not generated in significant quantities within the crystals because of the very low content of radioactive elements. (ii) The neutron flux produced due to (α, n) reactions and fissions is characterized by a comparatively long range ($\sim$10 cm) and penetrates deeper into the minerals. The reaction $^6Li+n\rightarrow\alpha+T\xrightarrow{\beta}{}^3He$ provides a continuous accumulation of ^{3}He which is (provided other things are equal) proportional to Li content. The ^{4}He content is provided by trapped helium and is approximately constant in a course of time. (iii) This model seems to be consistent with the data listed in Table 4. The high $^3He/^4He$ ratios are observed in minerals with high values of $Li/^4He$. It is noteworthy that in crystals free from trapped helium, extremely high ratios of $^3He/^4He$ may occur (up to 10^{-3} and higher).

It was found the $^3He/^4He$ ratios in microinclusions and crystallic lattice in quartz crystals are somewhat different as are the ratios in two samples selected from one large crystal (marked[1)] in Table 4). On the basis of this difference we calculated the helium diffusion coefficient (for crystalline quartz) and obtained a very low value of $\sim 10^{-20}$ cm^2/sec (TOLSTIKHIN *et al.*, 1974c).

Quartz and topaz crystals (with well preserved helium contents and which are poor in radioactive elements) are excellent detectors of a natural neutron flux. These detectors were set by nature itself and have been working for billions of years (!). Under the assumption that the neutron flux is identical for all parts of one large crystal, we calculated that a flux, $F=8$ neutrons cm^{-2} day^{-1}, which was irradiating samples 27 and 28 during 1.8 b.yr. This value is in excellent agreement with the average flux in granites (GORSHKOV *et al.*, 1966) and is evidence for the absence of any excess neutron flux in the crust in the past (Oklo-phenomenon is the only known exception).

This model is of principal significance for helium isotope geochemistry because it provides a mechanism capable of generating helium with a high $^3He/^4He$ ratio in minerals poor in lithium.

Another result of this investigation (TOLSTIKHIN *et al.*, 1974b) also seems to be interesting and instructive. While the ratios of $^3He/^4He$ in the trapped helium (in average 2.2×10^{-6}) are about one order of magnitude higher than those in enclosing granites (3×10^{-8}), they are very similar to the ratios in gabbro-labradorites which border the granitic massive. Very high ratios of $(^4He/^{40}Ar)_{rad}=10^2$ to 10^3 observed in included argon (compare with the average value ~ 10) also argue against the idea that the granites (with Clark's ratio of U/K) are the source of the trapped gases. We postulate that the source of the volatiles were rocks which had released volatiles under comparatively low-temperature metamorphism (see section 2.8). The gabbro-labradorites appear to be acceptable candidates. It is notable that the supposition above was independently confirmed by δC^{13} measurements (MAMCHUR *et al.*, 1975).

2.7 Some recent advances in the isotope geochemistry of neon

Some new results have recently been obtained in the study of neon isotope geochemistry. The pioneer investigations (of neon isotope composition in radioactive minerals) by WETHERILL (1954) have been extended. In particular it was shown that ^{22}Ne is produced in fluorine rich defects in minerals. The ^{22}Ne is released by minerals more readily than ^{21}Ne and the ratio of these isotopes may be changed due to migration processes (SHARIF-ZADE *et al.*, 1972; ASHKINADZE *et al.*, 1973).

Isotope analyses of neon from old granitic rocks confirm a radiogenic origin of some $^{21,22}Ne$ when measured $^{21,22}Ne/^{20}Ne$ ratios are above the atmospheric values (VERKHOVSKII and SHUKOLYUKOV, 1976b).

Large excess of ^{21}Ne and ^{22}Ne were observed in minerals which trap rare gases (beryls, cordierites, amphiboles and others). The ratios of $^{21}Ne/^{20}Ne$ attain 900% of the atmospheric ratio and the ratios of $^{22}Ne/^{20}Ne$ reach 160%.

The authors (VERKHOVSKII *et al.*, 1976a) note that some ratios ($^4He/^{21}Ne_{rad}$; $^4He/$

$^{40}Ar_{rad}$) in such minerals and terrestrial gases are similar.

So recent investigations confirm the traditional concept that nuclear reactions of $^{18}O(\alpha, n)^{21}Ne$; $^{19}F(\alpha, p)^{22}Ne$ and $^{19}F(\alpha, n)^{22}Na \xrightarrow{\beta^-} {}^{22}Ne$ are the main natural processes producing the radiogenic neon isotopes (WETHERILL, 1954; GORSHKOV *et al.*, 1966; SHUKOLYUKOV *et al.*, 1974).

In addition one very interesting and unexpected result was described. In some samples the ratio of $^{20}Ne/^{22}Ne$ was found to be as high as 10.6—well above the atmospheric value of 9.85. It is notable that excess ^{20}Ne is observed in the samples which vary in age, origin and location. The ancient metamorphic rocks sudberites (the Kola Peninsula, VERKHOVSKII and SHUKOLYUKOV, 1975), hot-springs of Iceland and Kurils-Kamchatka volcanic zone (ANUFRIEV *et al.*, 1976), Kilauea gas and young submarine basalts (CRAIG and LUPTON, 1976) are among these samples (Fig. 6).

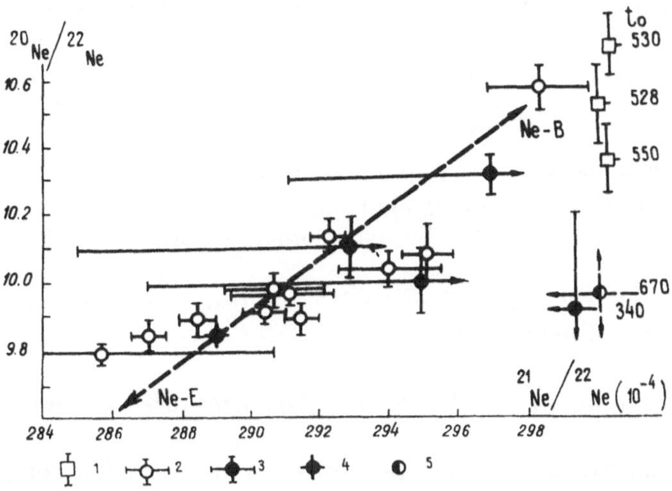

Fig. 6. The isotopic anomalies of terrestrial neon. 1-ancient rocks (VERKHOVSKII and SHUKOLUKOV, 1975); 2-thermal fluids (ANUFRIEV *et al.*, 1976); 3-Kilauea gas and submarine basalts (CRAIG and LUPTON, 1976); 4-earth's atmosphere; 5-amphibole (SAITO *et al.*, 1978).

The excess of ^{20}Ne is not always accompanied by high $^3He/^4He$ ratios. In sudberites these ratios are about 0.2×10^{-6}. On the other hand some samples very rich in ^3He do not show excess ^{20}Ne (SAITO *et al.*, 1978). Other isotopic ratios ($^4He/^{20}Ne$, $^3He/^{20}Ne$, $^{40}Ar/^{36}Ar$, $^4He/^{40}Ar$) also seem to vary widely.

In discussing these facts, the authors tend to favor interpreting the excess ^{20}Ne as evidence for primordial neon in the earth, but some of them do it too categorically. Without doubt, the problem is clear not yet. A brief account of the discovery of "mantle" argon seems to be to the point. CHERDYNTSEV *et al.* (1967a, b) observed excess ^{36}Ar in thermal fluids of the Kurils-Kamchatka volcanic zone and attributed it to the occurrence of juvenile mantle argon. They thought that the inequality of $(^{40}Ar/^{36}Ar)_{mantle} < (^{40}Ar/^{36}Ar)_{air}$ was evidence for the preservation of primordial argon (mainly ^{36}Ar) in the deep earth's interior. But subsequent investigations (KRUMMENACHER, 1970; PRASOLOV, 1976;

SMELOV, 1976; and others) show that the observed effect is due to isotope fractionation of atmospheric argon and variations of $^{40}Ar/^{36}Ar$ ratios go together with corresponding variations of $^{38}Ar/^{36}Ar$ values (see also the polemics of BROWN *et al.*, 1976 and ALEXANDER and SCHWARTZMAN, 1976). This example of course does not mean that isotope fractionation is the process producing the $^{20}Ne/^{22}Ne$ anomaly. The ratios of $^{38}Ar/^{36}Ar$ in the sudberites and in the atmosphere were found to be identical, and the quantity of gas analyzed was more than an order of magnitude higher than the apparatus background (VERKHOVSKII and SHUKOLYUKOV, 1975). On the other hand, note that nearly all the samples in Fig. 6 lie along the line Ne-E—Ne-B (which is close to the fractionation line) or to the right of the line (usually interpreted as an addition of $^{21}Ne_{rad}$). It is likely that the $^{20}Ne/^{22}Ne$ anomaly is associated with the trapping process but our knowledge of this process is limited and we know nothing about the possibility of isotopic fractionation (by trapping) of one rare gas (Ne?) while the others avoid this process. At first glance a comparison of the ratios listed below contradicts the idea of mantle sources of the excess ^{20}Ne:

$$(^{36}Ar/^{40}Ar)_{mantle} < (^{36}Ar/^{40}Ar)_{air}$$

$$(^{20}Ne/^{22}Ne)_{mantle} \; ? \; (^{20}Ne/^{22}Ne)_{air}.$$

The isotopes in the numerators are primordial and in the denominator are (partly) radiogenic: one is tempted to write the symbol "<" in place of "?" until one remembers that the production of $^{40}Ar_{rad}$ is higher by many orders of magnitude than the production of $^{22}Ne_{rad}$.

The problem of the origin of excess ^{20}Ne is one of the most mysterious in the isotope geochemistry of light rare gases.

2.8 Some general regularities in distribution of the light rare-gases isotopes in terrestrial gases

The earth's natural gases are complex mixtures of the volatile components released from different sources whose chemical and isotopic compositions have been averaged by the migration processes. This point of view indicates that investigations of terrestrial gases will allow the recognition of general regularities in the isotope geochemistry of rare gases as well as the major features of earth's sources of rare gases isotopes. Many authors have worked in this field during the last decade (MAMYRIN *et al.*, 1969a, b; KAMENSKII *et al.*, 1971, 1974, 1976; STROUD *et al.*, 1967; BENNET and MANUEL, 1970; MATSUO, 1970; TOLSTIKHIN *et al.*, 1969a, 1972a, 1977b, c; VORONOV *et al.*, 1974a, b; MAZOR, 1972; MAZOR and FOURNIER, 1973; GAVRILOV *et al.*, 1972; GAVRILOV and TEPLINSKII, 1973; KONONOV *et al.*, 1974; POLAK *et al.*, 1976; LOMONOSOV *et al.*, 1976; VERKHOVSKII *et al.*, 1977; and others). This section discusses only some of the data obtained. The constancy of $^{3}He/^{4}He$ ratios in terrestrial gases within the same regional tectonic structure appears to be the most important trend in helium isotopes.

The data available (about one thousand helium isotope analyses) confirm this conclusion which was first proposed by KAMENSKII *et al.*, in 1971. A representative collection of the available data is shown in Fig. 7. Gases from ancient plates are characterized by ratios of $^{3}He/^{4}He$ ($\sim 10^{-8}$) which are as low as in radiogenic helium from old granites (see also Fig. 3). Intermediate ratios (10^{-7} to 10^{-6}) are typical of gases

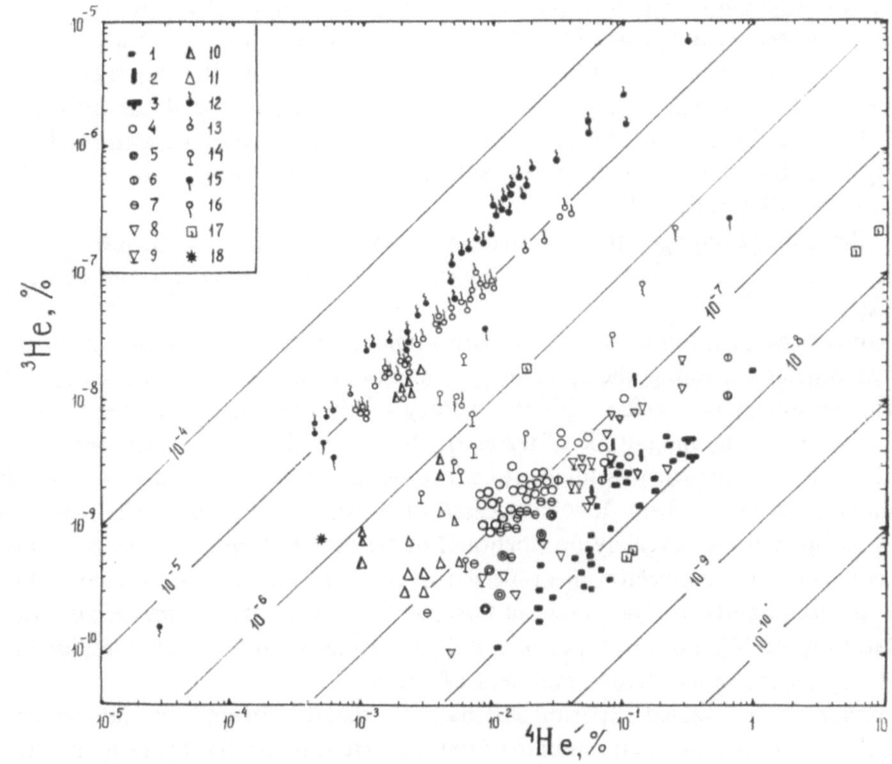

Fig. 7. The concentration of helium isotopes in terrestrial gases. Bed waters and gases of: Precambrian plates, 1-East Europian; 2-Dneprovo-Donezkaya basin; 3-East Siberian, Epihercynian platforms; 4-Turanian; 5-West Siberian; 6-West Kazakhstanian; 7-Skyphian. Springs, bed waters and gases of: intermountaineous areas, 8-Ferganian; 9-Tadjikian, geosyncline regions; 10-Sakhalin; 11-Turkmenian. Hot-springs of rift and volcanic zones: 12-Iceland; 13-East Kamchatka; 14-West Kamchatka (springs and well-gases); 15-the Caucasus; 16-Baikal region; 17-gases from mines of East Europian plate; 18-atmospheric helium. Sources of data: MAMYRIN *et al.*, 1969a, b, 1970, 1972a; KONONOV *et al.*, 1974; POLAK *et al.*, 1976; KAMENSKII *et al.*, 1971, 1974, 1976; TOLSTIKHIN *et al.*, 1969a, 1972a; DEVIRTZ *et al.*, 1971; and others.

from regions with younger (Epihercynian) basement. The nature of these intermediate ratios is not yet clear. Either it is "the effect of memory" of tectonic and igneous activity in the past or some flux of mantle helium exists up to the present time. High ratios (10^{-6} to 10^{-5} and higher) are normal for regions of modern igneous and tectonic activity, i.e., volcanic (the Kurils, Kamchatka, Japan) and rift (both oceanic and intra-continental—Iceland, Baikal) zones, Alpine orogens (the Caucasus), etc. A difference between $^3He/^4He$ ratios in thermal fluids of the Pacific circle and the oceanic rift zones is well-documented ($\sim$200 isotopic analyses were carried out for these regions).

The main source of thermal fluids for the Pacific circle may be oceanic basalts remelting in the Benioff zone (see section 2.3) while helium in the intraoceanic volcanic zones (both Atlantic and Pacific ones) probably represents "purer" samples from the earth's mantle. But the possibility of various contaminations of hot-springs by crustal, radiogenic helium cannot be eliminated at this time. It is noteworthy that within the limits of a single region marked differences in $^3He/^4He$ values are sometimes observed. As a rule, these differences are due to tectonic structure. Thus, the $^3He/^4He$ ratios in gases of the Dnieper-Don depression (rejuvenated in the Paleozoic) are about one order of magnitude higher than those of the surrounding Precambrian Russian platform (Fig. 7). Structurally complex regions (such as the Sakhaline) contain gases characterized by a wide range of $^3He/^4He$ ratios.

Thus, the isotope composition of helium in terrestrial gases is obviously related to tectonic structure. Ten years ago such a conclusion would have been viewed as pure fantasy.

The isotope composition of helium in natural gases therefore becomes a geological-historical parameter reflecting the influence of mantle volatiles in the past. Another independent parameter connected with the history of the geological development of a region is known to be the magnitude of terrestrial heat flow (POLAK and SMIRNOV, 1968). A comparison of both parameters might therefore be informative. Figure 8 shows that such a "materialization" of heat flow is successful (POLAK *et al.*, 1976). The magnitude of the flow is found to be inversely proportional to ^4He generation (!). The correlation obtained confirms that convective heat flow (in volcanic and rift zones) contributes greatly or predominantly to the total outflow of heat from the earth. Only within ancient plates are both helium isotopes of pure radiogenic origin. In these regions it is possible to neglect the convective component of the heat flow.

The numerous and varied experimental data discussed here and in the preceding sections (2.1, 2.3 and 2.4) allow an estimate of the $^3He/^4He$ ratio in the earth's mantle to be made. The following assumptions seem to be reasonable for this assessment: (i) results obtained from the youngest rocks and gases of probably mantle origin should be preferred; (ii) samples reflecting regional properties of the mantle are more representative (i.e., samples of gases as well as rocks and minerals rich in trapped gases); (iii) the estimate must include the majority of the highest values of $^3He/^4He$ found in such suitable samples. In the

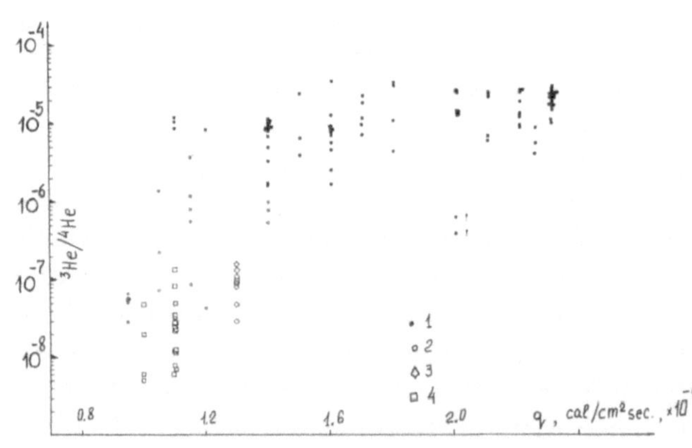

Fig. 8. The relationship between the ratio of $^3He/^4He$ in terrestrial waters and gases and regional heat flow (after POLAK *et al.*, 1976). 1-thermal fluids of volcanic and rift zones, bed waters and gases of 2-intermountaneous areas, 3-Epihercynian, and 4-Precambrian platforms.

light of the two former assumptions the thermal fluids (as well as very young rocks containing trapped gases) are the most suitable samples. This view is confirmed by a comparison of data obtained for rocks and hot-springs of Iceland and for an amphibole from New Zealand. The third assumption seems to correspond to the range of $^3He/^4He=(2\text{ to }4)\times 10^{-5}$ which was first proposed by KONONOV *et al.*, in 1974. The following studies support this value.

In the calculations by KAMENSKII *et al.* (1976) the $^3He/^4He$ values in mantle (3×10^{-5}), crustal (or radiogenic, 3×10^{-8}), and atmospheric (1.4×10^{-6}) types of helium were used together with some other parameters (^4He, ^{20}Ne, ^{36}Ar contents and $^4He/^{20}Ne$, $^4He/^{36}Ar$ ratios). These authors found that the contribution of each type of helium is equal to 9–38, 54–80, 0.4–26% (respectively) in thermal fluids of the east volcanic zone of Kamchatka; 0–3, 0–79, 19–100% in cold springs of the active water-exchange zone and 2–12, 79–98, 1–10% in well gases of the west regions of Kamchatka.

The progress in rare gases isotope geochemistry has also shed light on the location of the earth's degassing. Juvenile volatiles are degassed through deep faults located in volacnic, rift and orogenic zones on continents and islands as well as through the submarine crests of oceanic rifts.

The comparison of isotope analyses of helium and sulphur in Icelandic thermal fluids (KONONOV *et al.*, 1974; VINOGRADOV *et al.*, 1974) shows that both sulphur (in all its forms, $\delta S^{34}\sim 0$ as compared with troilitic sulphur) and helium [$^3He/^4He=(2\text{ to }4)\times 10^{-5}$] are mainly juvenile in many of the thermal features on the island. These authors confirmed that juvenile sulphur and helium are contaminated to different degrees. (i) The near-shore hot-springs are characterized by values of δS^{34} up to $+2.0‰$, which are similar to those in seawater and differ but strongly from juvenile sulfur. (ii) On the contrary, the $^3He/^4He$ ratios in all thermal fluids of Iceland are very high and approximately constant. The convincing illustrations of the ways of the earth's degassing may be found in the excellent paper by JENKINS and CLARKE (1976)***. The ^3He flows out mainly from the tops (!) of submarine ridges (see Fig. 4 of the paper cited).

Now we shall dwell upon some other regularities in light rare gases geochemistry established or developed during the last decade.

It was found that the ratios of $(^4He/^{40}Ar)_{rad}$ in various rocks (poor in trapped gases) are about two orders of magnitude lower than those which should have been formed by the K/U ratios of the rocks (TOLSTIKHIN and DRUBETSKOI, 1975c). This observation does not seem surprising. It reflects the well-known fact that radiogenic argon is more resistant to loss than is radiogenic helium. If such rocks (poor in helium) were strongly metamorphosed they would be a source of gases with low $(^4He/^{40}Ar)_{rad}$ ratios (in some cases as low as 0.01). Such low ratios can, therefore have been produced from common rocks with average Clark's contents of radioactive elements. Then one may explain the relation (e.g., VORONOV *et al.*, 1974b) between the value of $(^4He/^{40}Ar)_{rad}$ in gases and tectonic structure. The ancient platforms contain gases with higher ratios of $(^4He/^{40}Ar)_{rad}\sim 19$. The average value of these ratios in gases of young platforms is equal to 10. In the Alpine fold systems this ratio is ~ 4 and even lower ratios ~ 1 are found in volanic and rift zones. A proportion between ratios of $^3He/^4He$ and $(^4He/^{40}Ar)_{rad}$ in gases is also

*** There is also a very promising Fig. 6 in this paper which shows a strong correlation between ^3He and SiO_2 (!) in South Atlantic waters.

quite probable. If regions of higher heat flow values (and intense metamorphic processes) produced the main part of the flow of volatiles from the earth into the atmosphere, then the value of $(^4He/^{40}Ar)_{rad} \lesssim 10$ may be accepted as an upper limit for the total earth flux of volatiles. The theoretical ratio of $(^4He/^{40}Ar)_{rad}$ for average ("Clark's") rocks is about 3 to 7 and remains approximately constant in time (ZARTMAN *et al.*, 1961). These data will be used in the next section.

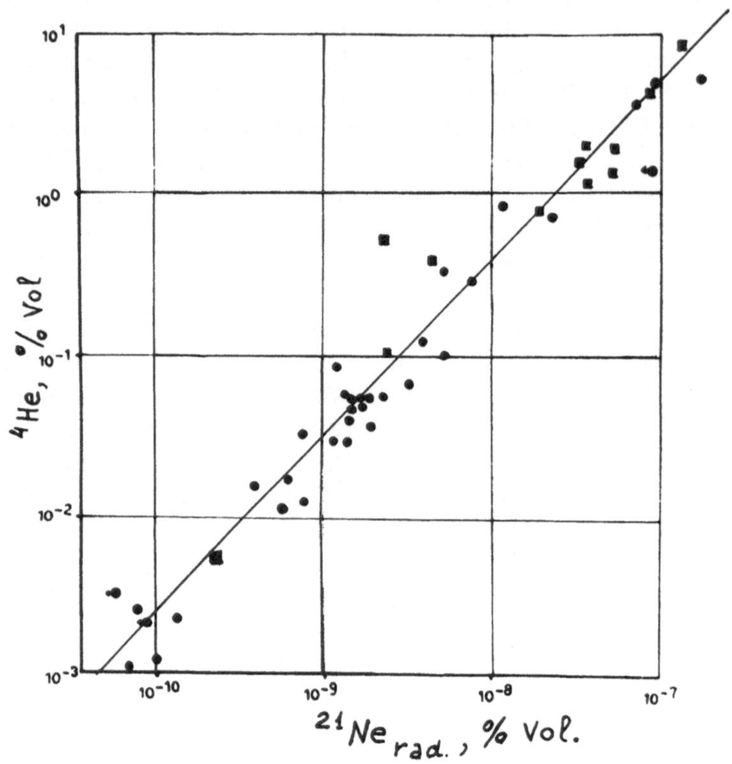

Fig. 9. The correlation between radiogenic ^{21}Ne and ^4He. Sources of data: circles-VERKHOVSKII *et al.*, 1977; squares-STROUD *et al.* (1967), and BENNET and MANUEL (1970).

A close correlation between the contents of $^4He_{rad}$ and $^{21}Ne_{rad}$ in terrestrial gases was obtained (Fig. 9) on the basis of numerous analyses carried out by STROUD *et al.* (1967), BENNET and MANUEL (1970), and VERKHOVSKY *et al.* (1977). The ratios of $(^4He/^{21}Ne)_{rad}$ are approximately constant and equal to $(2.7 \pm 1.5) \times 10^7$ $(\pm 2\sigma)$. The correlation is provided by a genetic affinity of both isotopes as well as the identical location of these isotopes in the crystalline structure of minerals. ^4He and ^{21}Ne are produced and released in similar proportion and their ratio does not depend on P-T conditions of degassing rocks (unlike $(^4He/^{20}Ar)_{rad}$ ratio). The constancy of the $(^4He/^{21}Ne)_{rad}$ ratio may be used successfully for an assessment of the abundance of one isotope, in case the abundance of another is known (TOLSTIKHIN *et al.*, 1977b).

2.9 The helium isotopes in "undissipated" atmosphere

The phenomenon of the helium loss from the atmosphere into space was established more than 20 years ago (NICOLET, 1957) but the mechanism of the loss is still being discussed (AXFORD, 1968; SHELDON and KERN, 1972; KOCKARTS, 1973; and others). The discovery (MAMYRIN *et al.*, 1969a, b) of the high $^3He/^4He$ ratios ($\sim 10^{-5}$ i.e., about one order of magnitude higher than the atmospheric ratio) in terrestrial helium requires a revision of some important initial parameters of the helium loss models. According to our estimate the solid earth is the main source of the atmospheric ^3He (TOLSTIKHIN *et al.*, 1975b, 1976b). The "restoration" of the helium isotopic abundance in an "undissipated" atmosphere seems to be useful for a revision of dissipation models and an assessment of uranium content in the earth. Such an estimation is independent of (i) data on cosmic abundance of radioactive elements; (ii) measurements of the U content in terrestrial materials; or (iii) the use of heat-flow models.

Some well known ratios of rare gas isotopes were used for the "restoration" (TOLSTIKHIN *et al.*, 1977c): $(^4He/^{20}Ne)_{prim} \sim 500$; $(^3He/^4He)_{prim} \sim 3 \times 10^{-4}$ (see section 2.5). These data along with the content of ^{20}Ne in the atmosphere (COOK, 1961) allow the computation of the abundance of primordial helium isotopes in an "undissipated" atmosphere:

$$^4He_{A,prim} \sim 0.33 \times 10^{23}\ cm^3$$
$$^3He_A \sim 1 \times 10^{19}\ cm^3.$$

An estimate of the upper limit of the total ^3He_A content is possible by analogy with the isotope geochemistry of argon. Atmospheric argon includes radiogenic ^{40}Ar and primordial ^{36}Ar isotopes introduced into the atmosphere by degassing. The crust is known to contain mainly radiogenic ^{40}Ar and the $(^{40}Ar/^{36}Ar)_M$ ratio is somewhat higher than that of the atmosphere. Argon and helium isotopic ratios are produced by similar processes of degassing and radiogenic isotope accumulation. Therefore the relations between the isotopic ratios of these elements are probably similar in the mantle, crust and atmosphere. In other words, if the inequality $(^{40}Ar/^{36}Ar)_M > (^{40}Ar/^{36}Ar)_A$ is correct (see section 2.5) it is reasonable to infer that the $(^4He/^3He)_M$ is higher than the $(^4He/^3He)_A$ of "undissipated" atmosphere. The ratio of $^3He/^4He$ $\sim 3 \times 10^{-5}$ in the mantle was adopted in the section 2.8; then

$$(^4He/^3He)_A < 3.3 \times 10^4 \quad \text{and} \quad ^4He_A < 3.3 \times 10^{23}\ cm^3;$$
$$^4He_{A,rad} < 3 \times 10^{23}\ cm^3.$$

WASSERBURG *et al.* (1963) used the $(^4He/^{40}Ar)_{rad}$ ratio and estimated the value of U/K in the earth as well as the $^4He_{rad}$ flux into the atmosphere. Assuming the average value of $(^4He/^{40}Ar)_{rad} < 10$ (see section 2.8) one may calculate an upper limit of the $^4He_{rad}$ content in "undissipated" atmosphere equal to 3.7×10^{23} cm^3.

The above mentioned limits are very similar and we shall adopt the following values of $^3He_A \sim 1 \times 10^{19}$ cm^3, $^4He_A < 3 \times 10^{23}$ cm^3 for further discussion. These values are correct only if almost all the isotopes of He, Ne and Ar have been released from the solid earth and have accumulated in "undissipated" atmosphere. Fortunately, both "continuous" and "catastrophic" models imply a high degree of earth degassing (the earth has lost more than 3/4 of its volatiles).

The similarity of the estimates of ^4He_A and ^3He_A obtained by two independent methods is encouraging and permit the following observations:

(i) The first estimate is based among other things, on the content of primordial ^{20}Ne released from solid earth mainly at the initial stage of its evolution. On the contrary, the second estimate uses the content of radiogenic ^{40}Ar emanated into the atmosphere continuously as it is produced by the radioactive decay of ^{40}K (FANALE, 1971). If the earth's atmosphere had lost gases in a hypothetical early event then the estimates would have been different (the loss of ^{20}Ne would predominate in this case). Therefore the similarity may be considered as an argument for a good preservation of the atmospheric volatiles.

(ii) The similarity of both estimates as well as the "helium" degassing models apparently argues for a primordial origin of the ^3He excess in the earth's interior.

(iii) The value of ^3He_A and the upper limit on ^4He_A allow a rough guess to be made about the concentrations of these isotopes in the "undissipated" atmosphere—$^4He_A < 10\%$ (probably 5%) and $^3He_A \sim 3 \times 10^{-4}\%$ vol. It is interesting to compare these results with the contents measured in thermal fluids from the Kurils-Kamchatka volcanic zone—$^4He \sim 10^{-2}\%$, $^3He \sim 10^{-7}\%$ and $^{40}Ar_{atm} > 0.1\%$ vol. These data allow an evaluation of the contamination of thermal fluids by atmospheric helium. In the case of an "undissipated" atmosphere the $^3He/^4He$ ratios in hydrothermal fluids would have been only a few percent higher than the "atmospheric" value. Such a situation is actually true for the argon isotope distribution in hot-springs (TOSTIKHIN *et al.*, 1972a; KAMENSKII *et al.*, 1976).

(iv) Using the ^{40}Ar content in the atmosphere GAST (1968) and HURLEY (1968) established independently a lower limit on the potassium content ($\sim$85 ppm) for the earth (see also LARIMER, 1971). Our estimates of $^4He_{A,rad} \sim 3 \times 10^{23}$ allows us to calculate in the same way the uranium content of the earth—which is lower than 3×10^{-8} g per g. This value is in good agreement with numerous literature data (WASSERBURG *et al.*, 1964; WAKITA *et al.*, 1967; FISHER, 1972; KOMAROV and JITKOV, 1973; and others).

3. Conclusions

Instead of a conclusion I will dwell on some important and promising items which go beyond the scope of this review.

The progress in the isotope geochemistry of light rare gases was permitted by advances in mass-spectrometry. The pioneer work of Prof. B. A. Mamyrin and his co-workers (MAMYRIN and SHUSTROV, 1957, 1962; MAMYRIN *et al.*, 1969b, 1972b) resulted in a high-resolution time-travel analytical mass-spectrometer which is in particularly good for measurements of $^3He/^4He$ ratios as low as $\sim 10^{-10}$. Prof. W. B. Clarke and colleagues worked out a static instrument of high resolution, sensitivity and accuracy (CLARKE *et al.*, 1969; KUGLER and CLARKE, 1971, 1972).

The review did not cover the very interesting studies of the helium isotope distribution in oceanic waters as well as in waters of midland basins (CLARKE *et al.*, 1969, 1970, 1976; CLARKE and KUGLER, 1973; CRAIG and CLARKE, 1970; CRAIG *et al.*, 1975; BIERI, 1971; BIERI *et al.*, 1967; BARNES and BIERI, 1976; and others).

The T-^3He method of dating young water seems to be very promising for scientific and practical purposes (TOLSTIKHIN and KAMENSKII, 1969b; DEVIRTZ *et al.*, 1971;

JENKINS *et al.*, 1972; JENKINS and CLARKE, 1976; and other works by W. B. Clarke and colleagues).

Finally, a very attractive scientific problem has not been discussed in this review: "evolution of the atmosphere: continuous or catastrophic?" (OZIMA, 1973)—in future this question undoubtedly will be repeatedly focused upon.

This review became possible only due to the efforts of many scientists in different countries. The author hopes that these efforts are reflected, at least partly, in the text and references. I would express my sincere gratitude to Prof. B. A. Mamyrin for his hopeful support of researches in a field of helium isotope geochemistry during many years. The courteous invitation by Prof. M. Ozima and Prof. C. Alexander to submit a review in this volume was of course very pleasant and an honor for the author.

REFERENCES

ALDRICH, L. T. and A. O. NIER, The occurrence of ^{3}He in natural sources of helium, *Phys. Rev.*, **74**, 1590–1593, 1948.

ALEXANDER, E. C. and D. W. SCHWARTZMAN, Argon isotopic evolution of upper mantle, *Nature*, **259**, 104–108, 1976.

ALVAREZ, L. W. and R. CORNOG, ^{3}He in helium, *Phys. Rev.*, **56**, 379, 1939.

ANDERS, E., D. HEYMANN and E. MAZOR, Isotopic composition of primordial helium in carbonaceous chondrites, *Geochim. Cosm. Acta*, **34**, 127–132, 1970.

ANUFRIEV, G. S., I. L. KAMENSKII, and V. P. PAVLOV, Anomalous isotopic composition of neon in hot-springs of modern volcanic zones, *Dokl. Akad. Nauk USSR*, **231**, 1454–1457, 1976 (in Russian).

ASHKINADZE, G. Sh., V. B. SHARIF-ZADE, and Yu. A. SHUKOLYUKOV, Migration of stable isotopes of neon in minerals, *Geokhimiya*, No. 11, 1704–1710, 1973 (in Russian).

AXFORD, W. I., The polar wind and the terrestrial helium budget, *J. Geophys. Res.*, **73**, 6855–6859, 1968.

BARNES, R. O. and R. H. BIERI, Helium flux through marine sediments of the northeast Pacific Ocean, *Earth Planet. Sci. Lett.*, **28**, 331–337, 1976.

BENNET, G. A. and O. K. MANUEL, Xenon in natural gases, *Geochim. Cosm. Acta*, **34**, 593–610, 1970.

BIERI, R. H., M. KOIDE, and E. D. GOLDBERG, Geophysical implications of the excess helium found in Pacific waters, *J. Geophys. Res.*, **72**, 2497–2511, 1967.

BIERI, R. H., Dissolved noble gases in marine waters, *Earth Planet. Sci. Lett.*, **10**, 329–333, 1971.

BROWN, J. F., C. T. HARPER, and A. L. ODOM, Petrogenic implications of argon isotopic evolution in the upper mantle, *Nature*, **250**, 130–133, 1974.

BROWN, J. F., A. L. ODOM, and C. T. HARPER, Argon isotopic evolution of upper mantle, *Nature*, **259**, 106–108, 1976

CHERDYNSTEV, V. V., Yu. V. SHITOV, and I. V. LIZARSKAYA, Isotope composition of argon from volcanic gases of USSR, *Dokl. Akad. Nauk USSR*, **172**, 1180–1181, 1967a (in Russian).

CHERDYNTSEV, V. V. and Y. V. SHITOV, Excess ^{36}Ar in volcanic and hydrothermal gases of the USSR, *Geokhimiya*, No. 5, 618–621, 1967b (in Russian).

CLARKE, W. B., M. A. BEG, and H. CRAIG, Excess ^{3}He in the sea: Evidence for terrestrial primordial helium, *Earth Planet. Sci. Lett.*, **6**, 213–220, 1969.

CLARKE, W. B., M. A. BEG, and H. CRAIG, Excess ^{3}He at the North Pacific geosecs station, *J. Geophys. Res.*, **75**, 7676–7678, 1970.

CLARKE, W. B. and G. KUGLER, Dissolved helium groundwater: A possible method for uranium and thorium prospecting, *Econ. Geol.*, **68**, 243–251, 1973.

COOK, G. A. (ed.), *Argon, Helium and the Rare Gas,* Vol. 1, p. 42, Interscience Pub., New York, London, 1961.

CRAIG, H and W. B. CLARKE, Oceanic ^{3}He: Contribution from cosmogenic tritium, *Earth Planet. Sci. Lett.*, **9**, 45–48, 1970.

CRAIG, H., W. B. CLARKE, and M. A. BEG, Excess ^{3}He in deep water on the east Pacific rise, *Earth Planet. Sci. Lett.*, **26**, 125–132, 1975.

CRAIG, H. and J. E. LUPTON, Primordial neon, helium and hydrogen in oceanic basalts, *Earth Planet. Sci. Lett.*, **31**, 369–385, 1976.

DEVIRTZ, A. L., I. L. KAMENSKII, and I. N. TOLSTIKHIN, Helium isotopes and tritium in volcanic hot-springs, *Dokl. Akad. Nauk USSR*, **197**, 450–452, 1971 (in Russian).

DOWNING, R. G., E. M. HENNECKE, and O. K. MANUEL, Josephinite: A terrestrial alloy with radiogenic ^{129}Xe and nobe gas imprint of iron meteorites, *Geochem. J.*, **11**, 4, 1977.

FANALE, F. P., A case for catastrophic early degassing of the Earth, *Chem. Geol.*, **8**, 79–105, 1971.

FAURE, G. and J. L. POWELL, *Strontium Isotope Geology*, edited by W. Engelhardt *et al.*, 188 pp., Springer Verlag, Berlin, Haidelberg, New York, 1972.

FISHER, D. E., Search for ^{3}He in deep-sea basalts, *Earth Planet. Sci. Lett.*, **8**, 77–78, 1970.

FISHER, D. E., Uranium content and radiogenic ages of hypersthene, bronzite, amphoterite and carbonaceous chondrites, *Geochim. Cosm. Acta*, **36**, 15–33, 1972.

FISHER, D. E., Trapped helium and argon and the formation of the atmosphere by degassing, *Nature*, **256**, 113–114, 1975.

FUNKHOUSER, J. G. and J. J. NAUGHTON, Radiogenic He and Ar in ultramafic inclusions from Hawaii, *J. Geophys. Res.*, **73**, 4601–4607, 1968.

GAST, P. W., Upper mantle chemistry and the evolution of the earth's crust, in *The History of the Earth's Crust*, edited by R. A. Phiney, pp. 15–27, Princeton Univ. Press, 1968.

GAVRILOV, E. Ya., Yu. A. JUROV, and G. I. TEPLINSKII, On the connection between isotope composition of argon and carbon in natural gases, *Dokl. Akad. Nauk USSR*, **206**, 448–452, 1972 (in Russian).

GAVRILOV, E. Ya. and G. I. TEPLINSKII, The distribution of argon isotopes in carbon-hydrogen gases, *Geokhimiya*, No. 4, 558–569, 1973 (in Russian).

GERLING, E. K., Yu. A. SHUKOLYUKOV, T. V. KOLCOVA, I. I. MATVEEVA, and S. Z. YAKOVLEVA, Age determination of basic rocks according to the K/Ar-method, *Geokhimiya*, No. 11, 931–938, 1962 (in Russian).

GERLING, E. K., I. N. TOLSTIKHIN, Yu. A. SHUKOLYUKOV, Z. N. NESMELOVA, and I. Ya. AZBEL, Argon isotopes and helium in natural carbohydrogen gases, *Geokhimiya*, No. 5, 608–617, 1967 (in Russian).

GERLING, E. K., I. M. MOROZOVA, and V. D. SPRINZSON, On the nature of excess argon in some minerals, in *The Problems of Geokhimii and Kosmologii, Trudy XXIII Mezdunarodn. Geol. Congr.*, pp. 76–81, Nauka, Moskva, 1968.

GERLING, E. K., B. A. MAMYRIN, I. N. TOLSTIKHIN, and S. S. YAKOVLEVA, Helium isotope composition in some rocks, *Geochem. Int.*, **8**, 755–761, 1971.

GERLING, E. K., I. N. TOLSTIKHIN, B. A. MAMYRIN, G. S. ANUFRIEV, I. L. KAMENSKII, and E. M. PRASOLOV, New investigations in helium isotope geochemistry, in *Proc. of the First Int. Geochem. Congr.*, edited by A. I. Tugarinov, Vol. 1, pp. 200–216, Moskva, Nauka, 1972 (in Russian).

GERLING, E. K., I. N. TOLSTIKHIN, E. R. DRUBETSKOI, R. Z. LEVKOBSKII, E. V. SHARKOV, and I. K. KOZAKOV, He and Ar isotopes in rock-forming minerals, *Geokhimiya*, No. 11, 1603–1610, 1976 (in Russian).

GORSHKOV, G. V., V. A. ZYABKIN, N. M. LYATKOVSKAYA, and O. S. ZVETKOV, *Natural Neutron Background of the Atmosphere and the Earth's Crust*, pp. 410, Atomizdat, Moscow, 1966 (in Russian).

HARPER, C. T. and S. SCHAMEL, Note on the isotopic composition of argon in quartz veins, *Earth Planet. Sci. Lett.*, **12**, 129–134, 1971.

HORIBE, Y. and H. CRAIG, Helium-3 in Hakone hot springs, in *Abstracts of U.S.-Japan Seminar on Rare-Gas Abundance and Isotopic Constraints on the Origin and Evolution of the Atmosphere*, p. 25, Hakone, Japan, 1977.

HURLEY, P. M., Absolute abundances and distribution of Rb, K and Sr in the earth, *Geochim. Cosmochim. Acta*, **32**, 273–284, 1968.

JENKINS, W. I., M. A. BEG, W. R. CLARKE, P. I. WENGERSKY, and H. CRAIG, Excess ^{3}He in the Atlantic Ocean, *Earth Planet. Sci. Lett.*, **16**, 122–126, 1972.

JENKINS, W. J. and W. B. CLARKE, The distribution of ^{3}He in the western Atlantic Ocean, *Deep Sea Res.*, **23**, 481–494, 1976.

KAMENSKII, I .L., V. P. YAKUTSENY, B. A. MAMYRIN, G. S. ANUFRIEV, and I. N. TOLSTIKHIN, Helium isotopes in nature, *Geokhimiya*, **8**, 914–931, 1971.

KAMENSKII, I. L., E. M. PRASOLOV, and V. V. TIKHOMIROV, On the juvenile components in natural gases of the Sakhalin, *Geokhimia*, **8**, 1220–1225, 1974.

KAMENSKII, I. L., V. A. LOBKOV, E. M. PRASOLOV, N. S. BESKROVNY, E. I. KUDRYAVTSEVA, G. S. ANUFRIEV, and V. B. PAVLOV, The components of the upper mantle of the Earth in gases of Kamchatka, *Geokhimiya*, No. 5, 682–695, 1976 (in Russian).

KANEOKA, I., Investigation of excess argon in ultramafic rocks from the Kola Peninsula by the $^{40}Ar/^{39}Ar$ method, *Earth Planet. Sci. Lett.*, **22**, 145–156, 1974.

KHLOPIN, V. G. and E. K. GERLING, New results in geochemistry of rare gases, *Dokl. Akad. Nauk USSR*, **61**, 297–299, 1958 (in Russian).

KOCKARTS, G., Helium in the terrestrial atmosphere, *Space Sci. Rev.*, **14**, 723–757, 1973.

KOMAROV, A. N. and A. S. JITKOV, Uranium in ultrabasic xenoliths from basalts, *Izv. Akad. Nauk USSR, Ser. Geol.*, No. 10, 79–85, 1973 (in Russian).

KONONOV, V. I., B. A. MAMYRIN, B. G. POLAK, and L. V. KHABARIN, The helium isotopes in Icelandic hydrothermal fluids, *Dokl. Akad. Nauk USSR*, **217**, 172–175, 1974 (in Russian).

KRUMMENACHER, D., Isotope composition of argon in modern surface volcanic rocks, *Earth Planet. Sci. Lett.*, **8**, 109–117, 1970.

KRYLOV, A. Ya., B. A. MAMYRIN, L. V. KHABARIN, T. I. MAZINA, and Y. A. SILIN, Helium isotopes in the basic rocks of the ocean floor, *Geokhimiya*, No. 8, 1221–1226, 1974 (in Russian).

KUGLER, G. and W. B. CLARKE, Mass-spectrometric search for Ne and Ar isotopes in ternary fission of ^{235}U, *Phys. Rev., C.*, **3**, 849–853, 1971.

KUGLER, G. and W. B. CLARKE, Mass-spectrometric measurements of ^{3}H, ^{3}He, and ^{4}He produced in thermal neutron ternary fission of ^{235}U: Evidence for short-range ^{4}He, *Phys. Rev. C.*, **5**, 551–560, 1972.

LARIMER, J. W., Composition of the Earth: Chondritic or achondritic, *Geochim. Cosm. Acta*, **35**, 769–771, 1971.

LEUTWEIN, F. G. and G. KAPLAN, Quelques recherches sur l'aptitudo de certains cristaux de ne'oformetion à capture argon radiogenic, *C. R. Acad. Sci., Paris*, **257**, 1315–1317, 1963.

LIPPOLT, H. Y. and W. GENTNER, K-Ar dating of some limestones and fluorites, in *Radioactive Dating*, pp. 239–242, 1963.

LIPPOLT, H. Y., Die zusammensetrung des übershub-argon in Schwarz-walder flubspaten, *Zs. Naturforsch. Bd.*, **21a**, 1162–1167, 1966.

LOMONOSOV, I. S., B. A. MAMYRIN, E. M. PRASOLOV, and I. N. TOLSTIKHIN, Isotope composition of helium and argon in some hot-springs of the Baikal rift, *Geokhimiya*, No. 11, 1743–1746, 1976 (in Russian).

LUPTON, I. E. and H. CRAIG, Excess ^{3}He in oceanic basalts: Evidence for terrestrial primordial helium, *Earth Planet. Sci. Lett.*, **26**, 133–139, 1975.

MAMCHUR, G. P., D. D. MATVIENKO, and O. A. YARYNYG, The genesis of chambered pegmatite in the light of δ ^{13}C data, *Geol. J.*, **35**, 91–97, 1975 (in Russian).

MAMYRIN, B. A. and B. N. SHUSTROV, Mass-spectrometer with resolution about some thousands, *J. Tekh. Phys.*, **27**, 1347–1350, 1957 (in Russian).

MAMYRIN, B. A. and B. N. SHUSTROV, High-resolution mass-spectrometer with two-cascade time-travel ions separation, *Prib. Tekh. Eksp.*, No. 5, 135–139, 1962 (in Russian).

MAMYRIN, B. A., I. N. TOLSTIKHIN, G. S. ANUFRIEV, and I. L. KAMENSKII, Anomalous isotopic composition of helium in volcanic gases, *Dokl. Akad. Nauk USSR*, **184**, 1197–1199, 1969a (in Russian).

MAMYRIN, B. A., I. N. TOLSTIKHIN, G. S. ANUFRIEV, and I. L. KAMENSKII, Isotopic analysis of terrestrial helium on a magnetic resonance mass spectrometer, *Geochem. Int.*, **6**, 517–524, 1969b.

MAMYRIN, B. A., G. S. ANUFRIEV, I. L. KAMENSKII, and I. N. TOLSTIKHIN, Determination of the isotopic composition of atmospheric helium, *Geochem. Int.*, **7**, 498–505, 1970.

MAMYRIN, B. A., I. N. TOLSTIKHIN, G. S. ANUFRIEV, and I. L. KAMENSKII, Isotope composition of helium of Iceland hot-springs, *Geokhimiya*, No. 11, 1396, 1972a (in Russian).

MAMYRIN, B. A., B. H. SHUSTROV, G. S. ANUFRIEV, B. S. BOLTENKOV, V. A. ZAGULIN, I. L. KAMENSKII, I. N. TOLSTIKHIN, and L. V. KHABARIN, The measurements of the helium isotope, composition on a

magnetic resonance mass-spectrometer, *J. Tech. Phys.*, **XLII**, 2577–2583, 1972b (in Russian).
MAMYRIN, B. A., V. I. GERASIMOVSKY, and L. V. KHABARIN, Helium isotopes in rocks of East Africa and Iceland rift zones, *Geokhimiya*, No. 5, 693–700, 1974 (in Russian).
MANUEL, O. K., A comparison of terrestrial and meteoritic noble gases, in *Abstracts of U.S.—Japan Seminar on Rare Gas Abundance and Isotopic Constraints on the Origin and Evolution of the Earth's Atmosphere*, pp. 28–29, Hakone, Japan, 1977.
MATSUO, S., Role of volatile components in Earth's evolution, *Mass Spectrosc.*, **18**, 1970.
MAZOR, E., D. HEYMANN, and E. ANDERS, Noble gases in carbonaceous chondrites, *Geochim. Cosm. Acta*, **34**, 781–824, 1970.
MAZOR, E., Paleotemperatures and other hydrological parameters deduced from noble gases dissolved in groundwaters, Jordan Rift Valley, Israel, *Geoch. Cosm. Acta*, **36**, 1321–1336, 1972.
MAZOR, E. and R. O. FOURNIER, More on noble gases in Yellowstone National Park hot waters, *Geochim. Cosm. Acta*, **37**, 515–525, 1973.
MORRISON, P. and J. PINE, Radiogenic origin of the helium isotopes in rock, *Ann. N. Y. Acad. Sci.*, **62**, art. 3, 69–92, 1955.
NAIDENOV, B. M., V. G. BOGOLEPOV, E. Ya. POLYVYANNYI, and A. I. ZAXARCHENKO, Argon isotopes in mineralogenetic fluids of pegmatite, *Geokhimiya*, No. 6, 734–737, 1972 (in Russian).
NAUGHTON, J. J., J. B. FUNKHOUSER, and I. L. BARNES, Fluid inclusions in K-Ar age anomales and related inert gasesstudies, *Trans. Am. Geophys. Union*, **47**, 197, 1966.
NICOLET, M., The aeronomic problem of helium, *Ann. Geophys.*, **13**, 1–21, 1957.
OZIMA, M. and K. KUDO, Excess Ar in submarine basalts and the Earth—Atmosphere evolution model, *Nature, Phys. Sci.*, **239**, 23–24, 1972.
OZIMA, M., Evolution of the atmosphere: Continuous or catastrophic?, *Rock Magn. Paleogeophys.*, **1**, 111–113, 1973.
OZIMA, M., Ar isotopes and Earth—Atmosphere evolution models, *Geochim. Cosm. Acta*, **39**, 1127–1133, 1975.
POLAK, B. G. and Ya. B. SMIRNOV, The connection between thermal flux and tectonic pattern of continents, *Geotectonika*, No. 4, 1968 (in Russian).
POLAK, B. G., V.I. KONONOV, I. N. TOLSTIKHIN, B. A. MAMYRIN, and L. V. KHABARIN, The helium isotopes in thermal fluids, in *Thermal and Chemical Problems of Thermal Waters*, edited by A. I. Johnson, pp. 15–29, Intern. Association of Hydrological Sci., Publication 119, 1976.
PRASOLOV, E. M. and I. N. TOLSTIKHIN, Isotopic composition of helium and argon from microinclusion in amphibole (rithchorrite), *Geokhimiya*, No. 2, 231–234, 1969 (in Russian).
PRASOLOV, E. M. and I. N. TOLSTIKHIN, Isotope composition of helium and argon from micro-inclusions in quartz, *Dokl. Akad. Nauk USSR*, **191**, 653–655, 1970 (in Russian).
PRASOLOV, E. M. and I. N. TOLSTIKHIN, On the genesis of ^{3}He in micro-inclusions in honeycomb quartz, *Geokhimiya*, No. 6, 727–730, 1972 (in Russian).
PRASOLOV, E. M., Excess Ar in gas-liquid inclusions in minerals and rocks, in *The Development and Application of Nuclear Geochronology Methods*, edited by Yu. A. Shukolykov and I. M. Morozova, pp. 153–176, Nauka, Leningrad, 1976 (in Russian).
RAMA, S. N. I., S. R. HART, and E. ROEDDER, Excess radiogenic argon in fluid inclusions, *J. Geophys. Res.*, **70**, 509–511, 1965.
SAITO, K., A. R. BASU, and E. C. ALEXANDER, Jr., Planetary rare gas in a mantle derived amphibole, *Earth Planet. Sci. Lett.*, **39**, 274–280, 1978.
SHARIF-ZADE, V. B., Yu. A. SHUKOLYUKOV, E. K. GERLING, and G. Sh. ASHKINADZE, The neon isotopes in radioactive minerals, *Geokhimiya*, No. 3, 314–319, 1972 (in Russian).
SCHWARTZMAN, D. W., Argon degassing models of the Earth, *Nature, Phys. Sci.*, **245**, 20–21, 1973a.
SCHWARTZMAN, D. W., Ar degassing and the origin of the sialic crust, *Geoch. Cosm. Acta*, **37**, 2479–2495, 1973b.
SCHELDON, W. R. and J. W. KERN, Atmospheric helium and geomagnetic field reversals, *Geophys. Res.*, **77**, 6194–6201, 1972.
SHUKOLYUKOV, Yu. A. and I. N. TOLSTIKHIN, Xe and Ar isotopes in the ancient rocks of the Earth, *Geokhimiya*, No. 10, 1179–1185, 1965.

SHUKOLYUKOV, Yu. A., Nuclear fission of uranium in nature, *Atomizdat, Moscow*, 270, 1970 (in Russian).

SHUKOLYUKOV, Yu. A. and L. K. LEVSKII, Geochemistry and cosmochemistry of rare gases isotopes, *Atomizdat, Moskow*, 335, 1972.

SHUKOLYUKOV, Yu. A., G. Sh. ASHKINADZE, V. B. VERKHOVSKII, and V. B. SHARIF-ZADE, The geochemistry of neon isotopes, in *Geochemistry of Radioactive and Radiogenic Isotopes*, edited by E. K. Gerling and Yu. A. Shukolyukov, pp. 79–90, Nauka, Leningrad, 1974.

SMELOV, S. B., V. I. VINOGRADOV, V. I. KONONOV, and B. G. POLAK, Argon isotope composition in Icelandic thermal fluids, *Dokl. Akad. Nauk USSR*, **222**, 429–432, 1975 (in Russian).

SMELOV, S. B., Argon isotope composition in Iceland's hydrothermal and in volcanic gases from i. Kynashir, in *Abstracts of VI Vses. Symposium po Stabilnym Isotopam v Geokhimii*, p. 91, Moskva, 1976 (in Russian).

SOKOLOV, V. A., Geochemistry of gases of the earth's crust and atmosphere, in *Nedra*, Moscow, 1966 (in Russian).

STROUD, L., T. O. MEYER, and D. E. EMERSON, Isotopic abundance of Ne, Ar and N in natural gases relationship to helium genesis, *Report of Investigations 6936*, US Department of the Interior, Bureau of Mines, 1967.

TAKAOKA, N. and M. OZIMA, Rare gas isotopic compositions in diamonds, in *Terrestrial Rare Gases*, edited by E. C. Alexander, Jr. and M. Ozima, pp. 65–70, Cent. Acad. Publ. Japan, Tokyo, 1978.

TOLSTIKHIN, I. N., I. L. KAMENSKII, and B. A. MAMYRIN, Isotopic criterion for investigation of terrestrial helium origin, *Geokhimiya*, No. 2, 201–204, 1969a (in Russian).

TOLSTIKHIN, I. N. and I. L. KAMENSKII, Determination of ground water ages by T-^{3}He method, *Geokhimiya*, No. 8, 1027–1029, 1969b (translated in *Geochem. Int.*, **6**, 810–811, 1969).

TOLSTIKHIN, I. N. and E. M. PRASOLOV, The methods of investigation of rare gases isotopes in microinclusion in rocks and minerals, in *The Investigations of Solutions and Melts by Analyses of Microinclusions in Minerals*, edited by N. P. Ermakov, L. N. Khetchikov, Proc. VNIISIMS, Aleksandrov, XIV, pp. 86–97, 1971 (in Russian).

TOLSTIKHIN, I. N., B. A. MAMYRIN, E. A. BASKOV, I. L. KAMENSKII, G. S. ANUFRIEV, and S. N. SURIKOV, Helium isotopes in gases of hot springs of the Kuril-Kamchatka volcanic zone, in *Ocherki Sovremennoj Geokhimii i Analiticheskoi Khimii*, edited by A. I. Tugarinov, pp. 405–414, Nauka, Moskva, 1972a (in Russian).

TOLSTIKHIN, I. N., B. A. MAMYRIN, and L. V. KHABARIN, Anomalous isotopic composition of helium in some xenoliths, *Geochem. Int.*, **9**, 407–409, 1972b.

TOLSTIKHIN, I. N., B. A. MAMYRIN, L. V. KHABARIN, and E. N. ERLIKH, Isotope composition of helium in ultrabasic xenoliths from volcanic rocks of Kamchatka, *Earth Planet. Sci. Lett.*, **22**, 75–84, 1974a.

TOLSTIKHIN, I. N., E. M. PRASOLOV, and S. S. YAKOVLEVA, The origin of the helium and argon isotopes in minerals of the chambered pegmatite, in *Zapiski Vses. Mineral. Obshestva*, CIII, i. 1, pp. 1–14, 1974b (in Russian).

TOLSTIKHIN, I. N., E. M. PRASOLOV, L .V. KHABARIN, and I. Ya. AZBEL, The estimation of diffusion coefficient in crystalline quartz, in *Geokhimiya Radioactive i Radiogenic Isotopov*, edited by E. K. Gerling, Yu. A. Shukolukov, pp. 79–90, Nauka, Leningrad, 1974c.

TOLSTIKHIN, I. N., Helium isotopes in the Earth's interior and in the atmosphere: A degassing model of the Earth, *Earth Planet. Sci. Lett.*, **26**, 88–96, 1975a.

TOLSTIKHIN, I. N., I. Ya. AZBEL, and L. V. KHABARIN, Isotopes of light noble gases in the mantle, crust and atmosphere, *Geokhimiya*, No. 5, 653–666, 1975b (translated in *Geochem. Int.*, **12**, 10–20, 1975b).

TOLSTIKHIN, I. N. and E. R. DRUBETSKOI, The isotopic ratios of the $^{3}He/^{4}He$ and $^{4}He/^{40}Ar$ in the rocks of the Earth's crust, *Geokhimiya*, No. 8, 1123–1136, 1975c (in Russian).

TOLSTIKHIN, I. N. and L. V. KHABARIN, Light noble gases isotopes in meteorites in the earth and in the atmosphere, I. He, Ne, and Ar in meteorites, in *The Development and Application of Nuclear Geochronology Methods*, edited by Y. A. Shukolukov and I. M. Morozova, pp. 78–102, Nauka, Leningrad, 1976a.

TOLSTIKHIN, I. N., L. V. KHABARIN, and I. Ya. AZBEL, Light noble gases isotopes in meteorites, in the earth and in the atmosphere, II. He, Ne and Ar isotopes in the earth and in the atmosphere (degassing model of the earth), in *The Development and Application of Nuclear Geochronology Methods*, edited by Yu. A. Shukolukov and I. M. Morozova, pp. 103–122, Nauka, Leningrad, 1976b.

TOLSTIKHIN, I. N. and E. R. DRUBETSKOI, The helium isotopes in rocks and minerals of the earth's crust, in *The Problems of Age Determination of Precambrian Rocks*, edited by Yu. A. Shukolukov, pp. 172–197, Nauka, Leningrad, 1976c (in Russian).

TOLSTIKHIN, I. N., E. K. DRUBETSKOI, E. N. ERLIKH, and B. A. MAMYRIN, On the origin of the acid volcanic rocks from Kamchatka, *Geokhimiya*, No. 7, 997–1004, 1976d (in Russian).

TOLSTIKHIN, I. N., E. R. DRUBETSKOI, F. P. MITROFANOV, and I. K. KOZAKOV, The helium isotopes in rocks of Sangilen massive (Tuva, Sayany), *Geokhimiya*, No. 4, 495–501, 1977a (in Russian).

TOLSTIKHIN, I. N., A. B. VERKHOVSKII, and YU. A. SHUKOLYUKOV, On the connection between isotopic composition of primordial neon and abundance of radiogenic He, Ne and Ar in natural gases and in the atmosphere, *Geokhimiya*, No. 5, 793–797, 1977b (in Russian).

TOLSTIKHIN, I. N., A. B. VERKHOVSKII, and E. R. DRUBETSKOI, The helium isotopes abundance in the "undissipated" atmosphere, *Geokhimiya*, No. 8, 1107–1110, 1977c.

TOLSTIKHIN, I. N., E. R. DRUBETSKAI, A. YA. SHARASKIN, About the isotope composition of argon in the earth's mantle, *Geokhimiya*, No. 4, S 14–S 20, 1978.

VERKHOVSKII, A. B. and YU. A. SHUKOLYUKOV, On the possible occurrence of primordial neon in sudberite of Monchegorsk Pluton, *Dokl. Akad. Nauk USSR*, **224**, 685–688, 1975 (in Russian).

VERKHOVSKII, A. B. YU. A. SHUKOLYUKOV, and G. SH. ASHKINADZE, Neon isotopes in minerals contained excess of helium and argon, *Geokhimiya*, No. 3, 315–322, 1976a (in Russian).

VERKHOVSKII, A. B. and YU. A. SHUKOLYUKOV, Isotope composition of neon in rocks of the earth's crust and the genesis of $^{21}Ne_{rad.}$ in natural gases, *Geokhimiya*, No. 5, 778–781, 1976b (in Russian).

VERKHOVSKII, A. B., YU. A. SHUKOLYUKOV, and G. SH. ASHKINADZE, Radiogenic helium, neon and argon in natural gases, in *The Problems of Age Determinations of Precambrian Rocks*, edited by Yu. A. Shukolukov, pp. 152–170, Nauka, Leningrad, 1977 (in Russian).

VINOGRADOV, V. I., V. I. KONONOV, and B. G. POLAK, Isotope composition of sulphur in Icelandic hydrothermae, *Dokl. Akad. Nauk USSR*, **217**, 1149–1152, 1974 (in Russian).

VORONOV, A. N. and E. M. PRASOLOV, Radiogenic argon in gas deposits of Nord-East region of the Volga-Urals oil-gas-producing province, *Geokhimiya*, No. 11, 1700–1710, 1974a (in Russian).

VORONOV, A. N., E. M. PRASOLOV, and V. V. TIKHOMIROV, Radiogenic argon in gas deposits, *Geokhimiya*, No. 12, 1842–1855, 1974b (in Russian).

WAKITA, H., H. NAGASAWA, S. UYEDA, and H. KUNO, Uranium, thorium and potassium contents of possible mantle materials, *Geochem. J.*, **1**, 183–198, 1967.

WASSERBURG, G. I., E. MAZOR, and R. E. ZARTMAN, Isotopic and chemical composition of some terrestrial natural gases, in *Earth Science and Meteorites*, pp. 219–240, North-Holland Publ. Co., Amsterdam, 1963.

WASSERBURG, G. I., G. I. F. MACDONALD, F. HOYLE, and W. A. FOWLER, Relative contributions of U, Th, and K to heat productions in the Earth, *Science*, **143**, 465–467, 1964.

WETHERILL, G. W., Variations in the isotopic abundances of Ne and Ar extracted from radioactive minerals, *Phys. Rev.*, **96**, 679–683, 1954.

ZARTMAN, R. E., G. I. WASSERBURG, and I. H. REYNOLDS, Helium, argon, and carbon in some natural gases, *J. Geophys. Res.*, **66**, 277–306, 1961.

ZVEREV, B. P., YU. F. SIMAKHIN, and A. G. DUTOV, The determination of lithium in solids by $^{6}Li\ (n, \alpha)\ ^{3}H$ reaction, *At. Energ.*, **32**, 39–42, 1972 (in Russian).

Abundances and Isotopic Compositions of Rare Gases in Granites and Thucholites

P. K. KURODA and R. D. SHERRILL†

Department of Chemistry, University of Arkansas
Fayetteville, Arkansas 72701, U. S. A.

It was felt to be highly desirable to carry out a systematic investigation on the abundances and isotopic compositions of rare gases in terrestrial rocks, especially in granites, in order to see if the Oklo phenomenon (NEUILLY *et al.*, 1972; see also BODU *et al.*, 1972 and KURODA, 1975) has left a mark on the rare gases in the rocks, which may have been once situated close to some of the 'extinct' natural reactors.

The contents and the isotopic compositions have been measured mass spectrometrically for neon, argon, krypton and xenon released from four specimens of granites in stepwise-heating experiment: (1) Carney Lake (Michigan), (2) Red Rock (Ontario), (3) Wausau (Wisconsin) and (4) Westerly (Rhode Isalnd) (KURODA *et al.*, 1977). These rocks contained appreciable amounts of ^{40}Ar from the decay of ^{40}K and $^{131-136}$Xe from the ^{238}U spontaneous fission. As shown in Table 1, the contents of neon, argon (excluding ^{40}Ar), krypton and xenon in these granites were considerably lower than the estimated rare gas contents of rocks reported by BROWN (1949), except for the Red Rock granite, which contained an abnormally large amount of xenon. The Ne/Xe and Kr/Xe ratios in the Red Rock granite were abnormal in that they resembled the relative abundances in meteorites. Moreover, a large fraction of the total xenon released from the Red Rock granite appeared to be a mixture of AVCC and atmospheric xenon. The Red Rock granite is from the Sudbury structure, Ontario, Canada, which is believed to have been formed by the impact of an asteroid in an event that formed a crater 30 miles in diameter 1,720 million years ago (DIETZ, 1964; FRENCH, 1967, 1970, 1972).

Table 1. Rare gas contents of granites.

	Ne	Ar	(^{36}Ar)	Kr	Xe
(1) Carney Lake	48,000	2,310,000,000	(32, 050)	104	21
(2) Red Rock	30,100	141,000,000	(32,700)	458	1,303
(3) Wausau	—	—	—	781	51
(4) Westerly	18,000	51,800,000	(10,160)	448	26
(5) Estimated rare-gas contents of rocks (BROWN, 1949)	77,000	22,000,000	(74,200)	1,200	340

The data are expressed in 10^{-12} cc STP/gram.

The isotopic compositions have been also measured mass spectrometrically for xenon and krypton fractions released from a sample of thucholite from Besner Mine, Parry Sound, Ontario, Canada, in stepwise heating experiments (KURODA and SHERRILL, 1977;

† Present address: Babcock and Wilcox Lynchburg Research Center, P.O. Box 1260, Lynchburg, Virginia 24505, U.S.A.

see also BOGARD *et al.*, 1965). The relative abundances of the fissiogenic xenon and krypton isotopes released from the thucholite indicated that they are primarily the product of ^{238}U spontaneous fission. A xenon fraction with an isotopic composition resembling that of the AVCC xenon was released at 400°C. In the case of krypton, the isotopic composition of a very large fraction of the total gas released at 250°C was closer to the AVCC krypton than to the atmospheric krypton. Small excesses of cosmic-ray-produced ^{124}Xe and ^{126}Xe appeared to be present in xenon fractions released at higher temperatures. A possible connection between the occurrence of a meteoritic rare gas component in terrestrial materials and the process of meteorite impact is discussed.

This investigation was supported by the National Science Foundation under Grant NSF EAR 76-00285.

REFERENCES

BODU, R., H. BOUZIGUES, N. MORIN, and J. P. PFIFFELMANN, On the existence of isotopic anomalies encountered in uranium from Gabon, Memorandum dated 25 Sept. 1972, ACE-tr-7393, U.S. Atomic Energy Commission, Technical Information Center, 22–24, 1972.

BOGARD, D. D., M. W. ROWE, O. K. MANUEL, and P. K. KURODA, Noble gas anomalies in the mineral thucholite, *J. Geophys. Res.*, **70**, 703–708, 1965.

BROWN, H., Rare gases and the formation of the earth's atmosphere, in *The Atmospheres of the Earth and Planets*, edited by G. P. Kuiper, pp. 258–266, The University of Chicago Press, Chicago, Illinois, 1949.

DIETZ, R. S., Sudbury structure as an astrobleme. *J. Geol.*, **72**, 412–434, 1964.

FRENCH, B. M., Sudbury structure, Ontario: Some petrographic evidence for origin by meteorite impact, *Science*, **156**, 1094–1098, 1967.

FRENCH, B. M., Possible relations between meteorite impact and igneous petrogenesis, as indicated by the Sudbury structure, Ontario, Canada, *Bull. Volcanol.*, **34**, No. 2, 466–517, 1970.

FRENCH, B. M., Shock-metamorphic features in the Sudbury structure, Ontario: A review, The Geol. Assoc. Canada, Special Number 10, 19–28, 1972.

KURODA, P. K., Fossil nuclear reactor and plutonium-244 in the early history of the solar system, in *The Oklo Phenomenon, Proceedings of a Symposium, Libreville, 23–27 June 1975, IAEA, Vienna, 1975*, pp. 479–487, 1975.

KURODA, P. K. and R. D. SHERRILL, Xenon and krypton anomalies in the Besner Mine, Ontario, thucholite, *Geochem. J.*, **11**, 9–19, 1977.

KURODA, P. K., R. D. SHERRILL, and K. C. JACKSON, Abundances and isotopic compositions of rare gases in granites, *Geochem. J.*, **11**, 75–90, 1977.

NEUILLY, M., J. BUSSAC, C. FRÉJACQUES, G. NIEF, G. VENDRYES, and J. YVON, Sur l'existence dans un passé recule d'une réaction en chaîne naturelle de fissions, dans le gisement d'uranium d'Oklo (Gabon), *C. R. Acad. Sci. Paris*, **275**, 1847–1849, 1972.

Rare Gas Isotopic Compositions in Diamonds

Nobuo TAKAOKA* and Minoru OZIMA**

*Department of Physics, Osaka University, Toyonaka-shi 560, Japan
**Geophysical Institute, University of Tokyo, Tokyo 113, Japan

The rare gases have been extracted from natural diamonds in a three step temperature program. Most of the rare gases were degassed during graphitization of the diamonds above 2,000°C. The rare gas elemental abundance pattern is similar to type 1 as defined by OZIMA and ALEXANDER (1976), showing deficiency in Ne and enrichment in Xe relative to the earth's atmosphere. The $^{3}He/^{4}He$ ratio is about an order of magnitude larger than the atmospheric ratio. The $^{40}Ar/^{36}Ar$ ratio is more radiogenic than the atmospheric ratio. Other rare gas isotopic ratios are indistinguishable from atmospheric ratios.

1. Introduction

Rare gas isotopic ratios such as $^{40}Ar/^{36}Ar$, $^{3}He/^{4}He$ and $^{129}Xe/^{132}Xe$ from terrestrial samples have provided a powerful tool to understand the origin and evolution of the terrestrial atmosphere (OZIMA and KUDO, 1972; SCHWARTZMAN, 1973; TOLSTIKHIN, 1975; BOULOS and MANUEL, 1971). Such isotopic information may be obtained from rare gases trapped in mantle-derived materials such as volcanic rocks, volcanic xenoliths or volcanic gases. Among mantle-derived materials, diamond seems to be unique in its inertness to all known chemicals and in its high temperature stability. The existence of O_2, H_2, CH_4, H_2O, CO, N_2, Ar, and CO_2 in diamonds has been reported (e.g., MELTON and GIARDANI, 1974, 1976). However, no measurement has been made so far either of the elemental composition or of the isotopic ratios of rare gases in diamonds. This is a preliminary report of rare gas elemental composition and isotopic ratios in diamonds. We have found that diamonds contain He whose $^{3}He/^{4}He$ ratio is more than an order of magnitude higher than the atmospheric value and Ar whose $^{40}Ar/^{36}Ar$ ratio is significantly higher than that in the atmosphere.

2. Sample

We studied industrial diamonds which are believed to come from the Kimberley Mines, South Africa. However, we have no means to confirm this. The diamonds investigated in this work were about 4 mm in size. Some contained black inclusions. A fraction of the inclusions were ferromagnetic and appeared to be pyrrhotite under a microscope. The existence of pyrrhotite inclusions in diamonds is not unusual (SHARP, 1966). The black inclusions appeared to be totally contained within the diamonds. This observation suggests a primary origin of the black inclusions in the diamonds. As will be discussed later, the thermal release pattern of rare gases from the diamonds also favours a primary origin of the black inclusions.

The diamonds were divided by eye into two groups purely on the basis of the amount of black inclusions; batch 1 of diamonds has more black inclusions (less than 10%) than batch 2 (less than 1 percent). To facilitate degassing, both batches of diamonds were crushed in a stainless steel mortar with a stainless steel burin and a hammer. There was no difficulty in crushing them. The crushed samples range from a few hundred microns to a few mm. No attempt was made to separate the inclusions from the crushed diamonds. Rare gas extraction was made on the crushed diamonds.

3. Experimental Methods

The experimental techniques used for the rare gas measurements are essentially the same as described in TAKAOKA (1976). The crushed diamonds were wrapped with 10 μm thick Al foil. The weights of Al foil used were 65 and 57 mg for batch 1 and 2 respectively. The rare gas concentrations (in units of cm^3STP/g) in the Al foil are 6.6×10^{-10} for ^{4}He, 2.5×10^{-10} for ^{20}Ne, 7.0×10^{-11} for ^{36}Ar, 4.5×10^{-12} for ^{84}Kr and 5.9×10^{-13} for ^{132}Xe. Samples were mounted in a sample holder of an extraction furnace and were heated at about 100°C overnight in a vacuum. A Ta crucible was thoroughly degassed by heating above 2,000°C. Before the sample measurements, blanks were run at 800 and 2,000°C with the same conditions as in an extraction experiment. The concentrations of rare gases and the isotopic ratios of He, Ne and Ar were corrected for the blank. However, as blanks at 2,100°C were not run, the blank correction in the 2,100°C fraction was made on the basis of the 2,000°C blank. The isotopic ratios of Kr and Xe were not corrected for the blank. The isotopic ratios of Kr and Xe in the blank were atmospheric.

The rare gases were extracted from the two batches of diamonds at 800, 2,000, and 2,100°C. The samples were held at 800°C for 30 minutes. Thirty minutes were required to raise the temperature to 2,000 and 2,100°C, and then these temperatures were maintained for one hour. Since Al foil melts at 800°C, blank correction for the 800°C fraction included corrections for backgrounds both from the extraction line and from the Al foil. Except for Ne, the amount of rare gases derived from the Al foil was less than 10% of the 800°C fraction both for batch 1 and batch 2. In the case of ^{20}Ne the background from the Al foil was larger than the 800°C fraction of ^{20}Ne in both batch 1 and 2. Hence, no information was obtained for ^{20}Ne in the 800°C fractions.

The sensitivity and mass discrimination of the mass spectrometer and sample system were determined by measuring known amounts of air. The mass discrimination for the ^{3}He/^{4}He ratio was calibrated using the known ^{3}He/^{4}He value in the Bruderheim standard prepared at Berkeley (Reynolds, J. H., written communication, 1973). The mass spectrometer was tuned to a mass resolution of about 600 by adjusting the collector slit width. With this resolution power, the spectrometer can separate ^{3}He from H_3 and HD, ^{38}Ar and ^{40}Ar from C_3H_2 and C_3H_4 peaks, and Kr and Xe from hydrocarbon peaks, respectively. The contributions from ^{3}H and HD at ^{3}He were less than 1%. The separation between ^{20}Ne and $H_2{}^{18}O$ is also fairly good.

4. Results

Rare gas elemental abundance and their isotopic ratios are shown in Table 1. The

Table 1. Elemental and isotopic compositions of rare gases in diamonds (concentration in cm^{3}STP/g).

Sample	Batch 1			Blank 1 (cm^{3}STP)		Batch 2			Blank 2 (cm^3 STP)	
Weight (g)		1.98					2.61			
Temperature (°C)	800	2,000	2,100	800	2,000	800	2,000	2,100	800	2,000
Heating time (min)	30	60	5	30	60	30	60	60	30	60
^{3}He ($\times 10^{-12}$)	<1	28.6	<1	<1	<2	<1	17.9	<1	<1	<2
^{4}He ($\times 10^{-9}$)	1.34	3,480	0.09	0.8	1.1	1.56	918	0.37	1	2
^{20}Ne ($\times 10^{-12}$)		20.7		10	20		10.2		10	20
^{36}Ar ($\times 10^{-11}$)	3.12	36.7	0.1	1	2.4	1.39	9.37	1.90	2	2
^{84}Kr ($\times 10^{-12}$)	1.6	3.6	<0.02	0.6	0.6	0.9	2.9	0.26	0.8	0.8
^{132}Xe ($\times 10^{-13}$)	3.3	9.1	<2	1	2	4.3	7.3	1.0	3	2
^{3}He/^{4}He		$(8.23\pm0.35)\times10^{-6}$					$(1.95\pm0.07)\times10^{-5}$			
^{40}Ar/^{36}Ar	359 ± 2	436 ± 2				574 ± 14	1121 ± 8			
^{128}Xe/^{132}Xe		0.0754 ± 0.0011					0.0718 ± 0.0037			
^{129}Xe/^{132}Xe		0.996 ± 0.019					0.978 ± 0.004			
^{130}Xe/^{132}Xe		0.161 ± 0.077					0.149 ± 0.006			
^{131}Xe/^{132}Xe		0.786 ± 0.023					0.779 ± 0.014			
^{134}Xe/^{132}Xe		0.391 ± 0.012					0.386 ± 0.009			
^{136}Xe/^{132}Xe		0.328 ± 0.012					0.318 ± 0.018			

Other isotopic ratios are indistinguishable from atmospheric values.

blacker diamonds of batch 1 gave almost three times as much rare gas as did the diamonds in batch 2. The isotopic ratios $^{40}Ar/^{36}Ar$ and $^{3}He/^{4}He$ are quite different between the two batches. It appears that the difference is mainly due to a difference in the amount of black inclusions. In both cases, major degassing is observed at 2,000°C heating. A third batch of samples (a fraction of the batch 1 diamond), which were heated at 1,500°C, produced only about 1×10^{-8} ccSTP/g of ^{40}Ar which is less than 10% of the rare gases degassed in batch 1 and 2 at 2,000°C. X-ray analyses on heated samples showed that 1,500°C heating for 1 hour (batch 3) did not result in graphitization, but 2,000°C heating for 1 hour (batch 1 and batch 2) completely converted diamond to graphite. From these experimental results we conclude that graphitization which involves a density change of more than 30% must be responsible for the major degassing of diamonds.

Figure 1 shows the rare gas elemental abundance patterns in a $^{m}F_{A}$ diagram, in which $^{m}F_{A}$ is defined as $(m/^{36}Ar)_{sample}/(m/^{36}Ar)_{atmosphere}$ and m represents the various rare gas isotopes. The patterns observed in both batch 1 and 2 are similar to the Type 1 defined by OZIMA and ALEXANDER (1976) for the pattern produced by the solubility of atmospheric rare gases in water. As will be discussed below, however, the observed $^{40}Ar/^{36}Ar$ ratios are distinctly higher than the atmospheric value and it is unlikely that there is any significant amount of water-absorbed rare gas contamination in the diamonds. Hence, the similarity between these abundance patterns is more likely to be accidental.

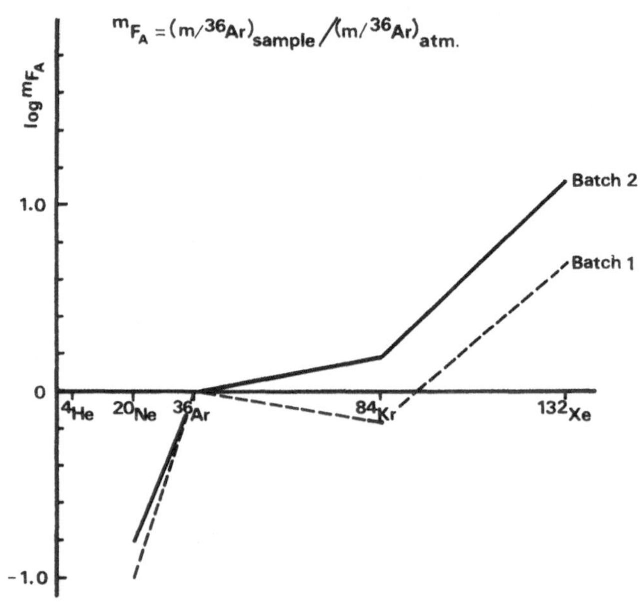

Fig. 1. Rare gas elemental abundance patterns in diamonds.

Except for He, other rare gases show similar thermal release patterns (Fig. 2). In the case of He, more than 99% was degassed at 2,000°C suggesting that most of the He residing in low temperature sites had already been degassed prior to the experiment, perhaps under high temperature mantle conditions. Although the 2,100°C extraction

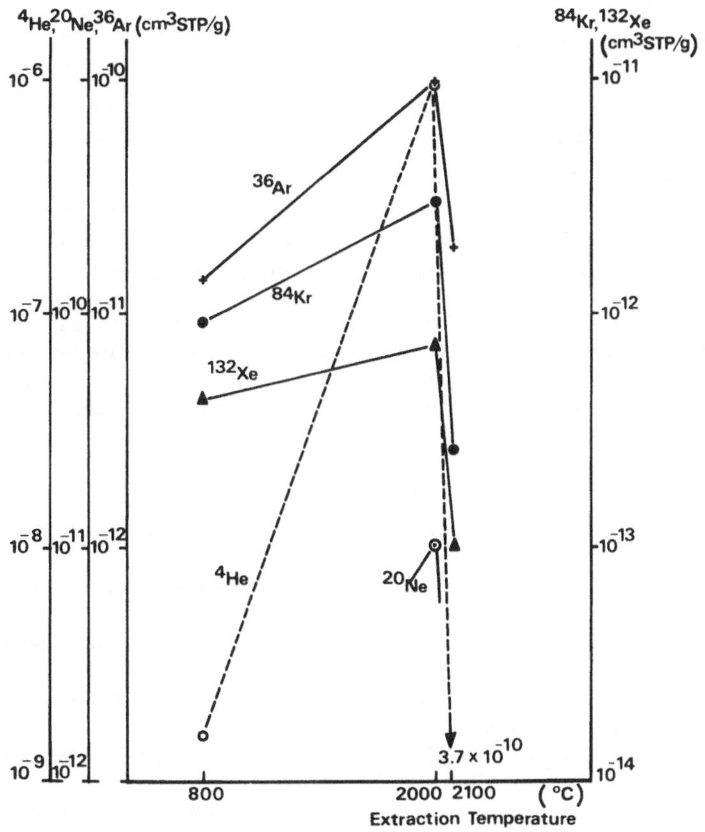

Fig. 2. Thermal release pattern of rare gases from diamonds (batch 2). Note that major degassing occurred at 2,000°C, at which diamonds were rapidly converted to graphite.

for batch 1 may not have been as complete as for batch 2 because of the damage in the furnace, the very similar thermal release patterns between the two batches, i. e., a dominant rare gas release at the 2,000°C step (Table 1), suggests that rare gases in the black inclusions reside at the same sites as those in the diamonds, favoring syngenetic origin of the black inclusions with the diamonds.

Potassium contents were measured on evaporates which were deposited on the inner wall of a high vacuum extraction furnace during the graphitization. The deposits were washed by dil. HNO_3 and the potassium content in the washed solution was measured with an isotope dilution method. The amounts of potassium were 8.22 ppm and 1.97 ppm for batch 1 and batch 2 respectively. Since potassium distillation from the diamonds might not be perfect, the potassium content thus estimated may give only the minimum values. Radiogenic ^{40}Ar which would have been produced in batch 1 diamonds by this amount of K will exceed the total amount of ^{40}Ar, if the age of the diamonds is older than 2.8 b.y. If an age of 2.8 b.y. is assumed, this will give a minimum $^{40}Ar/^{36}Ar$ ratio of 670 for the trapped Ar in the batch 2 diamonds. The observed $^{3}He/^{4}He$ ratio is a minimum since the diamonds must contain some U as well as K. Hence, $^{3}He/^{4}He$ and $^{40}Ar/^{36}Ar$ in the diamonds are likely to be larger than 1.95×10^{-5} and 670.

The $^3He/^4He$ ratio in the batch 2 diamonds is even higher than those commonly observed in submarine basaltic glasses (CRAIG and LUPTON, 1976). Ridge basalts are generally regarded to be derived from oceanic upper mantle or asthenosphere (GAST, 1968). The high $^3He/^4He$ ratio may suggest that the diamonds were derived from a region different from source regions for oceanic ridge basalts, most likely from a deeper region in the mantle. In this connection it is interesting to note that similar high $^3He/^4He$ ratios (2.09×10^{-5}) are observed in Hawaiian volcanic gases (CRAIG and LUPTON, 1976), which may be derived from the deeper mantle through a hot spot.

We are very grateful to Professor N. Kawai who generously offered us diamonds. We also benefited from discussions with Professors N. Kawai and I. Sunakawa. Isotope-dilution analyses of potassium were made by Mr. S. Zashu, Geophysical Institute, University of Tokyo. Dr. I. Kaneoka helped us in rare gas extraction experiments during an early stage in the present study.

REFERENCES

BOULOS, M. S. and O. K. MANUEL, The xenon record of extinct radioactivities in the earth, *Science*, **174**, 1334–1336, 1971.

CRAIG, H. and J. E. LUPTON, Primordial neon, helium and hydrogen in oceanic basalts, *Earth Planet. Sci. Lett.*, **31**, 369–385, 1976.

GAST, P. W., Trace element fractionation and the origin of tholeiitic and alkaline magma types, *Geochim. Cosmochim. Acta*, **32**, 1057–1086, 1968.

MELTON, C. E. and A. A. GIARDANI, The composition and significance of gases released from natural diamonds from Africa and Brazil, *Am. Mineral.*, **59**, 775–782, 1974.

MELTON, C. E. and A. A. GIARDANI, Experimental evidence that oxygen is the principal impurity in natural diamonds, *Nature*, **263**, 309–310, 1976.

OZIMA, M. and E. C. ALEXANDER, Jr., Rare gas fractionation patterns in terrestrial samples and earth-atmosphere evolution model, *Rev. Geophys. Space Phys.*, **4**, 385–390, 1976.

OZIMA, M. and K. KUDO, Excess argon in submarine basalts and an earth-atmosphere evolution model, *Nature Phys. Sci.*, **239**, 23–24, 1972.

SCHWARTZMAN, D. W., On argon degassing models of the earth, *Nature Phys. Sci.*, **245**, 20–21, 1973.

SHARP, W. E., Pyrrhotite: A common inclusion in South African diamonds, *Nature*, **211**, 402–403, 1966.

TAKAOKA, N., A low-blank, metal system for rare-gas analysis, *Mass Spectros.*, **24**, 73–86, 1976.

TOLSTIKHIN, I. N., Helium isotopes in the earth's interior and in the atmosphere: A degassing model of the earth, *Earth Planet. Sci. Lett.*, **26**, 88–96, 1975.

Rare Gases in Mantle-Derived Rocks and Minerals

Ichiro KANEOKA,* Nobuo TAKAOKA,** and Ken-Ichiro AOKI***

*Geophysical Institute, University of Tokyo, Tokyo 113, Japan
**Department of Physics, Osaka University, Toyonaka-shi 560, Japan
***Institute of Mineralogy, Tohoku University, Sendai 980, Japan

Rare gas analyses for mantle-derived rocks and minerals from South Africa, Hawaii and Oki Islands have revealed the heterogeneity of the rare gas state in the earth's deep interior both horizontally and vertically, at least in the lithosphere.

The occurrence of excess ^{129}Xe has been confirmed in some olivine-rich samples from South Africa and Hawaii together with a higher ^{3}He/^{4}He ratio than that of the atmosphere. However it is not yet clear whether excess ^{129}Xe exists rather ubiquitously in the earth's deep interior. Such rare gas isotopic anomalies can be used to discriminate the rocks and/or minerals of different magma sources.

Rare gas abundance patterns have been shown to be controlled by both the original patterns in the magma source and by the secondary processes involved. Hence we should be very careful in using such data to evaluate the original rare gas state in the earth's deep interior.

1. Introduction

The rare gas state in the earth's deep interior is closely related to the degassing history of the earth. The degassing of the earth has controlled the evolution of both the earth and the terrestrial atmosphere. Several lines of evidence suggest that the elemental and isotopic compositions of rare gases in the earth's deep interior are different from those of the atmosphere. The relatively high ^{3}He/^{4}He ratio observed in sea water (e. g., CLARKE *et al.*, 1969), submarine pillow basalts (e. g., LUPTON and CRAIG, 1975), volcanic gases (e.g., MAMYRIN *et al.*, 1969) and some mantle materials (TOLSTIKHIN *et al.*, 1974) is generally taken to indicate the presence of primordial ^{3}He in the earth's deep interior. Excess ^{129}Xe (compared to the atmospheric value) has also been found in CO_2 well gases (BOULOS and MANUEL, 1971; BUTLER *et al.*, 1963) and in a Hawaiian xenolith (HENNECKE and MANUEL, 1975) giving a very important constraint on the degassing history of the earth. Since the Hawaiian xenolith contains CO_2-rich inclusions, such excess ^{129}Xe may be accompanied by mantle-derived CO_2. CRAIG and LUPTON (1976) reported the occurrence (in Kilauea fumarolic gases and in some submarine pillow basalts) of excess ^{20}Ne compared to the atmospheric value. However, the relationships among these rare gas isotopic anomalies are not yet clear.

The ^{40}Ar/^{36}Ar ratio in the earth's interior gives some limits on the degassing history of the earth (OZIMA, 1975). From the correlation between excess ^{40}Ar and ^{36}Ar in ultramafic xenoliths from the Kola Peninsula, KANEOKA (1974, 1975) estimated this value to be very high, greater than 10,000, though he considered it to be variable from site to site. A similar high value has also been suggested from the excess Ar in Stillwater complex rocks (SCHWARTZMAN, 1973). FISHER (1975) inferred a degassing model of the earth from the $^4He/^{40}Ar$ ratio observed in submarine pillow basalts. As suggested by ALEXANDER (1976), however, selective transfer of He may have caused some change in the observed ratio from the original one. In this respect, the elemental abundance patterns of rare gasses classified by OZIMA and ALEXANDER (1976) probably reflect both the original abundance patterns and fractionation processes among each element.

To get information on the rare gas state in the earth's deep interior, mantle-derived rocks and minerals such as ultramafic nodules in alkali basalts are very significant. Since they solidified in the upper mantle, they will trap in-situ rare gases without any atmospheric contamination. For these samples, adsorbed atmospheric contamination, which may occur after the emplacement on the surface, can be removed by stepwise degassing technique. Pillow basalts can retain mantle-rare gases, but we have no guarantee that they are completely free from atmospheric contamination through sea water. Volcanic gases or sea water are more easily contaminated with atmospheric components. Hence, in principle, mantle-derived materials will give the most direct information concerning the rare gas state in the earth's deep interior, provided the sample is fresh enough to retain its primary rare gases.

In the present study, we intended to study the rare gas state in the earth's deep interior by analysing mantle-derived rocks and minerals. Our main concerns are as follows: (1) To check the homogeneity of the rare gas state in the earth's deep interior. (2) To check the occurrence of excess ^{129}Xe and its relationship with the $^3He/^4He$ ratio. (3) To evaluate the significance of rare gas abundance patterns. (4) Application of rare gas data to estimate the condition of magma formation. A part of this study was reported earlier (KANEOKA *et al.*, 1977b) and the complete data presentation shall be given elsewhere.

2. Samples

Samples were selected from three different regions: the South African kimberlite region, the Hawaiian Islands, and Oki-Dōgo Island, southwest of Japan. They represent the continental, oceanic and ocean-continent margin regions respectively. Sample descriptions and extrusion ages are shown in Table 1. Ultramafic nodules in kimberlites are well known to have been derived from the deep interior of the earth. Several lines of evidence indicate the existence of abundant CO_2 in this region, including the occurrence of igneous carbonatite. If excess ^{129}Xe is associated with CO_2 of deep origin, excess ^{129}Xe should be found in such ultramafic nodules. The nodules investigated came from pipes in the Kimberley region, whose eruption ages are about 86 to 90 m.y. by Rb-Sr (ALLSOPP and BARRETT, 1975) or ^{40}Ar-^{39}Ar analyses (KANEOKA and AOKI, 1978) on phlogopites. The carbonatite from the Premier mine was extruded about 1,250 m.y. ago (ALLSOPP, 1974). Hawaiian samples were extruded in historic eruptions, while the spinel lherzolite nodule from the Oki-Dōgo Island was extruded about 3.6 m.y. ago, as estimated from K-Ar ages of associated alkaline basalts (KANEOKA *et al.*, 1977a).

Samples from the South African kimberlite regions have been selected to represent different depths of deviation. Hawaiian and Oki-Dōgo samples are also mantle-derived rocks and minerals. Their formation depths have been estimated petrologically (Table 1). These samples were chosen after microscopic examination to exclude obvious con-

Table 1. Samples for rare gas analyses.

No.	Sample	Locality	Extrusion age	Estimated original depth (km)
South Africa				
1.	DU-02 Phlogopite nodule	Du Toitspan, Kimberley	86 m.y.	Lower crust or upper mantle
2.	BF-03 Phlogopite-bearing Peridotite	Bultfontein, Kimberley	90 m.y.	$\lesssim 100$
3.	MO-61 Augite-ilmenite intergrowth	Monastery, Orange Free State	90 m.y.	$100 \sim 150$
4.	MO-32 Olivine megacryst (unseparated)	Monastery, Orange Free State	90 m.y.	$150 \sim 200$
5.	MO-36 Olivine megacryst (separated pure)	Monastery, Orange Free State	90 m.y.	$150 \sim 200$
6.	Kimberlite, sill	Middle Sill, Benfontein, Kimberley	90 m.y.	$\sim 200 \lesssim$
7.	Carbonatite, dyke	Premier Mine, Transvaal	1, 250 m.y.	$\sim 200 \lesssim$
Hawaiian Islands				
8.	Hualalai 1801 Dunite	Hualalai, Hawaii	1801 A.D.	$30 \sim 60$
9.	Hualalai 1801b Dunite	Hualalai, Hawaii	1801 A.D.	$30 \sim 60$
10.	Olivine phenocrysts separated from Kapuho lava	Kilauea, Hawaii	1960 A.D.	$30 \sim 60$
11.	Glassy spatter from Kilauea Caldera	Kilauea, Hawaii	1975 A.D.	—
Oki Island				
12.	10B-13 Spinel lherzolite	Oki-Dōgo Is., SW of Japan	3.6 m.y.	$30 \sim 40$

tamination by crustal materials.

Hualalai dunites are similar to those studied by HENNECKE and MANUEL (1975), who found excess ^{129}Xe possibly associated with CO_2-rich inclusions. Hence we selected samples which were observed microscopically to contain larger numbers of CO_2 bubbles. Olivine phenocrysts from Kapuho lava, Kilauea, were separated to compare the results with those from nodules. Glassy spatter from Kilauea might be expected to contain mantle-ambient gases because it was formed rapidly like pillow basalts, though subaerially. However the microscopic observation suggests the incorporation of air bubbles.

Rare gases were analysed with a Nier-type mass spectrometer, installed with a secondary electron multiplier and having a high resolving power of about 600, which enables us to completely separate ^{3}He from ($HD+H_3$) and fully resolve Xe isotopes. More details on experimental procedures have been reported elsewhere (TAKAOKA, 1975).

3. Rare Gas Concentrations

Rare gas concentrations for samples from the South African kimberlite region and Hawaii and Oki-Dōgo Islands are summarized in Figs. 1 and 2 respectively. "Atmospheric" points in Figs. 1 and 2 correspond to the amounts of atmospheric rare gases divided by the mass of the earth. Hence if most rare gases have already been degassed

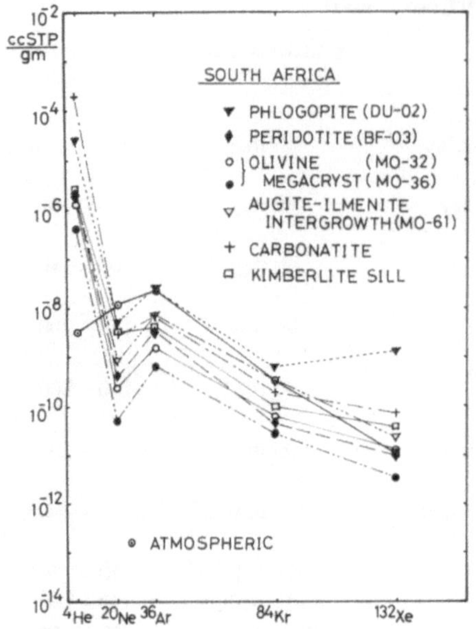

Fig. 1. Rare gas concentrations for South African samples. "Atmospheric" is defined as the amount of rare gases in the atmosphere divided by the mass of the earth. Errors in the measurement of rare gas concentrations are estimated to be about 10%.

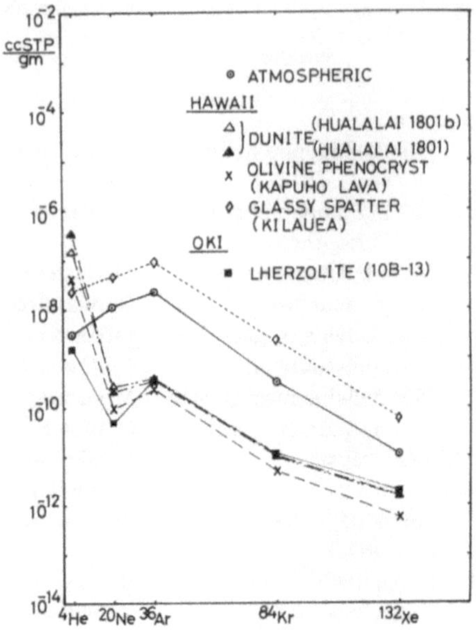

Fig. 2. Rare gas concentrations for Hawaii and Oki Island samples.

from the earth's interior into the atmosphere, the amount of each rare gas in a representative mantle-derived rock should be less than that of the "atmospheric" point, unless the rare gases are secondarily enriched into the sample.

As shown in Figs. 1 and 2, the rare gas concentrations of most samples are less than those of the "atmospheric" except for ^4He and Xe in some samples. The uranium contents required to generate the observed ^4He contents by in situ decay after extrusion are too large, $20 \sim 2{,}000$ ppb, for South African samples compared to the known uranium contents of similar rocks and minerals—except for the phlogopite nodule DU-02. Hence ^4He is probably of radiogenic origin for the phlogopite nodule, but is the mostly excess ^4He for the other samples. Similarly, the occurrence of excess ^{40}Ar in these samples is inferred—except for the sample MO-61. As shown in Fig. 1, the phlogopite nodule is enriched in rare gases, especially Xe. Since phlogopite is a hydrous mineral which is stable under high temperature and pressure, it will gather many volatiles in the upper mantle during its formation. Carbonatite is particularly enriched in ^4He. Its $^4He/^{40}Ar$ ratio is more than 50, larger than those of the other samples by one or two orders of magnitude. Since the K content of the carbonatite has been measured to be only 8.9 ppm, its U content is probably the order of ppb. Hence, to explain the amount of ^4He and the large $^4He/^{40}Ar$ ratio in the carbonatite, selective transfer of ^4He is required into the carbonatite magma. Since no such large enrichment of ^4He is observed in the kimberlite sill, it is more likely that the selective transfer of ^4He into the carbonatite magma occurred from the very old Precambrian wall rocks (more than 2,000 m.y., (CLIFFORD, 1968)) in this region.

Generally most olivine-rich samples show much lower rare gas concentrations (Figs.

1 and 2) than those of the other samples. Olivine phenocrysts contain a smaller amount of rare gases, reflecting its earlier formation in the magma reservoir together with the lower solubility of rare gases in olivine than in other phases. However rare gas concentrations are generally lower in Hawaiian dunites and olivine phenocrysts than those in South African olivine samples, which may reflect a difference in the properties of the upper mantle.

Glassy spatter contains larger amounts of rare gases than those of the "atmospheric" with a pattern similar to that of the atmosphere. Rare gas isotopes for the glassy spatter are also atmospheric, which supports the conjecture that the glassy spatter retained air bubbles during its solidification.

These results suggest that rare gas concentrations reflect a mixture of effects such as the properties of the upper mantle, mineral composition, conditions of magma formation and eruption, degassing history and so on.

4. Rare Gas Abundance Patterns

Rare gas abundance patterns are shown for samples from South Africa (Fig. 3) and the Hawaii and Oki-Dōgo Islands (Fig. 4). F^m is defined in Fig. 3.

All South African samples except for kimberlite show a depletion of Ne compared to the atmosphere. Dunites from Hawaii show an enrichment of Ne, while olivine phenocrysts are depleted in Ne. Oki lherzolite is also depleted in Ne.

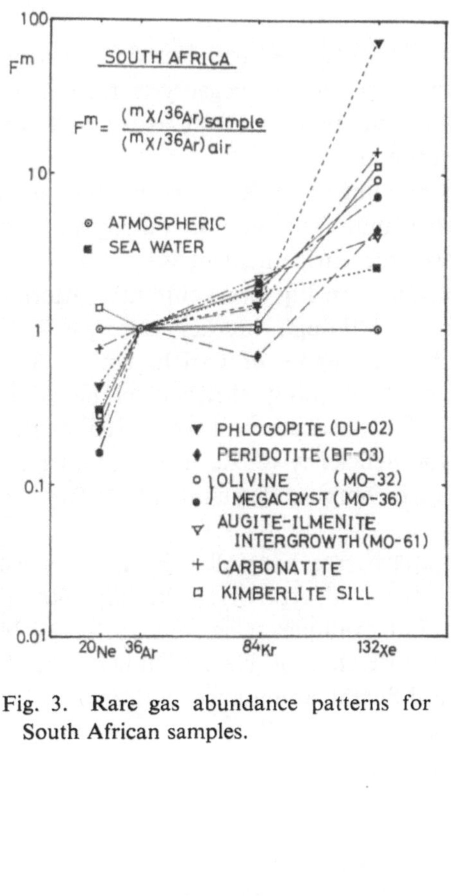

Fig. 3. Rare gas abundance patterns for South African samples.

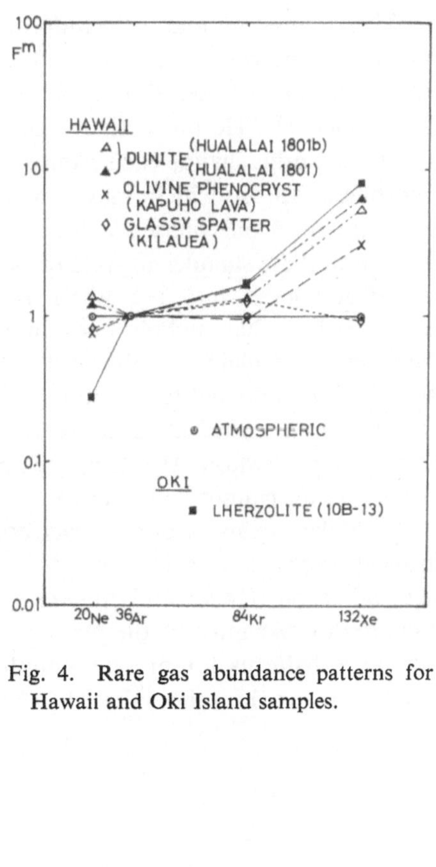

Fig. 4. Rare gas abundance patterns for Hawaii and Oki Island samples.

In Fig. 5, rare gas abundance patterns are compared among Hualalai nodules. Present dunite samples are more depleted in Ne, but more enriched in Kr and Xe than those investigated by HENNECKE and MANUEL (1975).

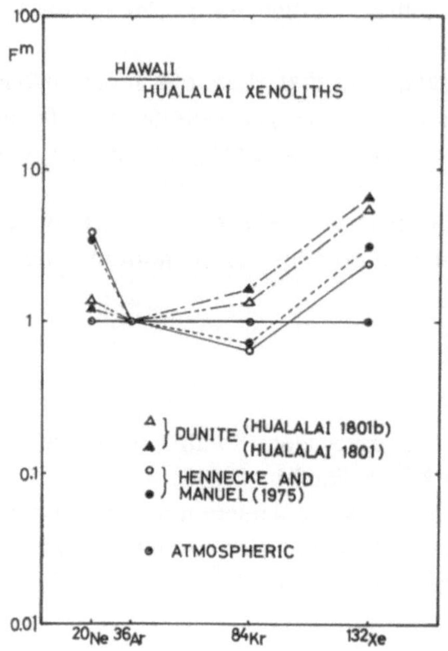

Fig. 5. Rare gas abundance patterns for Hualalai xenoliths. Heavier rare gases are more enriched in present samples than those examined by HENNECKE and MANUEL (1975), which suggests rare gas elemental fractionation for similar samples from the same site.

These dunite samples are composed of more than 95% olivine, but their rare gas abundance patterns are clearly different from those of olivine megacrysts from South Africa. In spite of such differences, samples from both regions contain excess ^{129}Xe and show higher $^{3}He/^{4}He$ ratios than that of the atmosphere.

On the other hand, each sample shows an enrichment of Xe as compared to the atmospheric value, which may partly reflect elemental fractionation during rare gas movement from one phase to the other. However the enrichment of Xe is commonly observed in these samples in spite of Ne characteristics. In Fig. 3, a solubility pattern of atmospheric gases into sea water is also drawn. Although heavier rare gases are more enriched than lighter ones in this pattern, the degree of enrichment of Xe is insufficient to explain the observed values. Hence Xe is surely enriched in the earth's deep interior compared to the atmospheric value. The adsorption of Xe on sedimentary rocks has been suggested as an explanation for the deficiency of Xe in the atmosphere (FANALE and CANNON, 1971), but the possibility of incomplete degassing of Xe from the earth's interior cannot be excluded.

Secondary changes of rare gas abundance patterns are controlled by the physicochemical properties of each rare gas such as diffusivity, adsorptivity, solubility into the melt and so on (OZIMA and ALEXANDER, 1976). For example, selective transfer of Ne compared to ^{36}Ar must be the main cause for the difference of Ne enrichments in the abundance patterns for present samples, because different abundance patterns are observed even for the same minerals, such as olivine.

Since the rare gases in Hualalai nodules are probably associated with CO_2 of deep origin, the difference in the abundance patterns might have occurred due to elemental fractionation during the incorporation of CO_2 and other volatiles into the source of olivine crystals. Secondary gas loss after extrusion cannot fully explain the observed pattern.

Several experiments have demonstrated that diffusion is capable of producing large but variable effects on apparent rare gas fractionation patterns. For example, KANEOKA *et al.* (1977b) have shown that light rare gases are degassed at lower temperatures than are heavy rare gases in phlogopite DU-02 (Fig. 6). For this sample, rare gas abundance patterns artificially based on each temperature fraction are different at each temperature (Fig. 7), which suggests that the partial gas loss will easily change the rare gas abundance patterns. Such patterns are clearly different among different minerals. For example, the results of stepwise degassing for Hawaiian dunites suggest that most rare gases (including He) are tightly trapped in olivine crystals. The difference in the retentivity of rare gases among different samples is also an important factor controling the rare gas state.

These results suggest that rare gas abundance patterns are controlled not only by the original pattern in the magma reservoir but also by any modifying processes involved. It is essential to discriminate these factors properly to evaluate the rare gas abundance patterns.

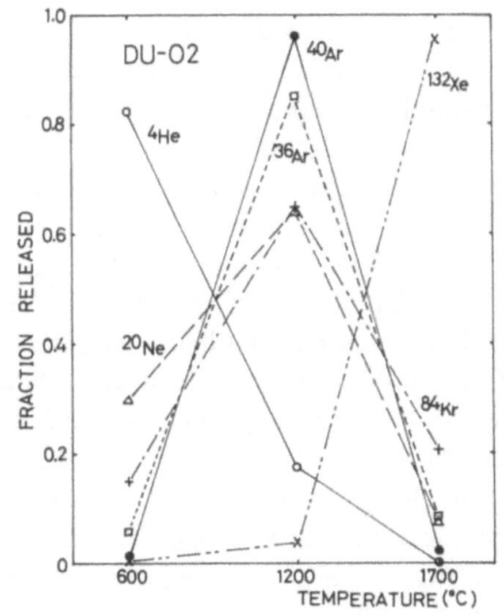

Fig. 6. Release patterns of rare gases for phlogopite DU-02 from South Africa. Lighter rare gases are degassed at low temperatures, while heavier ones degassed at high temperatures, suggesting different diffusion rates of rare gases in this mineral.

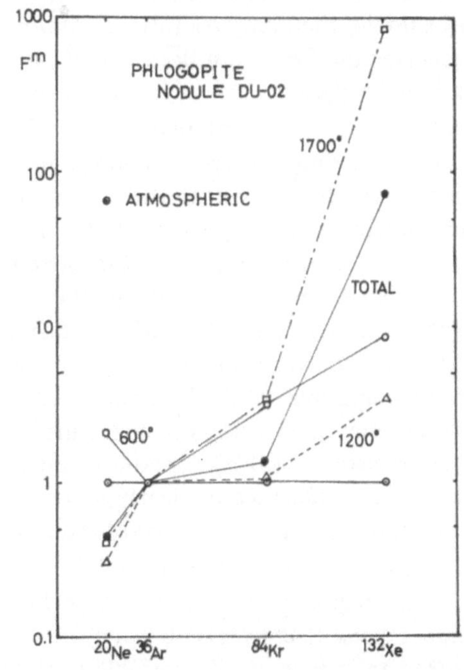

Fig. 7. Rare gas abundance patterns for DU-02, artificially made for each temperature fraction. Different diffusion rate of rare gases causes such different patterns from the same sample.

5. Rare Gas Isotopes

$^3He/^4He$, $^{20}Ne/^{22}Ne$, $^{40}Ar/^{36}Ar$, and $^{129}Xe/^{132}Xe$ ratios for the present samples are shown in Table 2. Other isotopic ratios are nearly atmospheric within the experimental error. The contamination of ^{22}Ne from doubly charged CO_2 is serious for some samples, and no definite $^{20}Ne/^{22}Ne$ ratio could be obtained for these samples.

MO-32 shows a higher $^3He/^4He$ ratio than the atmospheric value. This sample also contains excess ^{129}Xe. MO-36 contains excess ^{129}Xe, but no definite data concerning the $^3He/^4He$ ratio could be obtained from the sample due to a slight contamination from the previous measurement of meteorites. ^{21}Ne is also affected to a lesser degree, but no effects are observed for the other isotopes including ^{38}Ar. These samples are the most deep-seated nodules in this region. The $^{20}Ne/^{22}Ne$ ratio for kimberlite is possibly a little higher than the atmospheric value.

Hawaiian dunites and phenocrysts show high $^3He/^4He$ ratios and excess ^{129}Xe, but no excess ^{20}Ne, though the latter values are less precise. Isotopic ratios in the two Hualalai dunites are similar. However, the $^3He/^4He$ ratio for the olivine phenocrysts from Kilauea seems to be higher than those of Hualalai samples, whose significance shall be discussed in the next section.

The $^{40}Ar/^{36}Ar$ ratio is variable even if the aging effect is considered. Radiogenic ^{40}Ar loss and/or air contamination are possible explanations for the lowering of this ratio. For example, the $^{40}Ar/^{36}Ar$ ratio of MO-61 becomes lower than that of atmospheric Ar after correcting for in situ decay of ^{40}K with the assumption of no Ar loss. The He content is relatively low in this sample. Hence radiogenic ^{40}Ar loss after the extrusion of the sample seems to be most plausible in this case. However, a simple air contamination hypothesis cannot always explain the rare gas abundances. Furthermore, stepwise degassing results show that the $^{40}Ar/^{36}Ar$ ratio is low even at higher temperatures for some samples, especially for MO-36. Hence, the possibility of the occurrence of a low $^{40}Ar/^{36}Ar$ ratio in some part of the earth's interior cannot be excluded.

In MO-36, more than 90% of the Ar and ^4He are degassed at 1,700°C, whereas its $^{40}Ar/^{36}Ar$ ratio is low of nearly atmospheric. Although air contamination is a possibility, the occurrence and excess ^{129}Xe and ^4He and the high degassing temperature seem to be incompatible with this explanation. On the other hand, for MO-32 the $^{40}Ar/^{36}Ar$ ratio is about 1,000 at high temperature. Although both samples are olivine megacrysts from Monastery, MO-32 contains impurities. MO-36 is composed of purified olivine grains. This difference is clearly shown in their K contents (MO-32, 46 ppm, MO-36, 5 ppm) and the impurities probably contain much more excess ^{40}Ar than the pure olivines.

These results suggest that the $^{40}Ar/^{36}Ar$ ratio in the deeper part of the South African kimberlite region may not always be high, though the ratio in the upper part of the section represented by DU-02 and BF-03 is higher than 1,000. This implies that the $^{40}Ar/^{36}Ar$ ratio is vertically variable in the lithosphere.

Hawaiian dunites show $^{40}Ar/^{36}Ar$ ratios of more than 1,000, whereas the olivine phenocrysts have a lower $^{40}Ar/^{36}Ar$ ratio, reflecting the vertical inhomogeneity of the rare gas state in this region.

Oki lherzolite may be characterized by its relatively low $^{40}Ar/^{36}Ar$ ratio. This may reflect a different degassing history in this region than in the other regions.

Table 2. Rare gas isotopic ratios of mantle-derived rocks and minerals from South Africa, Hawaii and Oki Islands.

Sample		$^3He/^4He$ ($\times 10^{-6}$)	$^{20}Ne/^{22}Ne$	$^{40}Ar/^{36}Ar$		$^{129}Xe/^{132}Xe$
South Africa						
DU-02 Phlogopite nodule	600°C	<0.54	9.73±0.12	667.5±0.4		0.981±0.014
	1,200°C	<0.32	9.81±0.08	3,694 ±37		0.988±0.009
	1,700°C	<1,100	9.94±0.15	1,011 ±10		0.983±0.008
	Total	<0.687	9.80	3,282	2,054*	0.983
BF-03 Phlogopite-bearing peridotite						
	1,200°C	<5.9	9.74±0.15	4,981 ±59		0.990±0.014
	1,700°C	<4.3	9.87±0.07	1,898 ± 8		0.984±0.016
	Total	<5.4	9.79	4,275	3,396*	0.989
MO-61 Clinopyroxene-ilmenite intergrowth						
	1,700°C	<0.97	—	353.5±4.9	254.7*	0.983±0.008
MO-32 Olivine mega-cryst	700°C	<5.9	—	310.5±1.8		0.987±0.014
	1,700°C	6.55±0.27	9.79±0.27	1,183 ±11		1.014±0.005
	2,000°C	7.30±0.34	—	1,020 ± 8		1.009±0.006
	Total	6.25	—	808	797.5*	1.004
MO-36 Olivine mega-cryst	600°C	—	—	310.7±1.3		0.978±0.015
	1,700°C	—	—	297.5±0.5		1.005±0.007
	Total	—	—	299.5	296.8*	1.000
Kimberlite	1,700°C	<1.7	10.11±0.15	431.2±3.3	342.6*	0.983±0.024
Carbonatite	1,700°C	<0.26	—	498.0±8.9	491.7*	0.982±0.004
Hawaii						
Hualalai 1801						
Dunite	1,700°C	11.5 ±1.0	9.71±0.11	1,432 ±15		1.022±0.014
Hualalai 1801b						
Dunite	600°C	<44	9.63±0.22	298.9±4.2		0.975±0.006
	1,700°C	14.8±0.7	9.76±0.09	2,371 ±67		1.007±0.009
	Remelt	11.5±0.9	9.61±0.20	1,801 ±16		1.026±0.017
	Total	14.2	9.74	1,503		1.012
Olivine phenocryst Kapuho lava						
	1,700°C	25.1±3.4	9.64±0.19	732.0±2.2		1.014±0.009
Glassy spatter Kilauea	600°C	—	9.73±0.08	294.0±1.6		0.980±0.005
	1,700°C	—	9.70±0.05	294.0±1.2		0.977±0.003
	Total	—	9.72	294.0		0.978
Oki Island						
10B-13 Spinel lherzolite						
	600°C	<110	—	309.5±3.0		0.980±0.020
	1,700°C	< 49	9.5 ±0.5	375.4±1.6		0.985±0.013
	Total	< 73	—	353.9	345.0*	0.983
Air		1.4	9.81	295.5		0.983

N.B.: $^{40}Ar/^{36}Ar$ ratio with an asterisk (*) has been corrected for aging effect after the emplacement of a sample on the surface. ± means one standard deviation.

6. The Relationships among Rare Gas Isotopes and Concentrations

When atmospheric contamination mainly controls the $^{40}Ar/^{36}Ar$ ratio, an inverse correlation between the $^{40}Ar/^{36}Ar$ ratio and the ^{36}Ar content in each sample might be expected. However, the results for xenolithic rocks and minerals in the present study together with the data of Hawaiian samples determined by other investigators (FUNKHOUSER and NAUGHTON, 1968; GRAMLICH and NAUGHTON, 1972; HENNECKE and MANUEL, 1975) do not show such a correlation, but may rather show regional differences. Since ^{36}Ar concentration is affected by rock or mineral types, such a correlation might have been obscured. However even for similar samples with similar amount of ^{36}Ar, the $^{40}Ar/^{36}Ar$ ratio is also variable from site to site. Oki lherzolite and some Hawaiian peridotites, in which the $^{40}Ar/^{36}Ar$ ratio is about 350 for the former and more than 1,000 for the latter, are such an example. Hence, as suggested before, the $^{40}Ar/^{36}Ar$ ratio is probably heterogeneous in the upper mantle.

Concerning the $^{4}He/^{40}Ar$ ratio, the present samples show generally low values (0.1 to 3)—except for carbonatite. However we have no definite criteria to decide if they reliably represent the values of their sources. Some of them show secondary loss or addition of ^{4}He. Hence the discussion is not extended here.

In Fig. 8, the $^{3}He/^{4}He$ ratio is plotted against the ^{4}He content for present samples. Since most present samples show only the upper limit for the $^{3}He/^{4}He$ ratio, such data are shown with arrows. For present samples, there seems to exist a rough inverse correlation between the $^{3}He/^{4}He$ ratio and the ^{4}He content. If this is true, it may indicate that the apparent $^{3}He/^{4}He$ ratio is mainly controlled by the addition of radiogenic ^{4}He. The regional difference in the $^{3}He/^{4}He$ ratio may also reflect the difference in the $^{3}He/(U+Th)$ ratio of the region.

According to LUPTON *et al.* (1977), the $^{3}He/^{4}He$ ratios for pillow basalts have values around 1.4×10^{-5}, whereas the $^{4}He/Ne$ ratio is very variable from site to site by a factor

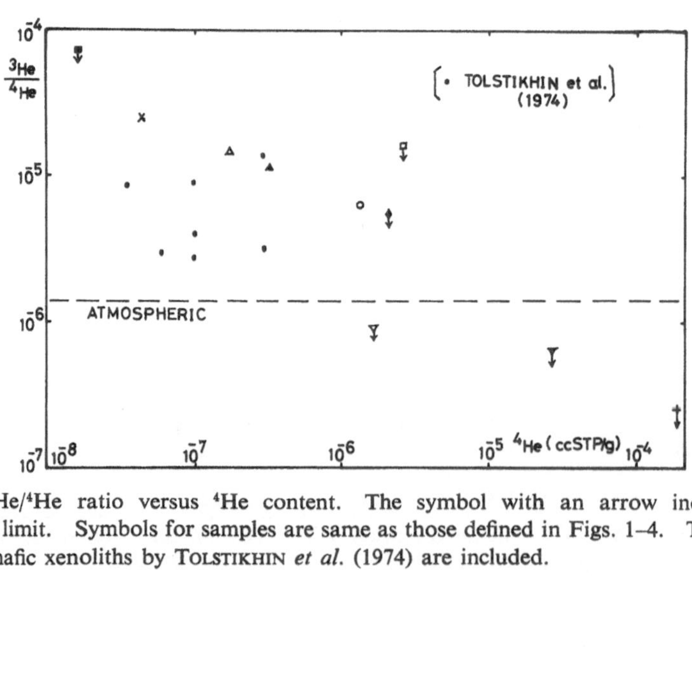

Fig. 8. $^{3}He/^{4}He$ ratio versus ^{4}He content. The symbol with an arrow indicates the upper limit. Symbols for samples are same as those defined in Figs. 1–4. The data of ultramafic xenoliths by TOLSTIKHIN *et al.* (1974) are included.

of more than 10^3. Although regional differences in the $^4He/Ne$ ratio may exist reflecting the heterogeneity of the upper mantle with respect to its U and Th contents and degassing history, different extrusion modes largely cause such an elemental fractionation. For example, Kilauea fumarolic gases show a higher $^4He/Ne$ ratio than that of olivine phenocrysts from Kapuho lava, Kilauea, reflecting the different mode of gas transfer in these samples.

Although the $^3He/^4He$ ratio is much lesser variable than the elemental ratio, such as $^4He/Ne$, differences in this value seems to exist. The $^3He/^4He$ ratio in the olivine phenocryst of Kapuho lava, Kilauea is similar to that of the Kilauea fumaroles, both of which have the value of $(2 \text{ to } 2.5)\times 10^{-5}$, and higher than those of Hualalai samples or pillow basalts from different regions. If the higher $^3He/^4He$ ratio for Kilauea samples is correct, we should also expect a vertical difference in the $^3He/^4He$ and the $^{40}Ar/^{36}Ar$ ratios under the island of Hawaii. Furthermore, the $^3He/^4He$ ratios for Hualalai samples are similar to those of submarine pillow basalts which were formed at the spreading center. Hence, the different isotopic ratios between Hualalai samples and olivine phenocrysts probably relate to the different magma sources from which they were formed. Since a hot spot has been suggested under the island of Hawaii (MORGAN, 1972), samples from Kilauea may reflect the characteristics of the hot spot, whereas Hualalai samples those of a spreading area. The occurrence of excess ^{129}Xe and a high $^3He/^4He$ ratio in the olivine phenocrysts of Kapuho lava also implies that they have not been formed at a shallow depth such as a few kilometers from the surface. Thus, we can discriminate rocks and/or minerals of different magma sources by comparing their rare gas isotopes to those in the earth's interior.

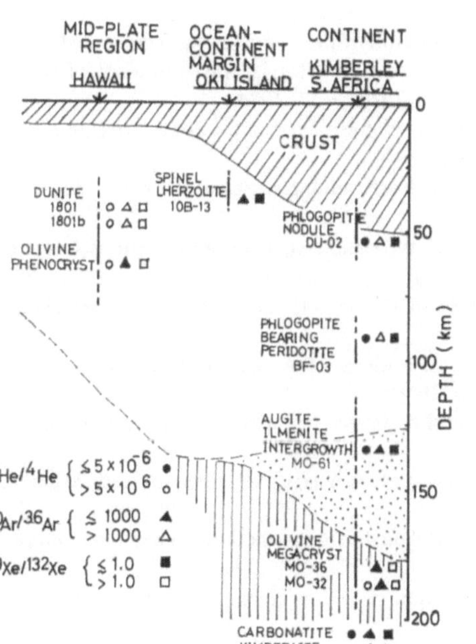

Fig. 9. Summary of the rare gas state in the earth's deep interior at the three regions investigated in this study. The schematic structure is modified after the figure by MACGREGOR and BASU (1974). ▨, Crust; □, Spinel peridotite; ▩, Spinel+garnet peridotites; ▥, Garnet peridotite. Open symbols represent ratios much higher than atmospheric ratios, whereas closed symbols represent ratios closer to atmospheric ratios.

7. Summary

Present results are summarized schematically in Fig. 9. (1) The rare gas state in the earth's interior is inhomogeneous both horizontally and vertically, at least in the lithosphere. (2) The occurrence of excess ^{129}Xe in Hawaiian xenoliths, olivine phenocrysts and olivine megacrysts from South African kimberlite region is confirmed. If excess ^{129}Xe exists rather ubiquitously in the deep interior of the earth, it should be found in submarine pillows too. However, so far no definite evidence for the occurrence of excess ^{129}Xe has been reported in these samples. If it is not found in such submarine glassy samples, excess ^{129}Xe may remain only in separated portions in the earth's deep interior, which implies an incomplete mixing of mantle materials through the earth's history and does not necessarily imply an early degassing of the whole earth. (3) Rare gas abundance patterns are controlled by both the original patterns and the secondary processes involved. The terrestrial samples are more or less affected by secondary processes. Hence, to evaluate the original rare gas state in the earth's deep interior from rare gas abundance patterns observed in rock or mineral samples, we should be very cautious in interpreting the data. (4) By applying rare gas isotopes, we can discriminate rocks and/or minerals of different magma sources, thus advancing our understanding of magma genesis.

We thank Prof. S. Aramaki and Mr. E. Takahashi of the University of Tokyo for providing us the Hawaii and Oki-Dōgo Island samples used in this study.

We appreciate the help of Mr. K. Nagao of Osaka University who assisted us in rare gas analyses and Mr. S. Zashu of the University of Tokyo who analysed the K contents of present samples.

This study was financially supported by the Matsunaga Science Foundation.

REFERENCES

Alexander, Jr., E. C., Trapped helium and argon and formation of the atmosphere, *Nature*, **261**, 77, 1976.

Allsopp, H. L., Referred to in H. J. Welke, H. L. Allsopp and J. W. Harris, Measurements of K, Rb, U, Sr, and Pb in diamonds containing inclusions, *Nature*, **252**, 35–37, 1974.

Allsopp, H. L. and D. S. Barrett, Rubidium-strontium age determinations on South African kimberlite pipes, *Phys. Chem. Earth*, **9**, 605–617, 1975.

Boulos, M. S. and O. K. Manuel, The xenon record of extinct radioactivities in the earth, *Science*, **174**, 1334–1336, 1971.

Butler, W. A., P. M. Jeffery, J. H. Reynolds, and G. J. Wasserburg, Isotopic variations in terrestrial xenon, *J. Geophys. Res.*, **68**, 3283–3291, 1963.

Clarke, W. B., M. A. Beg, and H. Craig, Excess ^{3}He in the sea: Evidence for terrestrial primordial helium, *Earth Planet. Sci. Lett.*, **6**, 213–220, 1969.

Clifford, T. N., Radiometric dating and the pre-Silurian geology of Africa, in *Radiometric Dating for Geologists*, pp. 299–416, Interscience Pub., London, 1968.

Craig, H. and J. E. Lupton, Primordial neon, helium, and hydrogen in oceanic basalts, *Earth Planet. Sci. Lett.*, **31**, 369–385, 1976.

Fanale, F. P. and W. A. Cannon, Physical adsorption of rare gas on terrigenous sediments, *Earth Planet. Sci. Lett.*, **11**, 362–368, 1971.

Fisher, D. E., Trapped helium and argon and the formation of the atmosphere by degassing, *Nature*, **256**, 113–114, 1975.

Funkhouser, J. G. and J. J. Naughton, Radiogenic helium and argon in ultramafic inclusions from Hawaii, *J. Geophys. Res.*, **73**, 4601–4608, 1968.

Gramlich, J. W. and J. J. Naughton, Nature of source material for ultramafic minerals from Salt Lake Crater, Hawaii, from measurement of helium and argon diffusion, *J. Geophys. Res.*, **77**, 3032–3042, 1972.

HENNECKE, E. W. and O. K. MANUEL, Noble gases in an Hawaiian xenolith, *Nature*, **257**, 778–780, 1975.

KANEOKA, I., Investigation of excess argon in ultramafic rocks from the Kola Peninsula by the $^{40}Ar/^{39}Ar$ method, *Earth Planet. Sci. Lett.*, **22**, 145–156, 1974.

KANEOKA, I., Non-radiogenic argon in terrestrial rocks, *Geochem. J.*, **9**, 113–124, 1975.

KANEOKA, I. and K. AOKI, $^{40}Ar/^{39}Ar$ analyses of phlogopite nodules and phlogopite-bearing peridotites in South African kimberlites, *Earth Planet. Sci. Lett.*, 1978 (in press).

KANEOKA, I., E. TAKAHASHI, and S. ZASHU, K-Ar ages of alkali basalts from the Oki-Dōgo Island, *J. Geol. Soc. Japan*, **83**, 187–189, 1977a.

KANEOKA, I., N. TAKAOKA, and K. AOKI, Rare gases in a phlogopite nodule and a phlogopite-bearing peridotite in South African kimberlites, *Earth Planet. Sci. Lett.*, **36**, 181–186, 1977b.

LUPTON, J. E. and H. CRAIG, Excess ^{3}He in oceanic basalts: Evidence for terrestrial primordial helium, *Earth Planet. Sci. Lett.*, **26**, 133–139, 1975.

LUPTON, J. E., R. F. WEISS, and H. CRAIG, Mantle helium in the Red Sea brines, *Nature*, **266**, 244–246, 1977.

MACGREGOR, I. D. and A. R. BASU, Thermal structure of the lithosphere: A petrologic model, *Science*, **185**, 1007–1011, 1974.

MAMYRIN, B. A., I. N. TOLSTIKHIN, G. S. ANUFRIEV, and I. L. KAMENSKIY, Anomalous isotopic composition of helium in volcanic gases, *Dokl. Akad. Nauk SSSR*, **184**, 1197, 1969 (in Russian).

MORGAN, W. J., Convection plumes and plate motions, *Am. Ass. Petrol. Geol. Bull.*, **56**, 203–213, 1972.

OZIMA, M., Ar isotopes and earth-atmosphere evolution models, *Geochim. Cosmochim. Acta*, **39**, 1127–1134, 1975.

OZIMA, M. and E. C. ALEXANDER, Jr., Rare gas fractionation patterns in terrestrial samples and the earth-atmosphere evolution model, *Rev. Geophys. Space Phys.*, **14**, 385–390, 1976.

SCHWARTZMAN, D. W., Argon degassing models of the earth, *Nature Phys. Sci.*, **245**, 20–21, 1973.

TAKAOKA, N., A low-blank, metal system for rare-gas analysis, *Mass Spectr.*, **24**, 73–86, 1975.

TOLSTIKHIN, I. N., B. A. MAMYRIN, L. B. KHABARIN, and E. N. ERLIKH, Isotope composition of helium in ultrabasic xenoliths from volcanic rocks of Kamchatka, *Earth Planet. Sci. Lett.*, **22**, 75–84, 1974.

A Comparison of Terrestrial and Meteoritic Noble Gases

O. K. MANUEL

Department of Chemistry, University of Missouri,
Rolla, Missouri 65401, U. S. A.

The isotopic compositions of Ar, Kr, and Xe in the mantle are distinct from those in carbonaceous chondrites. They show no evidence for the isotopically strange components found in carbon-rich residues of chondrites. Similarities in the isotopic compositions of noble gases in the mantle and in iron meteorites, some lunar breccias, and Ca-poor achondrites suggest that these bodies may have accumulated from common parent material. The $^{40}Ar/^{36}Ar$ ratio for ambient Ar in the mantle is similar to that in air. Atmospheric ^{40}Ar may be primordial rather than radiogenic.

1. Introduction

Recent reviews on the results of isotopic and elemental analyses of noble gases in meteoritic and lunar samples demonstrate the wealth of information which these elements have retained about the early history of our solar system (BOGARD, 1971; REYNOLDS, 1977; MANUEL and SABU, 1975). Our research group at the University of Missouri-Rolla has participated in several analyses on extraterrestrial samples, but an understanding of the noble gas record of the early history of the Earth and the evolution of its atmosphere has been the primary objective of our research for the past decade.

During this period there have been several observations which seemed inconsistent with the view that the Earth initially accumulated as a homogeneous body (UREY, 1952; RINGWOOD, 1966) from a primitive solar nebula containing condensable elements in approximately the abundances seen in C-1 carbonaceous chondrites (ANDERS, 1971). It seemed difficult to understand how subsequent differentiation into a core, mantle and crust could explain differences observed between the abundance pattern of nonradiogenic isotopes of noble gases in air and in carbonaceous chondrites (KRUMMENACHER *et al.*, 1962; EUGSTER *et al.*, 1967; MAZOR *et al.*, 1970) or the presence of radiogenic ^{129}Xe (BOULOS and MANUEL, 1971; HENNECKE and MANUEL, 1975a, b) and primordial ^{3}He (MAMYRIN *et al.*, 1969; CLARKE *et al.*, 1969) trapped inside the modern Earth. These observations seemed more compatible with the suggestion that the internal structures of the terrestrial planets were produced by inhomogeneous accumulation, with the iron-rich cores accumulating first (EUCKEN, 1944; TUREKIAN and CLARK, 1969, 1975; VINOGRADOV, 1975; ALFVÉN and ARRHENIUS, 1976). These models require neither identical isotopic compositions for elements in the Earth and the carbonaceous chondrites nor a stage in the Earth's history of drastic elemental partitioning.

Recently it has been suggested that many of the isotopic anomalies of noble gases and other elements in meteorites might be nucleogenetic (MANUEL, *et al.*, 1972; CLAYTON *et al.*, 1973, 1976; BLACK, 1975; CLAYTON, 1975, 1976; MANUEL and SABU, 1975, 1977a, b; SABU and MANUEL, 1976; JOVANOVIC and REED, 1976; BLAKE and SCHRAMM, 1976; CAMERON and TRURAN, 1977). Carbon-rich residues of carbonaceous chondrites contain isotopically anomalous components of Ar, Kr, and Xe (LEWIS *et al.*, 1975; FRICK and MONIOT, 1976) which correlate with the elemental abundances of the light noble gases, He and Ne, in the manner expected for condensate from supernova envelopes (MANUEL and SABU, 1975, 1977a, b). Iron meteorites trapped complementary isotopic components of Ar, Kr, and Xe (HENNECKE and MANUEL, 1977) and may represent condensate from the central region of a supernova. MANUEL and SABU (1975, 1977a, b) note that the correlations between elemental and isotopic heterogeneities in meteorites, the position of the iron-rich terrestrial planets in the inner part of the solar system, and the current low flux of solar neutrinos (BAHCALL and DAVIS, 1976) may indicate that the entire solar system formed primarily from the debris of a single supernova, concentric with the present sun.

In the present report the isotopic compositions of heavy noble gases in the atmosphere and the mantle will be compared with the isotopic compositions of these elements in meteorites to seek information on the early history of the Earth and the formation of its atmosphere. The two light noble gases are excluded from the present study because interference from spallation reactions makes it impossible to identify unambiguously the isotopic compositions of trapped He and Ne in some classes of meteorites, e.g., iron meteorites.

2. Isotopic Compositions of Argon, Krypton and Xenon

The isotopic compositions of Ar, Kr, and Xe in air are well known and are routinely used as standards to determine the mass discrimination of spectrometers used for isotopic analyses of noble gases. Values used in this report are given by NIER (1950a) for Ar, NIEF (1960) for Kr, and NIER (1950b) for Xe.

The isotopic compositions of these elements in crustal rocks frequently contain the decay products of ^{40}K and ^{238}U. Because of the high concentrations of K and U in surface rocks and their ability to alter the isotopic compositions of Ar, Kr, and Xe by spontaneous decay, vigilance for possible contamination from crustal material is necessary in selecting samples which might represent noble gases from the mantle. For this reason, we will use only the results of analyses on gas-rich samples which have been brought to the surface intact, as xenoliths, to represent noble gases in the mantle. The results of noble gas analyses on the following three gas-rich mantle samples will be used in this report, a kearsutitic amphibole inclusion in a brecciated nephelinitic basalt from Kakanni, New Zealand (SAITO *et al.*, 1978), a nickel-iron alloy-bearing rock, josephinite, which occurs as stream pebbles in an area of harzburgite, serpentinite and dioritic dikes in Josephine County, Oregon (DOWNING *et al.*, 1977), and an amphibole mineral separated from a peridotite xenolith from Dish Hill, California (THOMPSON *et al.*, 1978). Argon accounts for the bulk of the heavy noble gases in these three samples, and the concentrations of ^{36}Ar are 1.5×10^{-8}, 3.5×10^{-8}, and 2.0×10^{-8} ccSTP/g, respectively. Only the latter two mantle samples were included in our presentation at the Hakone Conference, but the authors of the report on noble gases in the New Zealand amphibole have kindly consented to our request to include their results in this discussion.

Noble gases in an olivine xenolith from Hawaii are not considered in this report for the following reasons. (i) The concentrations of heavy noble gases in the Hawaiian xenolith are almost two orders of magnitude lower than those in the above three samples, e.g., ^{36}Ar concentrations are about 0.035×10^{-8} cc STP/g in the Hawaiian xenolith. (ii) It is not possible to resolve the isotope spectra of Ar, Kr, and Xe clearly into simple mixtures of trapped plus radiogenic components because part of the radiogenic ^{129}Xe appears to have been carried into the Hawaiian xenolith with CO_2 (HENNECKE and MANUEL, 1975b), and the results of other studies show that radiogenic ^{40}Ar and fissiogenic $^{131-136}Xe$ accompany radiogenic ^{129}Xe in CO_2 gas (HENNECKE and MANUEL, 1975a).

To compare the isotopic compositions of Xe and Kr, we normalize the spectra to those observed in average carbonaceous chondrites, AVCC.

$$g_m^i = ({}^iX/{}^mX)_{sample}/({}^iX/{}^mX)_{AVCC} \quad (1)$$

Here m and i are the mass numbers of the reference isotope and any other isotope, respectively, of a given noble gas, X, and an average of the values given by KRUMMENACHER *et al.* (1962) and EUGSTER *et al.* (1967) is used for the isotopic composition in AVCC.

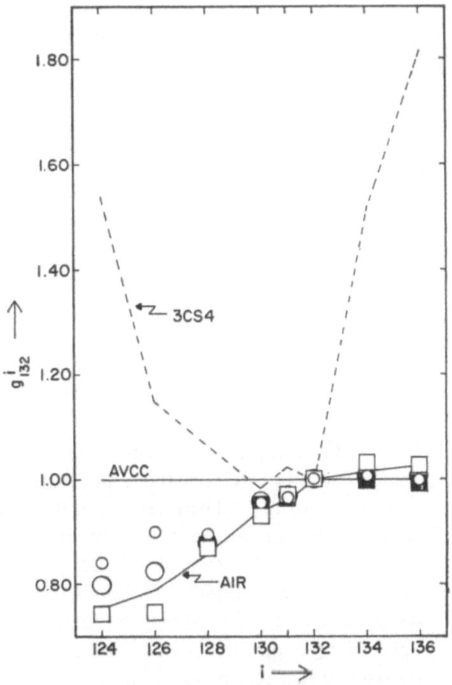

Fig. 1. The isotopic compositions of Xe in the mantle samples resemble those of Xe in air and in iron meteorites. Isotopic spectra of Xe in average carbonaceous chondrites, air and the carbon-rich residue (3CS4) of Allende are represented by the horizontal line, the other solid line and the broken line, respectively. Large open squares, New Zealand amphibole; large open circles, Oregon josephinite; large filled squares, California amphibole; small open circles, Cape York iron meteorite.

Figure 1 compares the isotopic spectra of Xe in the above three mantle samples, expressed as values of g_{132}^i, with Xe released from the Cape York iron meteorite at 1,500°C (HENNECKE and MANUEL, 1977), the Xe trapped in the carbon-rich residue, 3CS4, of the Allende carbonaceous chondrite (LEWIS *et al.*, 1975) and the Xe in air and bulk carbonaceous chondrites. Abundances of ^{129}Xe are not included in Fig. 1 because of possible interference from the decay of extinct ^{129}I. The Xe spectra in air and in meteorite samples are represented by lines or small symbols; large symbols are used to represent the isotopic compositions of Xe in the mantle samples.

As can be seen from Fig. 1, the isotopic composition of Xe in the mantle is similar to that in air and in iron meteorites. Xe in the mantle is isotopically distinct from AVCC Xe, and there is no evidence that Xe in the mantle contains a component of the strange Xe found in carbon-rich residues of Allende. Isotopic analyses on Xe in Ca-poor achondrites (ROWE and BOGARD, 1966) and in some breccias from the Apollo 16 mission (LIGHTNER and MARTI, 1974) also show atmosphere-like Xe, but in both cases it has been suggested that the Xe may represent terrestrial contamination rather than an indigenous component (ROWE and BOGARD, 1966; NIEMEYER and LEICH, 1976).

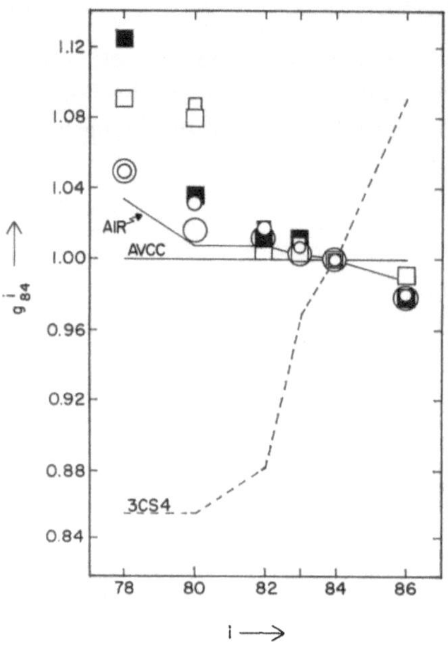

Fig. 2. The isotopic compositions of Kr in the mantle samples resemble those of Kr in air, Ca-poor achondrites and iron meteorites. Symbols and lines are as defined for Fig. 1, with the additional small open square used to represent Kr in the Peña Blacea Springs achondrite.

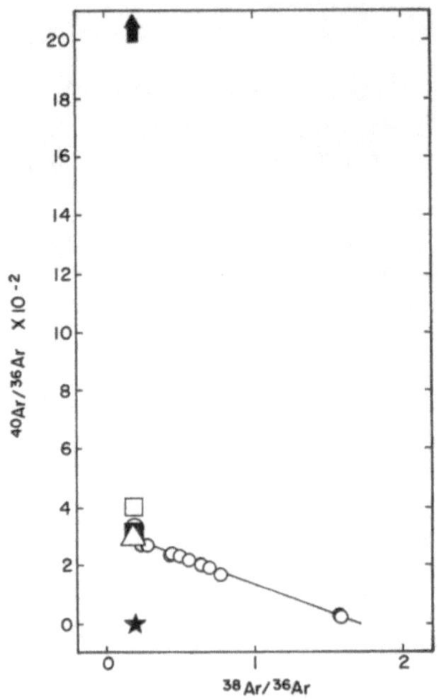

Fig. 3. The isotopic compositions of Ar in the mantle samples resemble those of Ar in air and in iron meteorites. The line defines the isotopic compositions of Ar generated by mixtures of cosmogenic and atmosphere-like Ar. Dark arrow, mantle estimate of $^{40}Ar/^{36}Ar > 2 \times 10^{3}$; dark star, carbon-rich, 3CS4, residue of Allende; open triangle, air; small open circles, Ar released from three iron meteorites.

The isotopic spectra of Kr in the three mantle samples are shown as values of g_{84}^{i} in Fig. 2. The same symbols have been used here as in Fig. 1, except that small open squares have been added to show the isotopic composition of Kr in the Ca-poor achondrite, Peña Blanca Springs (BOGARD, 1967). The isotopic spectrum of Kr in the mantle resembles that in air but it may deviate from atmospheric Kr in much the same manner as does Kr in iron meteorites and Ca-poor achondrites. Kr in the mantle is isotopically distinct from AVCC Kr and shows an opposite anomaly pattern to that seen in the carbon-rich residue of Allende.

The isotopic composition of Ar in the three mantle samples is shown in Fig. 3. The symbols represent the same samples as in Figs. 1 and 2, except that Ar released at all temperature fractions of three iron meteorite samples (HENNECKE and MANUEL, 1977) is shown by the small open circles, the large open triangle represents atmospheric Ar, the star represents the isotopic composition of Ar in the carbon-rich 3CS4 residue of Allende (LEWIS *et al.*, 1975), and the dark arrow near the top of Fig. 3 represents the minimum value of $^{40}Ar/^{36}Ar \geq 2000$ which OZIMA (1975) estimates for Ar in the mantle.

The solid line shows the isotopic compositions of Ar which could be produced by mixtures of atmospheric Ar with spallation products generated in iron meteorites during their exposure to cosmic rays. The alignment of Ar isotope ratios in iron meteorites, as shown in Fig. 3, has been interpreted as evidence that these sampled a component of primitive Ar similar to that found in the Earth's atmosphere (HENNECKE and MANUEL, 1977). It can be seen from Fig. 3 that the value of $^{40}Ar/^{36}Ar$ in all three mantle samples are far below the minimum estimate shown by the dark arrow. Approximate corrections for radiogenic ^{40}Ar in these three mantle samples yield values of $^{40}Ar/^{36}Ar = 298, 300$, and ≤ 230 for the Dish Hill amphibole (THOMPSON *et al.*, 1978), the josephinite (DOWNING *et al.*, 1977) and the New Zealand amphibole (SAITO *et al.*, 1978), respectively. It has been suggested that these results indicate that ambient Ar in the mantle is characterized by $^{38}Ar/^{36}Ar < 0.185$, $^{40}Ar/^{36}Ar = 3 \times 10^2$ (THOMPSON *et al.*, 1978; DOWNING *et al.*, 1977). Except for the calculated value of $^{40}Ar/^{36}Ar \leq 230$ in the New Zealand amphibole (SAITO *et al.*, 1978), there is no indication of an Ar component in the mantle similar to that seen in carbon-rich residues of Allende. Similarities between the values of $^{40}Ar/^{36}Ar$ in air, in iron meteorites and in the mantle suggest that the atmospheric inventory of ^{40}Ar may be primordial rather than radiogenic.

3. Conclusions

a) Similarities between the values of $^{40}Ar/^{36}Ar$ in the mantle and the atmosphere suggest that the atmospheric inventory of ^{40}Ar was generated very early, perhaps as the Earth accumulated (ALFVÉN and ARRHENIUS, 1976), rather than by later leakage through the K-rich crust. The atmospheric inventory of ^{40}Ar may be primordial rather than radiogenic.

b) The isotopic compositions of Ar, Kr, and Xe in the mantle are distinct from the compositions of these elements in carbonaceous chondrites. Noble gases in the mantle show no evidence for the isotopically strange components seen in carbon-rich residues of chondrites, but show similarities to the noble gases found in iron meteorites, in air, in Ca-poor achondrites and in some lunar breccias.

c) Differences in the isotopic compositions of noble gases in the mantle and in carbonaceous chondrites suggest that the latter constituted a very small part of the material which accumulated as the Earth's mantle.

d) The isotopic compositions of noble gases in the mantle are compatible with the suggestion of local element synthesis (MANUEL and SABU, 1975) and with models of inhomogeneous accumulation of the Earth (EUCKEN, 1944; TUREKIAN and CLARK, 1969, 1975; VINOGRADOV, 1975; ALFVÉN and ARRHENIUS, 1976).

The assistance of M.S. Phyllis Johnson in various stages of this manuscript is greatfully acknowledged. This work was supported by a grant from the National Science Foundation, NSF-EAR-76-03956, and by funds from the University of Missouri-Rolla.

REFERENCES

ALFVÉN, H. and G. ARRHENIUS, *Evolution of the Solar System*, pp. 499–500, NASA Publication, 1976.

ANDERS, E., How well do we know "cosmic" abundances?, *Geochim. Cosmochim. Acta*, **35**, 516–522, 1971.

BAHCALL, J. N. and R. DAVIS, Jr., Solar-neutrinos: A scientific puzzle, *Science*, **191**, 264–267, 1976.

BLACK, D. C., Alternative hypothesis for the origin of CCF xenon, *Nature*, **253**, 417–419, 1975.

BLAKE, J. B. and D. N. SCHRAMM, Nucleosynthesis and anomalous Xe and Kr in carbonaceous chondrites, *Nature*, **263**, 707–708, 1976.

BOGARD, D. D., Krypton anomalies in achondritic meteorites, *J. Geophys. Res.*, **72**, 1299–1309, 1967.

BOGARD, D. D., Noble gases in meteorites, *Trans. Am. Geophys. Union*, **52**, 429–435, 1971.

BOULOS, M. S. and O. K. MANUEL, The xenon record of extinct radioactivities in the earth, *Science*, **174**, 1334–1336, 1971.

CAMERON, A. G. W. and J. W. TRURAN, The supernova trigger for formation of the solar system, *Icarus*, **30**, 447–461, 1977.

CLARKE, W. B., M. A. BEG, and H. CRAIG, Excess ^{3}He in the sea: Evidence for terrestrial primordial helium, *Earth Planet. Sci. Lett.*, **6**, 213–220, 1969.

CLAYTON, D. D., Extinct radioactivities: Trapped residuals of presolar grains, *Astrophys. J.*, **199**, 765–769, 1975.

CLAYTON, D. D., Spectrum of carbonaceous-chondrite fission xenon, *Geochim. Cosmochim. Acta*, **40**, 563–565, 1976.

CLAYTON, R. N., L. GROSSMAN, and T. K. MAYEDA, A component of primitive nuclear composition in carbonaceous meteorites, *Science*, **182**, 485–488, 1973.

CLAYTON, R. N., N. ONUMA, and T. K. MAYEDA, A classification of meteorites based on oxygen isotopes, *Earth Planet. Sci. Lett.*, **30**, 10–18, 1976.

DOWNING, R. G., E. W. HENNECKE, and O. K. MANUEL, Josephinite: A terrestrial alloy with radiogenic xenon-129 and the noble gas imprint of iron meteorites, *Geochem. J.*, **11**, 219–229, 1977.

EUCKEN, A., Physikalisch-chemische Betrachtungen über die früheste Entwicklungsgeschichte der Erde: Nachr. d., *Akad. Wiss. Gottingen, Math.—Phys. Kl., Heft,* **1**, 1–25, 1944.

EUGSTER, O., P. EBERHARDT, and J. GEISS, The isotopic composition of krypton in unequilibrated and gas-rich chondrites, *Earth Planet. Sci. Lett.*, **2**, 385–393, 1967.

FRICK, U. and R. K. MONIOT, Noble gases in carbonaceous residues from the Orgueil and Murray meteorites, paper presented at the 39th Annual Meeting of the Meteoritical Society, Bethlehem, PA, Abstract No. 72, 1976.

HENNECKE, E. W. and O. K. MANUEL, Noble gases in CO_2 well gas, Harding County, New Mexico, *Earth Planet. Sci. Lett.*, **27**, 346–355, 1975a.

HENNECKE, E. W. and O. K. MANUEL, Noble gases in an Hawaiian xenolith, *Nature*, **257**, 778–780, 1975b.

HENNECKE, E. W. and O. K. MANUEL, Argon, krypton and xenon in iron meteorites, *Earth Planet. Sci. Lett.*, **36**, 29–43, 1977.

JOVANOVIC, S. and G. W. REED, ^{196}Hg and ^{202}Hg isotopic ratios in chondrites: Revisited, *Earth Planet. Sci. Lett.*, **31**, 95–100, 1976.

KRUMMENACHER, D., C. M. MERRIHUE, R. O. PEPIN, and J. H. REYNOLDS, Meteoritic krypton and barium versus the general isotopic anomalies in meteoritic xenon, *Geochim. Cosmochim. Acta*, **26**, 231–249, 1962.

LEWIS, R. S., B. SRINIVASAN, and E. ANDERS, Host phase of a strange xenon component in Allende, *Science*, **190**, 1251–1262, 1975.

LIGHTNER, B. D. and K. MARTI, Lunar trapped xenon, *Geochim. Cosmochim. Acta*, **2**, 2023–2031, 1974.

MAMYRIN, B. A., I. N. TOLSTIKHIN, G. S. ANUFRIER, and I. L. KAMENSKIY, Anomalous isotopic compositions of helium in volcanic gases, *Dokl. Akad. Nauk. USSR*, **184**, 1197–1200, 1969.

MANUEL, O. K. and D. D. SABU, Elemental and isotopic inhomogeneities in noble gases: The case for local synthesis of the chemical elements, *Trans. Mo. Acad. Sci.*, **9**, 104–122, 1975.

MANUEL, O. K. and D. D. SABU, Noble gas anomalies and synthesis of the chemical elements, Submitted to *Geochim. Cosmochim. Acta*, 1977a.

MANUEL, O. K. and D. D. SABU, Strange xenon, extinct superheavy elements and the solar neutrino puzzle, *Science*, **195**, 208–209, 1977b.

MANUEL, O. K., E. W. HENNECKE, and D. D. SABU, Xenon in carbonaceous chondrites, *Nature*, **240**, 99–101, 1972.

MAZOR, E., D. HEYMAN, and E. ANDERS, Noble gases in carbonaceous chondrites, *Geochim. Cosmochim. Acta*, **34**, 781–824, 1970.

NIEF, G., Reported in isotopic abundance ratios given for reference samples stocked by the National Bureau of Standards, edited by F. Mohler, NBS Tech. Note, 51, 1960.

NIEMEYER, S. and D. A. LEICH, Atmospheric rare gases in lunar rocks 60015, *Proc. Seventh Lunar Sci. Conf., Geochim. Cosmochim. Acta, Suppl. 7*, **2**, 587–597, 1976.

NIER, A. O., A redetermination of the relative abundances of the isotopes of carbon, nitrogen, oxygen, argon and potassium, *Phys. Rev.*, **77**, 789–793, 1950a.

NIER, A. O., A redetermination of the relative abundances of the isotopes of neon, krypton, rubidium, xenon and mercury, *Phys. Rev.*, **79**, 450–454, 1950b.

OZIMA, M., Ar isotopes and Earth-atmosphere evolution models, *Geochim. Cosmochim. Acta*, **39**, 1127–1134, 1975.

REYNOLDS, J. H., *Proc. Soviet American Conf. Cosmochemistry Moon, Planets, Moscow, June 4–8, 1974*, edited by J. H. Pomeroy and H. Hubbard, Part 2, pp. 771–780, NASA Publication, Washington, D.C., 1977.

RINGWOOD, A. E., Chemical evolution of the terrestrial planets, *Geochim. Cosmochim. Acta*, **30**, 41–104, 1966.

ROWE, M. W. and D. D. BOGARD, Isotopic composition of xenon from Ca-poor achondrites, *J. Geophys. Res.*, **71**, 4183–4188, 1966.

SABU, D. D. and O. K. MANUEL, Xenon record of the early solar system, *Nature*, **262**, 28–32, 1976.

SAITO, K., A. R. BASU, and E. C. ALEXANDER, Jr., Planetary type noble gases in an upper mantle-derived amphibole, *Earth Planet. Sci. Lett.*, **39**, 274–280, 1978.

THOMPSON, D. P., A. R. BASU, E. W. HENNECKE, and O. K. MANUEL, Noble gases in the Earth's mantle, *Phys. Earth Planet. Inter.*, 1978 (in press).

TUREKIAN, K. K. and S. P. CLARK, Jr., Inhomogeneous accumulation of the earth from the primitive solar nebula, *Earth Planet. Sci. Lett.*, **6**, 346–348, 1969.

TUREKIAN, K. K. and S. P. CLARK, Jr., The non-homogeneous accumulation model for terrestrial planet formation and the consequences for the atmosphere of Venus, *J. Atm. Sci.*, **32**, 1257–1261, 1975.

UREY, H. C., *The Planets* (*Their Origin and Development*), 245 pp., Yale University Press, New Haven, CT., 1952.

VINOGRADOV, A. P., Formation of the metal cores of the planets, *Geokhimiya*, No. 10, 1427–1431, 1975.

The Composition and History of the Martian Atmosphere

T. OWEN

Earth and Space Sciences, State University of New York,
Stony Brook, New York 11790, U. S. A.

The composition of the atmosphere of Mars near the surface has been determined by a series of measurements with the mass spectrometers on the two Viking Landers. The results agree with abundances for N_2 and ^{40}Ar determined by the upper atmosphere investigations and add Ne, ^{36}Ar, Kr, and Xe to the list of gases known to be in the Martian atmosphere. The abundance pattern of the noble gases is found to be similar to that in the Earth's atmosphere and the "planetary component" found in meteorites. This permits inferences about abundances of other volatiles which suggest that the atmosphere of Mars may once have been much more massive than it is at present, but never as massive as ours. Details of these arguments may be found in the series of papers referenced in the text.

The composition of the Martian atmosphere was determined by means of analyses of samples admitted through a special inlet system to the mass spectrometer of the GCMS on the two Viking Landers. The details of the instrument have been described by ANDERSON *et al.* (1972) and BIEMANN (1974). The first results provided a confirmation of the presence of N_2 and ^{40}Ar as reported by NIER *et al.* (1976a) and the discovery of ^{36}Ar (OWEN and BIEMANN, 1976). The ratio $^{36}Ar/^{40}Ar$ was found to be one-tenth the terrestrial value, while the total ^{40}Ar in the Martian atmosphere is much lower than scaling by the terrestrial argon abundance would predict. Subsequent investigations gained sensitivity by the cyclical use of chemical scrubbers in the gas inlet system to remove over 99.9% of the carbon dioxide in a given sample, thereby permitting the enrichment of minor and trace gases in the Martian atmosphere.

Using this procedure, we were able to confirm the discovery by NIER *et al.* (1976b) that $^{15}N/^{14}N$ on Mars is about 1.7 times the terrestrial ratio (BIEMANN *et al.*, 1976) and we were also able to discover the presence of krypton, xenon (OWEN *et al.*, 1976), and neon (OWEN *et al.*, 1977). A list of all the constituents presently known to exist in the lower atmosphere of Mars is given in Table 1. The uncertainties in these numbers are still being reviewed but are obviously largest for the least abundant species. A list of isotope ratios is given in Table 2.

The pattern of relative abundances exhibited by the noble gases in the Martian atmosphere is remarkably similar to the pattern found in the atmosphere of our own planet and in the "planetary component" of chondritic meteorites (Fig. 1). (The deviation shown by xenon in the Earth's atmosphere has been explained by preferential trapping of this gas in terrestrial shales (CANALAS *et al.*, 1968). It is not yet clear whether xenon is as deficient on Mars and if so, what the cause of this deficiency might be (FANALE *et al.*, 1978).) The general similarity in abundance patterns in these three cases provides support for the idea that the noble gases were fractionated prior to the accretion of the planets, and offers the possibility of trying to reconstruct the total volatile

Table 1. Composition of the lower atmosphere.

Carbon dioxide (CO_2)	95.32 percent
Nitrogen (N_2)[1]	2.7 percent
Argon (Ar)[1]	1.6 percent
Oxygen (O_2)	0.13 percent
Carbon monoxide (CO)	0.07 percent
Water vapor (H_2O)	0.03 percent[2]
Neon (Ne)[1]	2.5 ppm
Krypton (Kr)[1]	0.3 ppm
Xenon (Xe)[1]	0.08 ppm
Ozone (O_3)	0.03 ppm[2]

1) Discovered by Viking experiments.
2) Variable.

Table 2. Isotope ratios in atmospheric gases.

Ratio	Earth	Mars[1]
$^{12}C/^{13}C$	89	90
$^{16}O/^{18}O$	499	500
$^{14}N/^{15}N$	277	165
$^{40}Ar/^{36}Ar$	296	3,000
$^{129}Xe/^{132}Xe$	0.97	2.5

1) Uncertainties in these values are presently ±10% except for Ar and Xe (see text).

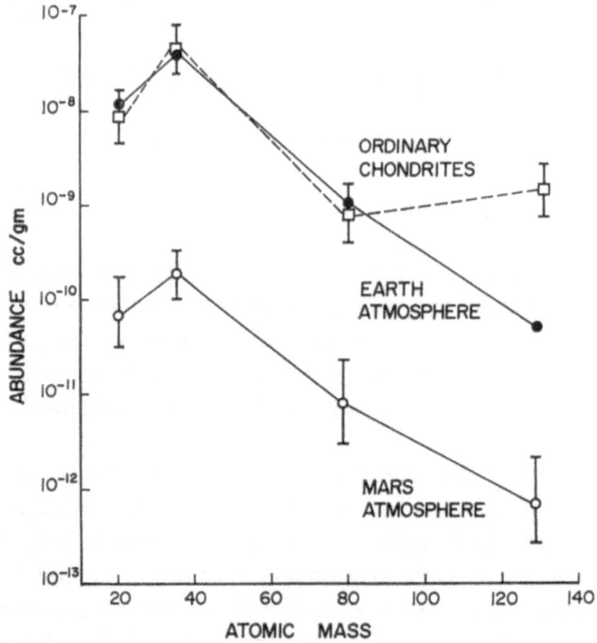

Fig. 1. The abundances of noble gases (per gram of planet or meteorite) in the atmospheres of Mars and the Earth compared with abundances in ordinary chondrites. Meteoritic abundances are from SIGNER (1964). Error bars are only approximate since systematic effects have not yet been fully determined.

inventory on Mars based on terrestrial and meteoritic abundances.

The most comprehensive model of this type to date has been developed by ANDERS and OWEN (1977) who use the C3V carbonaceous chondrites as the most likely example of a surviving remnant of the volatile-rich material that coated all of the inner planets. Adopting thallium as an index volatile element, and guessing a Tl abundance similar to that found on the moon, it is possible to reconstruct the entire Martian volatile inventory in a model that is found to be in good agreement with available observations.

This agreement requires that Mars has a smaller total inventory and that it has not outgassed as thoroughly as the Earth, conclusions that are substantiated by the value of $^{36}Ar/^{40}Ar$ in the atmosphere. Evidently small bodies in the inner solar system (Mars, the moon, the eucrite parent body) did not receive the same proportion of volatile elements per unit mass as did the Earth and Venus. The low volatile element abundance found on Mars is contrary to the prediction of homogeneous accretion models, but consistent with the theory of heterogeneous accretion. However, there is as yet no version of this theory that can give a quantitative explanation of the observed dependence of volatile content on planetary size.

Given the total inventory, our model allows us to predict that Mars has outgassed fifteen times more CO_2 and five times more N_2 than are presently found in the atmosphere, and a total amount of water equivalent to a 9-meter layer over the entire planet. The conclusions about the abundances of N_2 and H_2O agree well with minimum estimates based on a model for non-thermal escape that is calibrated by presently observed isotope ratios for N and O (MCELROY *et al.*, 1977). The CO_2 abundance is not estimated by that approach. The apparent enrichment of ^{129}Xe is also predicted by the C3V model, which suggests that the real anomaly may be a deficiency on Earth rather than an enrichment on Mars. ^{129}Xe is produced by the decay of ^{129}I (half-life 16.7 my), so the anomaly may arise from a difference in the manner in which the volatile rich veneer was accumulated by the two planets (ANDERS and OWEN, 1977).

If the atmosphere ever contained the full amount of outgassed CO_2, the surface pressure on Mars would have been on the order of 150 mb. Under these conditions, liquid water could exist if the mean temperature were elevated by some type of atmospheric greenhouse effect and the well known dendritic channels could have been carved at this time. But details of the climate modification and a time scale for the more clement

Table 3. Atmospheric abundances on Earth and Venus.

Gas	Atmospheres		
	Earth		Venus
	Now	Total[1]	Now
N_2	78%	1.5%	1.8%
O_2	21	Trace	Trace
Ar	0.9	190 ppm	200 ppm
CO_2	0.03	98%	98%
Water	3 km	3 km	Trace
Pressure	1 atm	~70 atm	88±3 atm

[1] No weathering, no life.

conditions have not yet been established.

The basic approach embodied by this model should also be applicable to Venus. Unfortunately, our current information about the atmosphere of Venus is very incomplete, but a preliminary report by SURKOV (1977) on abundances of CO_2, N_2, and ^{40}Ar determined by the Venera 9 and 10 missions indicates that they are similar to the values in the total terrestrial inventory (Table 3). This is exactly what heterogeneous accretion would predict (TUREKIAN and CLARK, 1975). Further tests will be provided by the NASA Pioneer mission to Venus, to be launched in May, 1978.

REFERENCES

ANDERS, E. and T. OWEN, Mars and Earth: Origin and abundance of volatiles, *Science*, **198**, 453–465, 1977.

ANDERSON, D. M., K. BIEMANN, L. E. ORGEL, J. ORO, T. OWEN, G. P. SHULMAN, P. TOULMIN, III, and H. C. UREY, Mass spectrometric analysis of organic compounds, water, and volatile constituents in the atmosphere and surface of Mars, *Icarus*, **16**, 111–138, 1972.

BIEMANN, K., Test results on the Viking gas chromatograph—mass spectrometer experiment, *Origins Life*, **5**, 417–430, 1974.

BIEMANN, K., T. OWEN, D. R. RUSHNECK, A. L. LAFLEUR, and D. W. HOWARTH, The atmosphere of Mars near the surface: Isotope ratios and upper limits on noble gases, *Science*, **194**, 76–78, 1976.

CANALAS, R. A., E. C. ALEXANDER, Jr., and O. K. MANUEL, Terrestrial abundance of noble gases, *J. Geophys. Res.*, **73**, 3331–3334, 1968.

FANALE, F. P., W. A. CANNON, and T. OWEN, Mars: Regolith adsorption and the relative concentrations of atmospheric rare gases, *Geophy. Res. Lett.*, **5**, 77–80, 1978.

MCELROY, M. B., T. Y. KONG, and Y. L. YUNG, Photochemistry and Evolution of Mars' atmosphere: A Viking perspective, *J. Geophys. Res.*, **82**, 4379–4388, 1977.

NIER, A. O., W. B. HANSON, A. SEIFF, M. B. MCELROY, N. W. SPENCER, R. J. DUCKETT, T. C. D. KNIGHT, and W. S. COOK, Composition and structure of the Martian atmosphere: Preliminary results from Viking 1, *Science*, **193**, 786–788, 1976a.

NIER, A. O., M. B. MCELROY, and Y. L. YUNG, Isotopic composition of the Martian atmosphere, *Science*, **194**, 68–70, 1976b.

OWEN, T. and K. BIEMANN, Composition of the atmosphere at the surface Mars: Detection of argon-36 and preliminary analysis, *Science*, **193**, 801–803, 1976.

OWEN, T., K. BIEMANN, D. R. RUSHNECK, J. E. BILLER, D. W. HOWARTH, and A. L. LAFLEUR, The atmosphere of Mars: Detection of krypton and xenon, *Science*, **194**, 1293–1295, 1976.

OWEN, T., K. BIEMANN, D. R. RUSHNECK, J. E. BILLER, D. W. HOWARTH, and A. L. LAFLEUR, The composition of the atmosphere at the surface of Mars, *J. Geophys. Res.*, **82**, 4635–4640, 1977.

SIGNER, P., Primordial rare gases in meteorites, in *The Origin and Evolution of Atmospheres and Oceans*, edited by P. J. Brancazio and A. G. W. Cameron, pp. 183–190, J. Wiley, New York, 1964.

SURKOV, Yu. A., Paper presented at NASA Lunar Science Conference, Houston, Texas, March, 1977 (in press).

TUREKIAN, K. K. and S. P. CLARK, Jr., The non-homogeneous accumulation model for terrestrial planet formation and the consequences for the atmosphere of Venus, *J. Atmos. Sci.*, **32**, 1257–1261, 1975.

THEORETICAL STUDIES

Nuclear Components in the Atmosphere

T. J. BERNATOWICZ and F. A. PODOSEK

Department of Earth and Planetary Sciences and McDonnell Center for the Space Sciences, Washington University, St. Louis, Missouri 63130, U. S. A.

A nuclear component is a contribution to the earth's composition which has been generated by nuclear-specific processes such as radioactive decay and nuclear particle reactions. Study of these components is useful in describing the origin and evolution of the earth.

The best known nuclear components in the earth's atmosphere are ^{40}Ar and ^{4}He, which dominate the earth's inventory of these isotopes and which originate in radioactive decay of ^{40}K and of ^{238}U, ^{235}U, and ^{232}Th, respectively. Smaller but still significant contributions to atmospheric ^{129}Xe and ^{136}Xe, from β decay of ^{129}I and spontaneous fission of ^{244}Pu and ^{238}U, respectively, are also reasonably well established. The possible presence of a variety of other components is conjectural: ^{21}Ne and ^{3}He from spallation in solids before accretion, ^{21}Ne from $^{18}O\ (\alpha, n)^{21}Ne$ reactions, ^{14}N from ^{14}C in the solar nebula.

The radiogenic components (^{4}He, ^{40}Ar, ^{129}Xe, ^{136}Xe) are particularly useful in study of the degassing of the earth, both the degree to which the earth is degassed and the history of this degassing. We discuss these topics within the framework of a simple first-order degassing model. We feel that even so simple a model is underconstrained by available data, however, and that even the basic question of whether or not the earth is only partially or nearly completely degassed of its primordial volatiles cannot be answered definitively at present. Nevertheless, we feel that the balance of evidence favors a relatively low degree of degassing. Calculations based on likely total inventories indicate that most of the radiogenic gases are still in the solid earth. Observations of isotopic ratio parameters—$^{40}Ar/^{36}Ar$ in some mantle materials, and xenon compositions in ancient sedimentary rocks—suggest the opposite conclusion: that degassing was rapid and extensive. A simple model consistent with both sets of data is that the present atmosphere originated by rapid degassing of only a small part of the earth, such as the upper mantle, but that most of the earth's supply of volatiles remains in relatively undegassed earth materials. We reemphasize, however, our feeling that any definitive statement of degassing history is premature.

1. Introduction

The presence of nuclear components in the earth affords opportunities to enhance our understanding of our planet, in terms of both the materials from which it was made and its subsequent evolution. In this review, we will use the term "nuclear component" to describe a source or reservoir of earth material which owes its identity to the nuclear properties of its constituents rather than their masses or chemical properties. In a sense, this distinction is rather arbitrary since ultimately the synthesis of all the elements is based on nuclear properties; it is possible, however, to make a distinction which is both unambiguous and useful. As a first approximation, let us suppose that at some time the solar system had a uniform composition, both elemental and isotopic. We will designate this composition by the term *primordial* (use of this term implicitly requires specification of an epoch, since the primordial composition changes with time). Accessible samples of solar system material, including samples of the earth, represent modifications of this primordial composition. For the most part, the first order modifications are those based on chemical properties. To a lesser extent, some processes in the history of different samples have also resulted in modifications of isotopic composition through mass-dependent fractionation. In addition, there have been modifications based on specific nuclear properties, and in this paper the term *nuclear component* will be used to designate such modifications generated after the establishment of a primordial composition. The best known nuclear components are those which originate in the decay of natural radionuclides, but other possibilities include nuclear reactions induced by galactic and solar cosmic rays and by particles emitted in radioactive decay.

We must now add the qualification that it appears that the solar system never was completely homogeneous, and that at the time when solids were condensing from the solar nebula, the time usually implicit in the designation of primordial composition, there were inhomogeneities on both micron and astronomical unit scales. The principal evidence for this is a variation of oxygen isotopic composition among various samples (CLAYTON *et al.*, 1973, 1976). In principle, these variations could be "nuclear" components, generated after a thorough homogenization, but this thesis is difficult to defend and the consensus is that these variations reflect nonhomogenization of components of different galactic nucleosynthetic origins. Isotopic variations in a few other elements may arise similarly, although in these cases a post-homogenization origin is not so easy to exclude. Whatever the case with other elements, however, the oxygen observations indicate that there is not a well-defined primordial composition applicable to all the parts of the solar system.

Nevertheless, individual samples of the solar system, hand-specimen size or planet size, have well-defined total compositions and it appears that, except for "high-temperature" inclusions in carbonaceous chondrites, the history of most accessible objects has included processes which homogenized any preexisting microscopic inhomogeneities. Among terrestrial samples in particular, there is no evidence for isotopic variations which have been preserved since before the origin of the solar system, so that it is possible to define a primordial composition for the earth. Because of the large scale inhomogeneities, the primordial composition of the earth may not be the same as that of any other object in the solar system. This may occasion some difficulties in attempting to determine primordial earth composition by comparison with extraterrestrial samples, but at the same time it offers the compensating possibility of deducing otherwise unobtainable information about the origin and formation of the earth.

Most of the attention of this paper is devoted to the noble gases. Although these gases constitute only a very minor part of the atmosphere, their chemical inertness is responsible for a much better established systematics in both the earth and in extraterrestrial samples. More important, their low abundances in both classes of samples render visible in their isotopic structures the presence of small additions of nuclear components which would be unobservable in elements present in cosmic proportions.

2. Planetary Gases

It is an important generalization that trapped noble gases (those not generated by in situ nuclear reactions) in extraterrestrial samples occur in two characteristic patterns (SIGNER and SUESS, 1963). The first pattern, designated *solar* and represented in Fig. 1 by the analysis of ilmenite from lunar soil, is characterized by relative noble gas abundances roughly comparable to solar proportions. It is believed that these gases in lunar soils originate primarily in solar wind implantation. Some meteorites also display solar relative abundances, and it is believed that they too originate in essentially the same way. In contrast, most meteorites are characterized by the *planetary* pattern, represented by Murray in Fig. 1, in which the heavier gases are progressively enriched, relative to solar

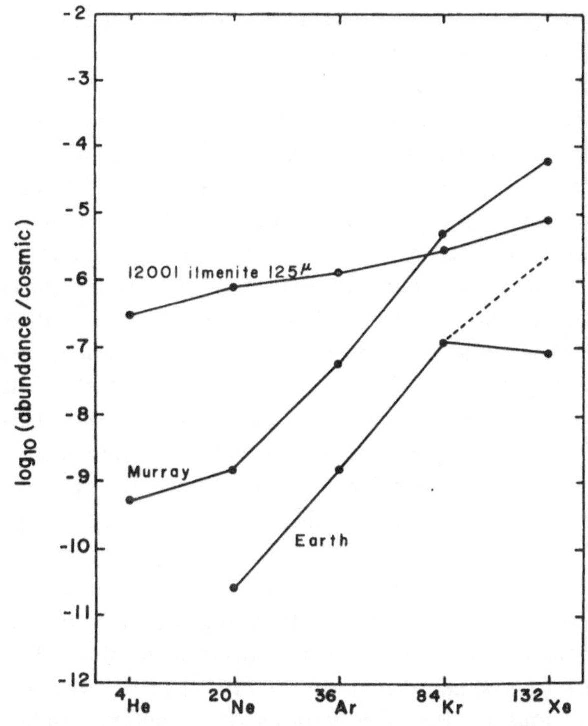

Fig. 1. Elemental abundance patterns of noble gases in lunar soil, the carbonaceous chondrite Murray, and the terrestrial atmosphere. The gases in the 125 μ ilmenite fraction of lunar soil 12001 (EBERHARDT *et al.*, 1972) exhibit the "solar" pattern, where gases occur in nearly cosmic proportion; in contrast, the Murray abundances (REYNOLDS, 1960) exhibit the "planetary" pattern, in which, relative to cosmic proportions, the heavier gases are progressively enhanced. The terrestrial abundances clearly more nearly resemble the planetary rather than the solar pattern, except for an underabundance of xenon relative to the amount expected in planetary gas (dashed segment). Data are plotted in terms of a depletion factor, the ratio of observed gas concentration to the value expected for cosmic proportions (CAMERON, 1973). The cosmic abundance is calculated assuming 17% silicon content for each sample. For the earth, concentration is calculated by dividing the atmospheric inventory by the mass of the earth.

proportions. Generation of this planetary pattern is not well understood, although several hypotheses have been advanced (JOKIPII, 1964; LANCET and ANDERS, 1973; FANALE and CANNON, 1974; SABU and MANUEL, 1976).

The abundances of noble gases in the earth's atmosphere are also shown in Fig. 1. The relative abundances of neon, argon and krypton are quite different from solar gas composition and essentially indistinguishable from planetary composition, and it is generally held that the gases in the earth's atmosphere are "planetary" rather than solar. Unfortunately, it is not possible to extend this comparison to helium, which escapes from the atmosphere, nor to xenon, which is underabundant compared to planetary composition (Fig. 1). It has been commonly accepted that the xenon underabundance is not a strong argument against identification of terrestrial gases with the meteoritic planetary pattern; plausible hypotheses for this underabundance stipulate that xenon, the heaviest stable noble gas, may have been less efficiently degassed from the solid earth than the other gases or that most of the earth's atmospheric xenon inventory is adsorbed on shales (FANALE and CANNON, 1971).

OWEN *et al.* (1977) have recently observed that the abundance pattern of noble gases in the atmosphere of Mars is strikingly similar to that in the atmosphere of the earth. This further strengthens the concept that planetary gas has a distinct identity of its own and is not the unique result of the particular evolution of a single planet. The match of Martian and terrestrial abundance patterns extends to the deficiency of xenon relative to the meteoritic planetary pattern. This suggests that at least the noble gas components of the earth and Mars are more closely related to each other than either is related to meteoritic gas. It further suggests that the xenon underabundance in both planetary atmospheres may be an intrinsic primordial characteristic rather than a modification peculiar to the earth (or Mars). Comparison of terrestrial noble gases with meteoritic planetary gases remains a very useful paradigm, since the meteoritic gases are still the closest accessible extraterrestrial analog to terrestrial gases, but the results of OWEN *et al.* (1977) underscore the precept that this analogy should not be accepted uncritically or extrapolated beyond its usefulness.

The earth's atmosphere is generally agreed to be *secondary*, i.e., it became part of the earth not through capture of a gas phase but by being incorporated in the solid particles that accreted to form the earth. This tenet is supported by the difficulties of producing the present atmosphere (in the sense of RUBEY's (1951) excess volatiles) from a primary gas phase atmosphere, and by the observations of planetary noble gas and plausibly appropriate contents of the major volatiles in meteorites, our best representative of early solar system solids. Thus, the present atmosphere must have been degassed from solid earth materials after the establishment of a gravitational field. A primary goal of much research is the deciphering of the history of this degassing, and in this paper we will consider the applicability of nuclear components to this task.

3. Isotopic Structures in Atmospheric Gases

There are two approaches available in attempting to characterize nuclear components in the atmosphere. One is to identify possible nuclear sources and make quantitative estimates of their likely contributions to the atmosphere; this approach is explored in the following three sections. A second approach, explored in this section, is to seek differences between observed and primordial terrestrial compositions. The principal basis for assessment of primordial composition is comparison with extraterrestrial compositions; because of the apparent large-scale inhomogeneities in the solar system noted earlier, terrestrial primordial composition may differ from that of extraterrestrial sources, a caveat which has become more apparent within the last few years. Because of the relative ease with which elemental composition may be changed in a variety of processes, we will concentrate on isotopic compositions.

3.1 Neon

The case for neon is a good example of the difficulty that occurs when there is not a unique extraterrestrial composition to which to compare terrestrial composition. Several "preferred" compositions have been identified in extraterrestrial samples, as illustrated in Fig. 2. These include cosmic ray induced spallation neon and several trapped components.

The composition denoted "neon-B" is defined by intersecting trends in bulk compositions and especially in compositions observed in stepwise heating analyses of lunar soils and meteorites (PEPIN, 1967; BLACK and PEPIN, 1969; BLACK, 1972a), most notably the "gas-rich" meteorites whose elemental abundances follow the solar pattern (Fig. 1). Neon-B is generally believed to result from direct implantation of the solar wind, and its composition is indeed close to (although distinct from) the solar wind composition directly measured during the Apollo missions (GEISS *et al.*, 1972). The difference between solar wind and neon-B could arise in secondary processes involved in solar wind implantation and redistribution. Alternatively, BLACK and PEPIN (1969) infer the existence of another component in lunar samples and solar gas rich meteorites, designed neon-C; BLACK (1972a) tentatively identified neon-C with solar flare neon and suggests that neon-B is a mixture, in relatively constant proportions, of neon-C and solar wind neon. Whatever its exact origin neon-B is the isotopic composition usually associated with the "solar" elemental abundance pattern.

The second component most commonly found in meteorites (but not lunar samples) is neon-A. It is identified by a number of carbonaceous chondrites which define a line A-S, interpreted as a mixture of spallation neon and a trapped component, neon-A (PEPIN, 1967). Because of its occurrence in carbonaceous chondrites whose elemental abundances define the "planetary" pattern, neon-A is usually considered to represent planetary neon.

More generally, carbonaceous chondrite neon compositions, both whole rock and stepwise release, fall within the triangular field B-A-S (Fig. 2) indicating mixtures in which neon-A, neon-B (and neon-C?) and spallation may be present. This indicates that many carbonaceous chondrites, in addition to containing trapped planetary gas, also contain solar gas due to exposure to solar wind; indeed, in some cases, the light-dark structure characteristic of other meteorites containing solar gases (in the dark portion) has been identified in carbonaceous chondrites. Because of the different elemental patterns (Fig. 1) it is possible in many cases that the heavier gases are essentially "planetary" while admixture of "solar" gases can be significant, or even dominant, in neon (and helium); a good example is Murray (Fig. 1), various samples of which have varying proportions of

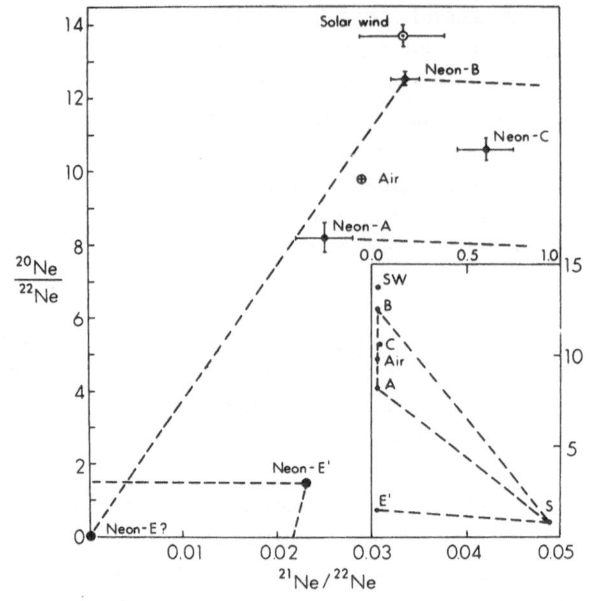

Fig. 2. A three-isotope correlation diagram for neon, illustrating the compositions of several observed or inferred components (described in text) and mixing relations among them. On this diagram the locus of all compositions formed by mixing of any two components is the straight line which joins their compositions. The inset at lower right shows, in smaller scale, the same compositions as the main diagram. Solar wind composition from GEISS *et al.* (1972); neon-B and neon-C from BLACK (1972a); neon-A from PEPIN (1967); neon-E′ from EBERHARDT (1975).

neon-A and neon-B (cf., BLACK and PEPIN, 1969; MAZOR *et al.*, 1970).

Another component of considerable interest is neon-E, characterized by a lower $^{20}Ne/^{22}Ne$ ratio than the other trapped components and initially identified by low $^{20}Ne/^{22}Ne$ ratios in stepwise heating analyses of carbonaceous chondrites (BLACK and PEPIN, 1969; BLACK, 1972b). The siting of neon-E is different from the other components, allowing the separation observed in the stepwise heating experiments; EBERHARDT (1974, 1975) also achieved a mineralogical separation, leading to the lowest $^{20}Ne/^{22}Ne$ ratio yet observed (plotted as neon-E′ in Fig. 2). Although neon-E is often thought of as being pure ^{22}Ne, its composition is rigorously constrained only to lie within the roughly rectangular area bounded by mixing lines through E′ and other available components (Fig. 2). Neon-E has obviously had a different nuclear history than the other components; hypotheses for its origin have included both extrasolar synthesis (in which case, it must have been imported in presolar grains which were never vaporized or heated enough to release neon) and particle reactions within the early solar system, either in the gas phase or by in situ production in solids (BLACK, 1972b; CLAYTON, 1975; HEYMANN and DZICZKANIEC, 1976; AUDOUZE *et al.*, 1976).

This multitude of neon components greatly complicates attempts to understand the origin of terrestrial neon. There are essentially two approaches to the problem (not necessarily mutually exclusive). One approach is to assume that all (or some) components were derived from a primordial composition by processes acting within the solar system. Especially noteworthy is the observation that the principal extraterrestrial compositions neon-B and neon-A, as well as air and solar wind, can reasonably be considered as related by mass-dependent fractionation, leading to the view that terrestrial neon is fractionated primordial neon and that meteoritic trapped neon is also fractionated (to varying degrees) primordial neon. The fractionations involved are large but not impossible so that this viewpoint is at least economical in number of hypotheses.

The second approach is to consider the various preferred compositions as independent components, whatever their origins, and account for observations in terms of mixing. We have already noted that, in general, carbonaceous chondrite neon seems to be a mixture of neon-B and neon-A and in this view terrestrial neon would also be a mixture of neon-B and neon-A. The origin of neon-A need not be specified, although neon-A might represent some reservoir ultimately derived by fractionation of primordial neon.

The multicomponent mixing approach has gained favor recently, primarily because of the discovery of neon-E, whose composition seems too extreme to be ascribed reasonably to fractionation of likely primordial composition, and because of the observations of correlated variations in helium, neon and argon compositions which cannot be obtained via fractionation (BLACK, 1970, 1971; ANDERS *et al.*, 1970; JEFFERY and ANDERS, 1970). We also note that if neon-C is indeed a well-defined component (an interpretation challenged by WALTON *et al.*, 1974) it cannot have been derived by fractionation of any other known composition and must be or contain a "nuclear" component.

In particular, it has been proposed that neon-A itself is composite (BLACK, 1972a), since stepwise heating analyses show no trends involving neon-A as they do neon-B and neon-C (BLACK 1972b; SMITH *et al.*, 1977). If neon-A is a mixture, one component must have lower $^{20}Ne/^{22}Ne$, presumably neon-E, another higher $^{20}Ne/^{22}Ne$, presumably a primordial solar composition near solar wind or neon-B. BLACK (1972b) has proposed that neon-A is a mixture of neon-E and neon-D, a presently rare component with $^{20}Ne/^{22}Ne \sim 14.5$ and $^{21}Ne/^{22}Ne$ unknown, identified as pre-main-sequence solar wind. If neon-A is a mixture, the well-defined composition inferred from whole rock carbonaceous chondrite compositions must be attributed to a relative constancy in the ratio of neon-E to trapped planetary primordial neon.

In this view, terrestrial neon would also be interpreted as a mixture of primordial neon and neon-E. The earth would have to have a smaller proportion of neon-E than carbonaceous chondrites; this result is thus analogous to the observation that the earth has different proportions of two oxygen components than do most meteorites, including carbonaceous chondrites (CLAYTON *et al.*, 1976).

HEYMANN *et al.* (1976) have noted that interpreting terrestrial neon as a mixture of neon-E with primordial neon suggests the possibility that the atmosphere contains an additional component, accounting for up to 20% of atmospheric ^{21}Ne. This follows from the position of air to the right of a possible mixing line such as that shown between neon-E (?) and neon-B in Fig. 2. Identification of any excess ^{21}Ne in air depends critically on assumed compositions for both neon-E and the primordial neon; for example, mixing neon-E′ with neon-B does not require any excess ^{21}Ne in air (similarly, obtaining terrestrial neon via fractionation of a composition like neon-B does not require excess ^{21}Ne). The likelihood of excess ^{21}Ne in the atmosphere is not unreasonable, and could be due to ^{18}O (α, n) ^{21}Ne or to cosmic ray spallation. The latter possibility is especially interesting and is considered in greater detail below.

3.2 *Helium*

In extraterrestrial samples helium, like neon, shows variations in trapped composition. The helium variations are less well characterized than the neon variations since in many samples calculation of trapped helium composition is subject to relatively large uncertainties because of the required corrections for radiogenic ^4He and spallation ^3He. The helium and neon variations are correlated (BLACK, 1970, 1972a, b) but correlation is observed only imperfectly, especially since variations in stepwise heating are of little utility (even identically sited helium and neon would be expected to be released at different temperatures).

"Solar" helium (helium-B or helium-C) is characterized by $^3He/^4He=4\times10^{-4}$ (cf., BLACK, 1972a). "Planetary" helium is more difficult to define because of the possibility that even in meteorites in which the heavier gases are clearly planetary the light gases might contain or be dominated by the solar component. It is, nevertheless, reasonably clear that planetary helium is characterized by a lower ^3He abundance, $^3He/^4He=1.5\times10^{-4}$ (JEFFERY and ANDERS, 1970; BLACK, 1972b). At least part of the difference between these two compositions is accounted for by nuclear burning of deuterium to ^3He in the surface of the sun.

Until relatively recently it was believed that the source of the ^3He in the earth's atmosphere was spallation induced by cosmic rays (in the atmosphere). CLARKE *et al.* (1969) discovered excess ^3He in sea water, however, and it is now generally believed that the principal source of atmospheric ^3He is a juvenile flux from oceanic spreading centers (CRAIG *et al.*, 1975). It is also generally believed that this juvenile ^3He is primordial. If so, ^3He is the only primordial atmospheric component for which a reasonably reliable juvenile flux estimate may be made (CRAIG *et al.*, 1975).

By analogy with the heavier gases, primordial terrestrial helium would be expected to have planetary isotopic composition. This is perhaps a moot point, however, since atmospheric helium, with $^3He/^4He=1.3\times10^{-6}$, contains far more ^4He than could be expected in any primordial composition. It is universally considered that atmospheric helium is dominated by radiogenic ^3He (from decay of ^{238}U, ^{235}U, and ^{232}Th). This follows not only from estimation of a primordial composition, as above, but also from quantitative considerations of how much radiogenic ^4He is available, as described below.

3.3 *Argon*

Argon isotopic compositions in extraterrestrial samples also show variations which correlate with those in neon and helium (BLACK, 1971, 1972a, b). Most of the variation in trapped $^{36}Ar/^{38}Ar$ occurs only in solar-gas-rich meteorites, however. The range of $^{36}Ar/^{38}Ar$ compositions in other samples, particularly including the carbonaceous chondrites which are the primary basis for the planetary gas pattern, are much more modest and essentially identical to the atmospheric value, $^{36}Ar/^{38}Ar=5.35$. There is thus no reason to suspect that the atmospheric $^{36}Ar/^{38}Ar$ ratio has been significantly modified from its primordial value either by fractionation or by addition of a nuclear component, nor is there any plausible candidate to produce such a nuclear component.

Primordial ^{40}Ar is essentially unobservable experimentally. Nucleosynthetic theory predicts that in primordial argon $^{40}Ar/^{36}Ar = 2 \times 10^{-4}$ (CAMERON, 1973); the lowest measured ratio in any sample is $^{40}Ar/^{36}Ar = 1.4 \times 10^{-3}$ (BEGEMANN *et al.*, 1976). In general, every accessible sample, even lunar soil exposed to solar wind, represents an environment in which primordial ^{40}Ar is overwhelmed by ^{40}Ar ultimately originating in decay of ^{40}K. As in the case of ^{4}He, comparison with extraterrestrial argon and estimation of the amount of ^{40}Ar available from decay of ^{40}K has led to the general recognition that atmospheric ^{40}Ar, with $^{40}Ar/^{36}Ar = 295.6$, is essentially all radiogenic. Indeed, the large ^{40}Ar abundance in the atmosphere led VON WEIZSÄCKER (1937) to suggest this identification even before this mode of decay of ^{40}K was observed experimentally.

3.4 Xenon

The case for xenon is certainly the most complicated of all the terrestrial gases. In part, this is because of the difficulty in identifying any well-defined primordial composition in extraterrestrial xenon; in part this is also because terrestrial xenon differs substantially from all samples of extraterrestrial xenon.*

Figure 3 illustrates the comparison between atmospheric xenon and three different extraterrestrial compositions which might serve as candidate models for primordial terrestrial xenon. Each composition will be discussed in turn, but two features are clear in all the comparisons.

All the extraterrestrial compositions have essentially identical relative abundances of the isotopes ^{124}Xe, ^{126}Xe, ^{128}Xe, and ^{130}Xe (they differ at the other five isotopes, presumably because of nuclear component additions). At these four isotopes and, to a first approximation also at the heavier isotopes, terrestrial xenon exhibits a clear mass dependent fractionation pattern of heavy isotope enrichment, a respect in which terrestrial xenon is unique (the Martian atmosphere data (OWEN *et al.*, 1977) are insufficiently precise to make the relevant comparison). The fractionation is very severe, approximately 3.5% per mass unit, and, in view of the large mass, more extreme than observed in any other element. The fractionation is sufficiently severe, and the mass range sufficiently great, that departures from an assumed linear dependence of fractionation on mass, usually applicable in less extreme cases, are likely to hamper identification of superposed nuclear components.

The second feature, evident in all three comparisons, is an excess of ^{129}Xe above the level predicted by interpolation of the fractionation patterns. Excess ^{129}Xe is a common feature of meteoritic xenon, where it has been demonstrated to be due to decay of now extinct ^{129}I (JEFFERY and REYNOLDS, 1961); it is usually accepted that excess terrestrial ^{129}Xe is also due to ^{129}I. It is not clear whether the ^{129}I was present when the earth accreted or had already decayed to ^{129}Xe and was imported in the accreting solids.

As seen in Fig. 3, the amount of ^{129}Xe excess is about 7%, as determined from the same considerations by, for example, MARTI (1969), and PEPIN and PHINNEY (1976). This should probably be viewed as a lower limit, since the extraterrestrial compositions used for

* A possible exception is the identity of terrestrial xenon and purported lunar trapped xenon (LIGHTNER and MARTI, 1974; LEICH and NIEMEYER, 1975), which would have far reaching consequences for the history of the earth and the moon. It appears, however, that the phenomenon is an artifact resulting from terrestrial contamination (NIEMEYER and LEICH, 1976).

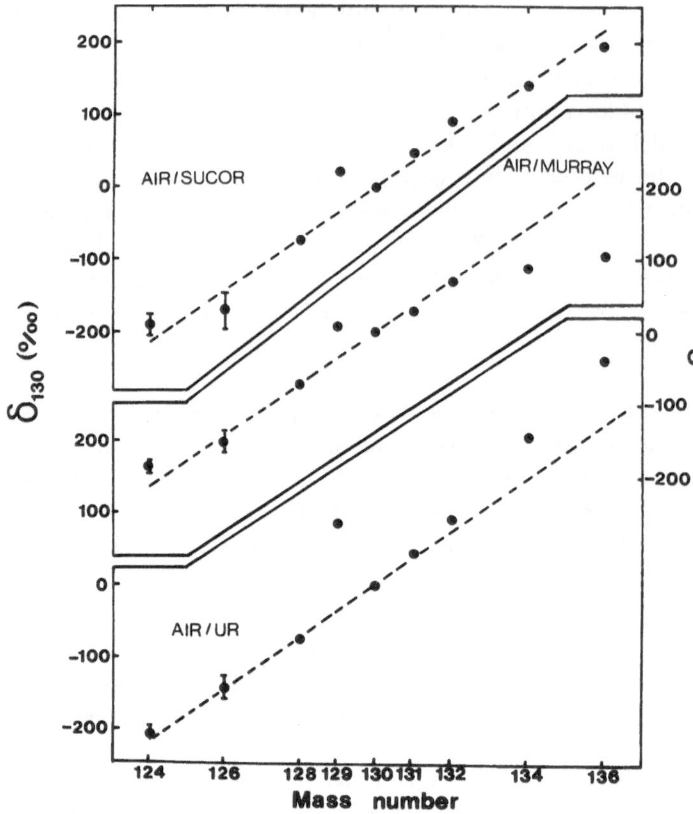

Fig. 3. A comparison of terrestrial xenon composition with three extraterrestrial compositions. The data are plotted as per mil differences: $\delta_{130}=1{,}000(R_a/R_s-1)$ where $R={}^{i}Xe/{}^{130}Xe$ ($i=124$, 126, ..., 136), a=air and s=SUCOR (PODOSEK *et al.*, 1971), s=Murray (PODOSEK *et al.*, 1971), or s=Ur (PEPIN and PHINNEY, 1976; PEPIN, 1976, and private communication). These compositions are described in the text (SUCOR designates surface-correlated gas in lunar soil). The abundance of ^{129}Xe is not specified in Ur-xenon; the figure is plotted assuming $(^{129}Xe/^{132}Xe)_{Ur}=0.99$, the "trapped" ratio in the Renazzo carbonaceous chondrite calculated by PODOSEK (1970). The dashed lines are straight lines drawn by eye through the data for isotopes ^{124}Xe, ^{126}Xe, ^{128}Xe, and ^{130}Xe; they illustrate the approximate degree of fractionation by which air xenon is believed to differ from these extraterrestrial compositions, but real fractionations of this magnitude are likely to be significantly nonlinear. The per mil scale at upper left is for the air-SUCOR comparison, that at lower left is for Air-Ur, that on the right is for Air-Murray.

comparison might also contain some radiogenic ^{129}Xe added to an unobserved primordial composition, although this effect is unlikely to be large (DROZD and PODOSEK, 1976). In principle, some of this ^{129}Xe excess could be attributed to the fission contributions discussed later, but this too is unlikely to be significant. One possibly serious ambiguity in the assessment of radiogenic ^{129}Xe in the atmosphere, particularly in view of the apparent large scale inhomogeneities in the early solar system, is the as yet unresolved suggestion that in some planetary objects (meteorite parent bodies) the abundance of primordial ^{129}Xe is substantially lower than would be inferred by comparison with the sun or carbonaceous chondrites (DROZD and PODOSEK, 1976; PODOSEK, 1970).

In addition to the fractionation and excess ^{129}Xe effects, terrestrial xenon differs from extraterrestrial xenon at the heavy isotopes in a manner suggesting the addition of fission components. The principal problem in assessing these effects, aside from resolving them from the fractionation, is that the extraterrestrial compositions also seem to include fission or fission-like components; the determination of a fission-free primordial xenon composition has been one of the long-standing problems in xenology.

Carbonaceous chondrites all have essentially the same xenon isotopic composition, usually designated AVCC for "average value carbonaceous chondrite" (KRUMMENACHER *et al.*, 1962). A comparison between xenon isotopic compositions of air and the carbonaceous chondrite Murray is shown in Fig. 3. We see that air is not AVCC, nor is it fractionated AVCC; relative to fractionated AVCC, air is deficient in the heavy isotopes.

AVCC is itself not a structureless component, as evidenced by substantial isotopic variations exhibited in stepwise heating analyses (first seen by REYNOLDS and TURNER, 1964). These variations correlate in such fashion that they may be interpreted as resulting from addition of a component containing only the heavy isotopes which lack more neutron-rich stable isobars (PEPIN, 1968), i.e., a fission component, usually dubbed CCF ("carbonaceous chondrite fission"). No known nuclide is a plausible parent of CCF; this has led to the suggestion that the parent is a superheavy element (ANDERS and HEYMANN, 1969; DAKOWSKI, 1969) and to a flourishing literature which has addressed this problem without resolving it. Separation of the components which comprise AVCC has been greatly enhanced by chemical techniques, as initially reported by LEWIS *et al.* (1975); such experiments have also revealed a large light-isotope enrichment as well as a heavy-isotope enrichment. It is not clear whether the light-isotope and heavy-isotope effects represent independent components or are coupled in a single component.

While the problem of the number and nature of independent components remains an open and vexing question, the hope has persisted that terrestrial xenon can be genetically related to meteoritic planetary xenon. In particular, the comparison between atmospheric and meteoritic xenon (Fig. 3) has sustained the hope that terrestrial primordial xenon can be identified as fractionated CCF-free (more generally, heavy-component-free) meteoritic xenon. This composition has proved elusive, however. The most comprehensive analysis has been performed by PEPIN (1976), and PEPIN and PHINNEY (1976) who, by isotopic correlations observed in a number of carbonaceous chondrites, have identified a heavy-component-free (and light-component-free) xenon component designated U- (Ur-) xenon. The comparison of atmospheric and U-xenon is shown in Fig. 3; the fractionation trend is obvious and relative to fractionated U-xenon air shows a marked heavy isotope excess. This is clearly suggestive of a fission component which has been added to a primordial composition identified as fractionated U-xenon; PEPIN and PHINNEY (1976) state that the pattern of heavy isotope enrichment in air, relative to a non-linear fractionation, indicates that the fissioning nuclide is ^{244}Pu.

Another useful comparison to atmospheric xenon is solar wind xenon, as observed in surface-correlated xenon in lunar soils (EBERHARDT *et al.*, 1970, 1972; PODOSEK *et al.*, 1971). As seen in Fig. 3, terrestrial xenon comes much closer to matching fractionated solar wind xenon than to matching fractionated Ur- or Murray-xenon. In detail, however, a precise interpretation of air xenon as fractionated solar wind xenon is impossible. Authors reporting solar wind compositions have pointed out that surface-correlated lunar soil gas compositions, loosely designated "solar wind", in general also incorporate components generated in the lunar environment, as seen most clearly in the case of high and variable $^{40}Ar/^{36}Ar$ ratios. In particular, PODOSEK *et al.* (1971) argue that comparison of air and "solar wind" xenon indicates that *both* contain admixtures of heavy isotope components, presumably fission. (Correspondingly, it is not possible to interpret AVCC as "solar wind" plus a heavy-isotope composition consistent with the isotopic correlations seen in stepwise heating.)

In summary, estimation of primordial terrestrial xenon, and thereby evaluation of a possible fission component added to it, is a difficult affair. Although there are plausible qualitative models there is no quantitatively unambiguous candidate for a primitive composition which, when subjected to a severe and unexplained fractionation, produces primordial terrestrial xenon. Several lines of reasoning suggest that air contains an added fission component; the suggested magnitude is a few percent (of ^{136}Xe) but it is difficult to be more precise. The only direct estimate follows from the U-xenon comparison, but this result should not be accepted uncritically; the derivation of the U-xenon analysis has not been described completely and this model has other, mostly unexplored ramifications (e.g., the suggestion that "solar wind" also contains a very large fission component coincidentally in nearly the same proportion as air). Quantitative estimates of the amount of available fission xenon are considered in the next section.

3.5 Krypton

In comparison with the other noble gases, the behavior of krypton is rather bland. AVCC and "solar wind" krypton are very similar to each other and to air (EBERHARDT *et al.*, 1970, 1972; BASFORD *et al.*, 1972); they differ principally in that relative to AVCC atmospheric krypton is slightly (0.5%/mass) but definitely fractionated, with the light isotopes enhanced (the opposite of the xenon trend).

An isotopic structure in meteoritic krypton comparable to that seen in xenon has not been observed. Relative to "solar wind" krypton, however, both AVCC and atmospheric krypton show an apparent excess of about 1% at ^{86}Kr, obviously suggestive of a fission component, but the existence of this excess is not as well-defined experimentally as other effects described above (EBERHARDT *et al.*, 1972; PODOSEK *et al.*, 1971).

3.6 Nitrogen

The preceding discussion has concerned instances in which nuclear components were fairly easy to identify, at least in principle. There may be others in which identification is not so easy, as in the cases of monisotopic elements or in elements with only two isotopes, where moderate effects cannot easily be distinguished from fractionation. A good illustration is the case for nitrogen. Substantial variations in $^{15}N/^{14}N$ have been observed among terrestrial, meteoritic, and lunar (solar wind) samples, and the composition of the Martian atmosphere is different from that of the earth (KERRIDGE, 1975; BECKER and CLAYTON, 1975; OWEN *et al.*, 1977). Differences among terrestrial samples are presumably due to fractionation; the differences between terrestrial and extraterrestrial samples are also usually attributed to fractionation, even though these differences are larger than the terrestrial range and fractionation hypotheses are sometimes somewhat strained. It is possible that these variations represent different admixtures of nuclear components rather than fractionation.

This suggestion is, of course, purely speculative, at least in the absence of any evidence to indicate that it is more plausible than fractionation, and particularly in the absence of any specific nuclear mechanism. It is always possible to suppose that nitrogen, like oxygen and perhaps neon, is characterized by residual inhomogeneity in the solar system, but it would be difficult to test this hypothesis unless extremely large effects were observed. The comparison with neon, specifically with neon-E, is interesting, however, since it suggests a more parochial mechanism. HEYMANN and DZICZKANIEC (1976) have proposed that neon-E could be generated by proton irradiation of a gas phase which would induce the reaction $^{22}Ne(p, n)^{22}Na$; formation of a condensate in which neon is depleted relative to sodium would generate a component enriched in ^{22}Ne upon subsequent decay of ^{22}Na. The same mechanism could be invoked to account for nitrogen isotopic variations: neutron irradiation of a gas phase could induce $^{14}N(n, p)^{14}C$, the reaction responsible for ^{14}C on the earth and the principal sink of cosmic ray secondary neutrons in the atmosphere. Condensation of a phase in which nitrogen is depleted relative to carbon would, upon decay of the ^{14}C, result in ^{14}N enrichment relative to ^{15}N.

This mechanism would require large neutron fluences to produce observable effects. For a (thermal) cross section of 1.8 barns the production of, say, a 10% enrichment of ^{14}N would require fluence of at least $6 \times 10^{22} \times f$ neutrons/cm^2, where f is the ratio of N/C in the condensate to N/C in the gas. In low grade meteorites, and in the earth's volatile inventory, nitrogen is depleted relative to carbon by factors of the order 10^{-1} to 10^{-2}. This suggested mechanism is thus similar to the well-known FGH theory (FOWLER *et al.*, 1962) not only in spirit but also in terms of the number of neutrons required, and is thus similarly liable to attack on both astrophysical and geochemical grounds, although in this case, irradiation of only a gas phase is required.

The conjecture should probably not be taken very seriously as it involves ad hoc assumptions of questionable plausibility. Nevertheless, it does provide a qualitative basis to account for a general nonuniformity in nitrogen composition among various planetary objects and the indication that their nitrogen might be lighter (richer in ^{14}N) than the sun's, and it may serve as a useful alternative hypothesis until the nature of nitrogen isotopic variations is better understood. In the present context, however, its more important function is to illustrate the possible presence of a greater roster of nuclear components than is usually recognized.

4. Radiogenic Components

In the previous section we considered the possibility of detecting nuclear components in the atmosphere by comparison of present and primordial compositions. Here and in the next two sections, we consider the complimentary approach: evaluating whether known or likely sources of nuclear components are quantitatively capable of accounting for a significant fraction of atmospheric gases. In this section, we will consider sources in naturally occurring radionuclides. Plausibly significant sources of radiogenic atmospheric components are summarized in Table 1.

4.1 ^{4}He production

As described above, comparison with extraterrestrial compositions indicates that atmospheric ^{4}He is predominantly a nuclear rather than a primordial component. The principal source is almost certainly α decay of uranium and thorium.

It is unfortunately rather difficult to make a quantitative description of the atmospheric ^{4}He economy. Since ^{4}He escapes from the atmosphere it is impossible to compare production with accumulation. The rate of escape is also not very well-known and may be dominated by nonthermal mechanisms (AXFORD, 1968) whose rate cannot be predicted accurately. The rate at which ^{4}He is added to the atmosphere by degassing of the earth is also not very well known. From their observations of ^{4}He in seawater in excess of solubility equilibrium with the atmosphere, CRAIG *et al.* (1975) calculate a total marine input flux of 3×10^5 atoms cm^{-2}sec^{-1} or 1.8×10^{12} (cm^{3}STP) yr^{-1} for the whole earth. The influx from continental and volcanic sources is unknown, however, and could predominate over the marine input.

Table 1. Radionuclides contributing to the atmosphere.

Nuclide	Half-life (10^9 yr)	Yield[1]	Daughter
^{232}Th	14.0	6	^{4}He
^{235}U	0.70	7	^{4}He
^{238}U	4.47	8	^{4}He
^{40}K	1.31	0.11	^{40}Ar
^{238}U[2]	4.47	2.8×10^{-8}	^{136}Xe
^{244}Pu[2]	0.082	6.9×10^{-5}	^{136}Xe
^{129}I	0.017	1	^{129}Xe

[1] Yield = number of daughter atoms per decay of parent atom.

[2] Although only ^{136}Xe is listed as daughter, spontaneous fission of ^{238}U and ^{244}Pu also produces other isotopes of xenon which are not shielded by isobars of higher neutron number (^{129}Xe, ^{131}Xe, ^{132}Xe, and ^{134}Xe as well as smaller amounts of ^{86}Kr and ^{84}Kr).

The present ^{4}He production rate from uranium and thorium combined (for Th/U = 3.8) is 2.3×10^{-7} (cm^{3}STP) yr^{-1} per gram of uranium (thorium accounts for about half of this). For a whole-earth uranium concentration of 32 ppb, which corresponds in steady state to the average heat flow (WILLIAMS and VON HERZEN, 1974; LANGSETH *et al.*, 1976), this results in a whole-earth ^{4}He production of 4.4×10^{13} (cm^{3}STP) yr^{-1}. This is more than adequate to account for the marine flux and is presumably adequate for the continental sources as well.

4.2 Radiogenic ^{40}Ar

As with ^{4}He, comparison with extraterrestrial compositions indicates that essentially all atmospheric ^{40}Ar is radiogenic. The atmospheric ^{40}Ar inventory could be produced in 4.5×10^9 yr by 4.9×10^{23} gm of potassium, establishing the well-known lower limit of around 80 ppm for the average potassium content of the earth. That the earth contains at least this much potassium is geochemically reasonable. Discussion of the important qeustion of whether or not the earth contains significantly more potassium will be taken up below.

4.3 *Radiogenic ^{129}Xe*

While not as prominent as in the cases of ^{4}He and ^{40}Ar, the existence of excess ^{129}Xe in atmospheric xenon is reasonably securely established, as is its attribution to decay of ^{129}I. The ^{129}I may have been present in the earth itself or may have decayed in the solids which later accreted to form the earth. The required amount of ^{129}I is not excessive in comparison with the amount known to be present when meteorites formed. Discussion of this subject, in fact, is usually framed in terms of an iodine-xenon age for the earth. WETHERILL (1975), for example, estimates $^{129}I/^{127}I=3\times10^{-6}$ from the ratio of excess atmospheric ^{129}Xe to the ^{127}I in the crust and ocean. This ratio is low compared to a typical value of $^{129}I/^{127}I=1\times10^{-4}$ for chondritic meteorites (HOHENBERG *et al.*, 1967; PODOSEK, 1970) and, at face value, leads to the surprising conclusion that the earth (or earth materials) did not form until about 80×10^6 yr after chondrites. If most of the inventory of xenon degassed from the interior of the earth is now adsorbed on shales, the amount of ^{129}Xe is correspondingly increased and the difference in formation ages reduced. More generally, this problem involves the total earth iodine inventory and the degree of ^{129}Xe liberation into the atmosphere, as discussed below in the context of degassing models.

4.4 *Fission*

As indicated earlier, there are strong indications to believe that atmospheric xenon contains a fission component, although it is difficult to estimate the quantity. Here we will examine the possible sources.

The only extant natural radionuclide which undergoes significant spontaneous fission is ^{238}U. In 4.5×10^9 yr one (present) gram of uranium will produce 2.7×10^{-6} cm^{3}STP of ^{136}Xe. At 32 ppb this corresponds to 5.1×10^{14} cm^{3}STP for the whole earth. The atmospheric ^{136}Xe inventory is 3.1×10^{16} cm^{3}STP. Thus, spontaneous fission of ^{238}U can account for about 1.7% of atmospheric ^{136}Xe.

The only other plausible source of spontaneous fission xenon is ^{244}Pu, a now-extinct radionuclide which, like ^{129}I, was present in the early solar system. For $^{244}Pu/^{238}U=0.015$ in the early solar system (PODOSEK, 1972), ^{244}Pu produces about a factor of 70 more fission ^{136}Xe than does ^{238}U. Thus, spontaneous fission of ^{244}Pu could account for slightly more than the total atmospheric inventory of ^{136}Xe; more realistically, it could account for a few percent if xenon is only partially degassed or if most degassed terrestrial xenon is adsorbed on shales.

Also as noted above, there is a suggestion of a fission component in ^{86}Kr. ^{238}U and ^{244}Pu are inadequate to produce any significant contribution. The larger contribution is made by ^{244}Pu; for a fission yield at ^{86}Kr which is about 2% of the ^{136}Xe yield (LEWIS, 1975), calculations made on the same basis as for xenon indicate that ^{244}Pu can account for at most 0.1% of the ^{86}Kr atmospheric inventory. This amount would be unobservable; if there is a fission or any other nuclear component in ^{86}Kr its source is unknown.

4.5 *Other radionuclides*

In addition to the sources already discussed (Table 1) there are a number of natural radionuclides which can contribute to atmospheric volatiles, for example α emission by rare earth nuclides or double β decay of ^{130}Te to ^{130}Xe. Interesting as these nuclides are in other contexts, they do not make significant contributions to the total atmospheric inventory. The largest known source not listed in Table 1 is probably ^{147}Sm which, for cosmic (CAMERON, 1973) abundance, produces ^{4}He at about 1% of the rate for ^{232}Th. A possible exception is a superheavy nuclide, as discussed in connection with CCF in meteoritic xenon; the existence of such a species is still hypothetical, however.

5. Spallation Components

The interaction of cosmic rays with samples of the solar system produces nuclear components through spallation reactions with high energy primaries and secondaries and through absorption of moderated secondary neutrons. Except in a few extreme cases, the effects of cosmic ray interactions are not large enough to produce observable depletions of the target nuclides, but when the background abundance of the product nuclides is low the addition of these cosmogenic components can be significant. In general, the noble gases are greatly depleted in solids, and spallation effects in the noble gases are commonplace in meteorites and lunar samples. The noble gases of the earth's atmosphere are similarly depleted and most of the present-day interaction of earth with cosmic rays occurs in the atmosphere, so that the atmosphere is the most reasonable place to look for these nuclear components.

5.1 Present-day cosmogenic components

Cosmic ray interactions in the atmosphere produce a number of nuclides, both radioactive and stable. In general, only the radionuclides are detectable (through low-level counting). These radionuclides are of considerable interest in a number of applications; the best known and most useful example is, of course, ^{14}C, but there are several others (cf., LAL, 1963). The only stable nuclide produced in significant or detectable quantity is ^{3}He. Until the discovery that oceanic spreading centers were a source of juvenile ^{3}He (CLARKE *et al.*, 1969; CRAIG *et al.*, 1975) it was believed that cosmic ray spallation was the only source of atmospheric ^{3}He; CRAIG *et al.* (1975) estimate that spallation accounts for about 25% of the atmospheric ^{3}He inventory. The ^{3}He does not accumulate in the atmosphere since it has a geologically short lifetime against escape.

The larger variety of target elements in the earth's crust allows a correspondingly wider variety of nuclear interaction products, but attenuation by the atmosphere effectively limits reactions to those induced by neutrons and muons. Reactions in the crust do not appear to account for any quantitatively significant fractions of the earth's inventory of noble gases.

The present-day interaction of cosmic rays with the earth, interesting as it is in a number of other contexts, thus apparently has little bearing on the problem of the evolution of the earth's atmosphere.

5.2 Ancient cosmogenic components

While there is now no significant accumulating input of cosmogenic components in the earth, the earth's atmosphere may still contain components from an earlier exposure. Such components could have been accumulated at the surface of the earth before the generation of an atmosphere sufficiently dense to shield the surface, or could have been brought to the earth by exposure of the material from which it accreted.

As we have seen, consideration of possible nuclear components in terrestrial neon leads to the possibility that a nontrivial fraction of atmospheric ^{21}Ne, perhaps up to 20%, is a nuclear component in excess of primordial composition. We will illustrate the orders of magnitude involved in assessing a cosmogenic contribution to ^{21}Ne. In solid (silicate) materials, the range of significant spallation production is of the order of a meter. For objects smaller than this size, i.e., 4π exposure geometry, the production rate of spallation ^{21}Ne in earth composition is approximately 4×10^{-15}(cm^{3}STP)gm^{-1} yr^{-1}; for larger objects spallation occurs only near the surface (2π geometry) and the relevant production rate is about 8×10^{-13}(cm^{3}STP)cm^{-2} yr^{-1} (both rates are based on the present intensity of galactic cosmic rays). For comparison, atmospheric ^{21}Ne is 2×10^{17}(cm^{3}STP) or 3×10^{-11}(cm^{3}STP)gm(earth)$^{-1}$.

^{21}Ne spallation production at the surface of an airless earth, at the rate above, is 4×10^6(cm^{3}STP)yr^{-1}. In 10^9 yr, this would come to 2% of the ^{21}Ne in air. Thus, irradiation of the earth is unlikely to contribute a significant fraction of ^{21}Ne unless the intensity of cosmic rays was substantially higher in the past.

The possibility of importing spallation ^{21}Ne in exposed material accreting to the earth is more interesting but also more difficult to evaluate. Presently, infalling meteorites have much larger ^{21}Ne concentrations than the earth, but even over the age of the earth their influx at the present rate is too small to account for a significant fraction of atmospheric ^{21}Ne (HEYMANN *et al.*, 1976). Nevertheless, 10% of atmospheric ^{21}Ne could be accounted for if the material which accreted to form the earth were exposed as small particles for an average of only 10^3 yr. This is a very short time but whether or not it is reasonable is critically dependent on the sequence and timing of condensation, accretion, and dispersal in the early solar system, and it would be treacherous to argue from this direction. An alternative, and better defined calculation has been performed by HEYMANN *et al.* (1976), who find that if earth materials were irradiated in bodies distributed in size as are present asteroids, an average irradiation time of the order of 10^8 yr would account for about 10% of a atmospheric ^{21}Ne.

These illustrative calculations are certainly not definitive, but they are sufficient to point out exciting possibilities. The generation of terrestrial neon involves several coupled uncertainties. If there are no nuclear components, either spallation ^{21}Ne or neon-E, we are left with the traditional problem of severely fractionated neon (and xenon) but (nearly) unfractionated argon and krypton in the atmosphere. If terrestrial neon is dominated by a mix of solar neon and neon-E, it may also contain non-trivial spallation neon; the calculations described above show that this idea may not be too far-fetched and illustrate the constraints this would place on the pre-accretion history of earth materials.

^{21}Ne is the most sensitive indicator of spallation components. This can be seen in Table 2 where we have shown the spallation contributions expected in gas of planetary or atmospheric composition in which 10% of the ^{21}Ne is spallogenic. We see that the corresponding spallation components in argon, krypton and xenon are negligible. This is not necessarily true for helium, however, and this point merits further discussion.

As shown in Table 2, if spallation gas (line 4) were added to planetary gas (line 2) in sufficient proportion to account for 10% of total ^{21}Ne, it would also account for about 8% of the ^3He. There are many uncertainties involved here, however, not the least of

Table 2. Comparison of noble gas compositions.

	^{3}He	^{4}He	^{20}Ne	^{21}Ne	^{36}Ar	^{38}Ar	^{78}Kr	^{84}Kr	^{126}Xe	^{130}Xe
(1) Air	—	—	25.4	0.075	48.4	9.0	0.0060	=1.0	0.00012	0.0054
(2) Planetary gas[1)]	0.62	4,928	22.4	0.068	80	14.8	0.0060	=1.0	0.0034	0.134
(3) Solar gas[2)]	3,000	8×10^{6}	3×10^{5}	86	1,560	310	0.0060	=1.0	0.00057	0.021
(4) Spallation[3)]	7.5	40	0.9	=1.0	0.12	0.19	5×10^{-5}	1×10^{-4}	2×10^{-6}	2×10^{-6}
(5) Spallation[4)] / Air	—	—	3×10^{-4}	=10%	2×10^{-5}	2×10^{-4}	6×10^{-5}	7×10^{-7}	1×10^{-4}	3×10^{-6}
(6) Spallation[4)] / Planetary	8%	6×10^{-5}	3×10^{-4}	=10%	1×10^{-5}	9×10^{-5}	6×10^{-5}	7×10^{-7}	4×10^{-6}	1×10^{-7}

1) Elemental ratios of planetary (type II carbonaceous chondrites) and helium-A and neon-A isotopic ratios from MAZOR *et al.* (1970).

2) Analysis of ilmenite grains (125 μ) from lunar soil 12001 by EBERHARDT *et al.* (1972); this composition is not intended to represent and is systematically different from the composition of the sun (cf., Fig. 1).

3) Helium, neon and argon production rates as calculated and used for type II carbonaceous chondrites by MAZOR *et al.* (1970); relative production rates for krypton and xenon in ordinary chondrites from EUGSTER *et al.* (1969).

4) Spallation ratios from row (4), renormalized to make ^{21}Ne 10% of the ^{21}Ne abundances in rows (1) and (2), respectively.

which is that there may not be any spallation ^{21}Ne in the earth at all. Additionally, the abundance of helium in planetary gas is not well known, or even well-defined. In general, meteoritic trapped gases can be regarded as a mix of planetary and solar contributions, and neon isotopic patterns in carbonaceous chondrites, the meteorites richest in planetary gas, commonly indicate nontrivial "solar" gas. It may then be that helium in carbonaceous chondrites, considered "planetary", is in some measure actually "solar". Earth materials might not have contained the same proportions of planetary and solar gases, might thus have contained more or less helium, and might thus have contained a lesser or greater fraction of spallation ^3He than indicated in Table 2. In particular, if the relative ^3He abundance in the earth's primordial gas is substantially less than in the usual conception of planetary gas, the earth's initial inventory of ^3He could have been dominated by a spallation rather than a primordial component.

The preceding paragraph illustrates the wide range of uncertainty involved in the estimation of the earth's initial ^3He inventory, and probably severely overinterprets generalizations based on presently accessible extraterrestrial samples. We wish only to point out the possibility that initial terrestrial ^3He might have been predominantly spallation rather than primordial; it seems impossible to make convincing arguments, pro or con, on this issue. This issue is potentially important however, since the juvenile ^3He observed in seawater (Clarke *et al.*, 1969; Craig *et al.*, 1975) and basaltic rocks (Lupton and Craig, 1975) is generally accepted as representing primordial gas. If this juvenile ^3He is spallation rather than primordial some of the conclusions based on this association will require modification, as discussed in greater detail below.

6. Other Sources of Nuclear Components

As likely sources of nuclear components in the atmosphere we have discussed inherited presolar inhomogenieties, daughters of natural radionuclides, and nuclear reactions induced by cosmic rays. There are a few more possible sources; these can be characterized as related to the natural radionuclides but otherwise constitute an untidy miscellany which, with one possible exception, are not important sources of atmospheric volatiles.

The possible exception is the reaction ^{18}O (α, n) ^{21}Ne, where the α particle comes from uranium-thorium series decays. This is the most important reaction of its type, primarily because of the low ^{21}Ne abundance. ^{21}Ne excesses presumably attributable to this reaction have been observed in terrestrial samples, for example by Wetherill (1954), who calculated that 1–2% of atmospheric ^{21}Ne could be due to this reaction. A more recent calculation by Heymann *et al.* (1976) indicates that as much as 4% of ^{21}Ne could be due to this reaction (if ^{21}Ne is all in the atmosphere). In view of uncertainties in the nuclear physics and the extent of degassing this reaction may or may not be a significant source of ^{21}Ne. If indeed there is any excess ^{21}Ne in the atmosphere it might plausibly be attributed to either the ^{18}O (α, n) ^{21}Ne reaction and/or imported spallation.

Human activity, in the form of bombs and nuclear power generation, has made significant changes in the normal concentration levels of short-lived radionuclides such as ^3H and ^{14}C. As yet, however, this activity has not been sufficiently intense to make nontrivial additions to permanent atmospheric constituents. As an example, for an estimate of about 10^{20} joules liberated in induced fission in commercial reactors, the production of fission ^{136}Xe is about 10^{10} cm^{3}STP, less than 10^{-6} of the atmospheric inventory.

Induced fission also occurs in natural reactors, for instance Oklo (cf., COWAN, 1976). The Oklo reactor produced about 5×10^{17} joules, corresponding to 4×10^7 cm^{3}STP fission ^{136}Xe, impressive but not a significant contribution to the atmosphere. Oklo is the only known phenomenon of its type, but there may have been others. While the possible number of Oklo-type reactors is completely unknown, DROZD (1974) has noted that if this number is large enough to have a nontrivial effect on the atmosphere we would expect to see large and wide-spread variations in the $^{235}U/^{238}U$ ratio in a major part of the earth's uranium. Since these variations have not been observed, this phenomenon has apparently been unimportant for the atmosphere.

More generally, uranium/thorium series decays produce a variety of α, β and γ particles, neutrons, neutrinos, and fission fragments. These will correspondingly induce a wide variety of nuclear reactions, some of which result in atmospheric species. In selected samples the products may be observable and interesting (e.g., excess ^{128}Xe from ^{127}I (n, γ) in iodyrite; SRINIVASAN *et al.*, 1971) but, except for ^{18}O (α, n) ^{21}Ne we are unaware of any cases of possibly appreciable effects in the atmosphere.

7. Degassing Models

Much of the study of the evolution of the terrestrial atmosphere has focused on models of the rate at which solid earth has degassed to form the atmosphere. Of paramount importance in such studies are the radiogenic noble gases; unlike the primordial gases they were not present at the time of formation of the earth, so that the rate at which they enter the atmosphere will be, in general, different from the rate for primordial gases, and it is this difference which offers the greatest promise for delineating the evolution of the atmosphere.

In this section we will describe the features of a very simple set of degassing models: first order degassing with two reservoirs, the solid earth and the atmosphere. Such simple models have been considered by many authors (a partial list: DAMON and KULP, 1958; TUREKIAN, 1959; OZIMA and KUDO, 1972; OZIMA, 1975). More sophisticated models involving different degassing behavior (e.g., coherent degassing models such as those presented by SCHWARTZMAN, 1973) and larger numbers of reservoirs (e.g., mantle, crust, and atmosphere, such as presented by HAMANO and OZIMA, 1978) have also been considered. It is unlikely that the actual evolution of the earth and its atmosphere followed very closely any such simple course. Nevertheless, such models exhibit qualitative features which may approximate the real behavior of the earth. In particular, special cases of the first order models correspond to either very rapid or very slow degassing. We feel that the question of whether the earth degassed rapidly or is degassing slowly has not yet been answered satisfactorily. Until this basic question is settled, first order models are as profitable a vehicle for its attack as are the more sophisticated models.

7.1 First order degassing

In this model it is assumed that there are two reservoirs of noble gases involved in the generation of the atmosphere: one is the atmosphere itself, hereafter designated by subscript a, the other is the solid earth, hereafter designated by subscript m. It is assumed that the atmosphere is a permanent reservoir, from which there is no escape (except for He) or return to the mantle. First order degassing means that there is a continuous transfer from reservoir m to reservoir a at a rate proportional to the content or reservoir m. Finally, it is assumed that at time $t=0$ there is no atmosphere.

With these assumptions the behavior of a stable primordial isotopic species S is governed by

$$\frac{\mathrm{d}}{\mathrm{d}t}S_a=-\frac{\mathrm{d}}{\mathrm{d}t}S_m=\alpha S_m, \tag{1}$$

where α is the degassing rate, assumed to be a constant. With the boundary condition $S_a(0)=0$ the solution of (1) is

$$S_m=S_{m_0}\mathrm{e}^{-\alpha t}, \tag{2}$$

$$S_a=S_{m_0}(1-\mathrm{e}^{-\alpha t}), \tag{3}$$

where S_{m_0} is the amount of S originally present in the earth.

In the remainder of this paper we will use the terms *rapid* and *slow* degassing to designate $\alpha T \gg 1$ and $\alpha T \ll 1$, respectively (taking $T=4.55\times10^9$ yr for the age of the earth). Within the confines of the first order model rapid degassing indicates that the earth is now essentially completely degassed ($S_m \ll S_a$) and slow degassing indicates that most of the original volatile inventory is still in the solid earth ($S_m \gg S_a$).

For an isotope D which is produced by decay of a radioactive parent P with yield y (Table 1) the behavior is governed by

$$\frac{\mathrm{d}}{\mathrm{d}t}D_m=\lambda y P_{m_0}\mathrm{e}^{-\lambda t}-\alpha D_m, \tag{4}$$

$$\frac{\mathrm{d}}{\mathrm{d}t}D_a=\alpha D_m, \tag{5}$$

where λ is the decay constant of the parent and P_{m_0} is the amount of the parent initially present in the earth. The solution to these equations is

$$D_m=D_{m_0}\mathrm{e}^{-\alpha t}+\frac{\lambda y P_{m_0}}{\lambda-\alpha}(\mathrm{e}^{-\alpha t}-\mathrm{e}^{-\lambda t}), \tag{6}$$

$$D_a=D_{m_0}(1-\mathrm{e}^{-\alpha t})+yP_{m_0}\left(1-\frac{\lambda\mathrm{e}^{-\alpha t}-\alpha\mathrm{e}^{-\lambda t}}{\lambda-\alpha}\right), \tag{7}$$

where D_{m_0} is the amount of any primordial daughter isotope.

It is often convenient to represent this system by the ratio of a daughter isotope to a suitable stable isotope. Dividing (6) by (2) and (7) by (3) we obtain:

$$\left(\frac{D}{S}\right)_m=\left(\frac{D}{S}\right)_{m_0}+\frac{\lambda y}{\lambda-\alpha}\left(\frac{P}{S}\right)_{m_0}(1-\mathrm{e}^{-(\lambda-\alpha)t}), \tag{8}$$

$$\left(\frac{D}{S}\right)_a=\left(\frac{D}{S}\right)_{m_0}+\frac{y}{\lambda-\alpha}\left(\frac{P}{S}\right)_{m_0}\left(\lambda-\alpha\frac{1-\mathrm{e}^{-\lambda t}}{1-\mathrm{e}^{-\alpha t}}\right). \tag{9}$$

This model is characterized by two free parameters. The first is α, the degassing rate. The second is essentially a geochemical parameter. If we consider the cace for S separately, the second parameter is S_{m_o}, the initial content of S. Considering separately also the case for D, there are actually two geochemical parameters, D_{m_o} and P_{m_o}; this is usually reduced to one by knowledge or assumption of one of these parameters, typically D_{m_o}, so that the "free" parameter is P_{m_o}. In the case of the ratios, Eqs. (8) and (9), the geochemical parameter is $(P/S)_{m_o}$, where $(D/S)_{m_o}$ is usually presumed known on the basis of comparison with extraterrestrial primordial compositions. Usually the ratio characterization, Eqs. (8) and (9), is more convenient than the single isotope characterization, (2) and (3), or (6) and (7), since estimates or measurements of D/S may be feasible where D or S separately are not.

One constraint on this two parameter system is provided by a measurement of the present atmosphere, S_a, D_a, or $(D/S)_a$, at time $t=T=4.55\times10^9$ yr. The model may then be completely defined by one other datum. One approach has been an independent estimation of the geochemical parameter, e.g., $(P/S)_{m_o}$. An alternative approach, independent of geochemical arguments, may be based solely on rare gas measurements: the system can be defined either by a measurement of $(D/S)_m$, the present-day composition of the source reservoir, or by a measurement of $(D/S)_a$ at a suitable time t other than the present, i.e., measurement of the composition of the ancient atmosphere. These various avenues have been pursued vigorously, although so far without unambiguous success. Ultimately, the goal of such studies is to provide sufficient data to overconstrain the first order model, which would allow a test for its validity and provide meaningful constraints or more sophisticated models with more parameters.

7.2 Argon

The isotopic system which has attracted the most attention is atmospheric argon. In terms of the model above, $S={}^{36}Ar$, $P={}^{40}K$, and $D={}^{40}Ar$; with reasonable assurance we may take $({}^{40}Ar/{}^{36}Ar)_{m_o}=0$.

As indicated above, the choice or determination of one additional parameter (besides knowing ${}^{40}Ar$ and ${}^{36}Ar$ in the present atmosphere) completely specifies the model. For illustration, if we arbitrarily select a degassing rate α, we may calculate $({}^{40}K/{}^{36}Ar)_{m_o}$ (9), $({}^{40}Ar/{}^{36}Ar)_m$ (8), and $({}^{40}K)_{m_o}$ (7). The results are shown in Fig. 4. If the degassing is rapid ($\alpha T\gg1$), then most of the ${}^{40}Ar$ ever produced is now in the atmosphere, the potassium content of the earth is close to the lower limit (80 ppm) set by the ${}^{40}Ar$ content of the atmosphere, and the mantle contains only the radiogenic ${}^{40}Ar$ produced during the last $1/\alpha$ yr and so has a very low gas content (compared to the atmosphere divided by the mass of the earth) with $({}^{40}Ar/{}^{36}Ar)_m\gg296$. Conversely, for slow degassing ($\alpha T\ll1$), the mantle has a high gas content with $({}^{40}Ar/{}^{36}Ar)_m=432$ (the lowest value possible for continuous degassing) and $(K)_{m_o}\gg80$ ppm. A given value of α also dictates $({}^{40}Ar/{}^{36}Ar)_a$ as a function of time (9), as illustrated in Fig. 5.

A geochemical approach to the problem would be specification of the potassium content of the earth (or, less likely, the ratio of potassium to primordial ${}^{36}Ar$). For example, if we assume a chondritic potassium content of 880 ppm we must have $\alpha=3.1\times10^{-11}$ yr^{-1} (Fig. 4a); this was the approach taken and conclusion reached by TUREKIAN (1959). For chondritic potassium in a first order degassing model we have $\alpha T=0.14$, so that the earth is now only $1-e^{-\alpha t}=13\%$ degassed and the atmosphere has grown essentially uniformly over geologic time.

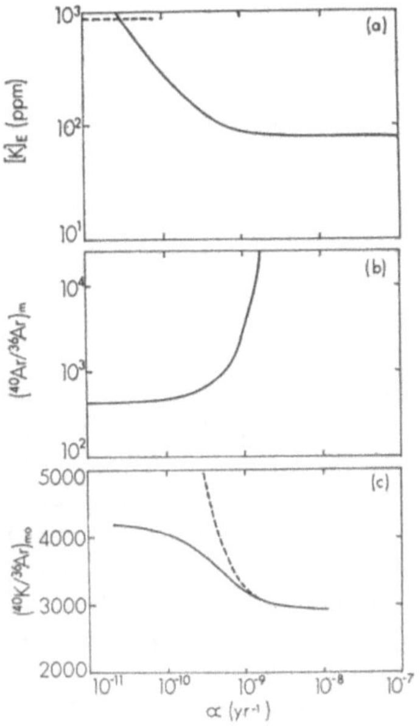

Fig. 4. Parameters in a first order degassing model for argon. Such models contain one free parameter; this figure illustrates required values (ordinate) when α, the degassing rate, is assumed (abscissa). (a) Potassium content of the earth $[K]_E$(ppm); for chondritic K-content of 880 ppm (dashed line), $\alpha=3.1\times10^{-11}$ yr^{-1}; if degassing is rapid ($\alpha T\gg1$), most of radiogenic ^{40}Ar is in the atmosphere, and a lower limit $[K]_E\approx80$ ppm is approached. (b) (^{40}Ar/^{36}Ar) presently in the mantle; for $\alpha T\ll1$ a lower limit $(^{40}Ar/^{36}Ar)_m\approx430$ is obtained. For rapid degassing most of the ^{36}Ar is lost from the mantle, and $(^{40}Ar/^{36}Ar)_m$ approaches infinity. (c) Initial (^{40}K/^{36}Ar) in the mantle; for a continuous degassing model (solid line), $(^{40}K/^{36}Ar)_{m_0}$ is restricted to a finite range of values; for catastrophic degassing at $t=0$ (dashed line) values are identical to those of continuous degassing for large α, but diverge for $\alpha\lesssim10^{-9}$ yr^{-1}. $(^{40}K/^{36}Ar)_{m_0}$ is not restricted to a finite range. Calculations from Eqs. (7)–(9) and (15) using: ^{36}Ar in atmosphere $=1.25\times10^{20}$cm^{3}STP, ^{40}Ar/^{36}Ar in atmosphere$=$ 295.5; $\lambda=5.31\times10^{-10}$ yr^{-1}; $\lambda_e/\lambda=0.112$; ^{40}K/K$=$ 1.19×10^{-4}.

In principle, specification of the terrestrial potassium content is the simplest and most promising way to define the characteristics of a first order degassing model or, more generally and in a model-independent fashion, to determine the degree of degassing of the earth. It is, therefore, very unfortunate that the problem of potassium abundance, quite apart from its importance in this context, is itself a source of considerable controversy. Crustal rocks have K/U ratios lower than chondritic (WASSERBURG *et al.*, 1964) and the upper mantle source region of oceanic basalts apparently has a low potassium content (cf., WYLLIE, 1971). Geochemical evidence thus suggests that the earth may indeed be depleted in potassium, relative to chondritic abundance, by as much as an order of magnitude; if so, the earth must be nearly completely degassed. Contrarily, however, it has been argued that it may be only the upper mantle which is depleted in potassium (which has been extracted into continental crust) and that the lower mantle, with chondritic potassium, remains undifferentiated, or that in a fully differentiated earth most of the chondritic potassium is sequestered in the core (cf., GOETTEL, 1976).

Another approach is direct measurement of ancient atmospheric argon composition. This is experimentally rather difficult (cf., ALEXANDER, 1975), however, and only one datum (Fig. 5) purporting to be a measure of the paleoatmosphere has been published (CADOGAN, 1977). We are skeptical about the validity of this datum; even if this result is valid, however, CADOGAN's (1977) datum is too recent to provide much discrimination among models. More generally, as noted by ALEXANDER (1975), the field of atmospheric ^{40}Ar/^{36}Ar variation (Fig. 5) for first order degassing is relatively narrow, so that paleoatmospheric composition data, while possibly able to exclude first order models if outside this field, would not be very sensitive indices of the degassing rate α if within this field.

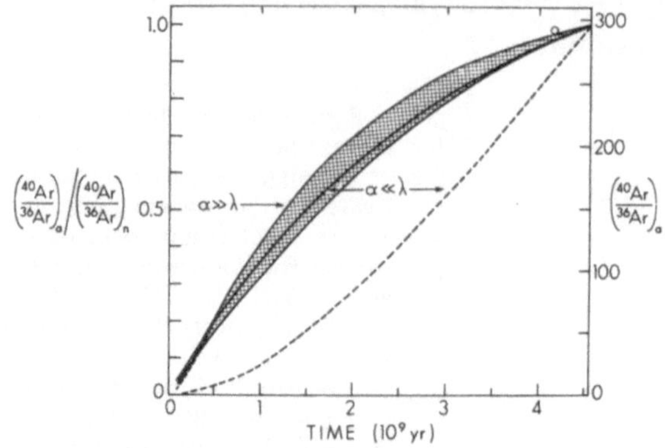

Fig. 5. Evolution of atmospheric argon composition for continuous and catastrophic degassing models. Continuous model (solid lines) narrowly restricts the possible range of $(^{40}Ar/^{36}Ar)_a$ (shaded area). Curves reach maximum concavity for $\alpha \gg \lambda$, minimum concavity at an intermediate value; for $\alpha \ll \lambda$, the evolution curve lies between these two extremes. Catastrophic degassing (at $t=0$, followed by continuous degassing) model admits larger differences in evolution curves. The $\alpha \gg \lambda$ curve is identical to that for continuous degassing except for $t \lesssim 0.5 \times 10^9$ yr; when $\alpha \ll \lambda$ (dashed line), values of $(^{40}Ar/^{36}Ar)_a$ are substantially lower for all t. All curves are constrained to pass through the present ($T=4.55 \times 10^9$ yr) atmospheric value $(^{40}Ar/^{36}Ar)_a=295.5$. Values on the left-hand side of the graph are $(^{40}Ar/^{36}Ar)_a(t)$ normalized to the present atmospheric ratio, assuming that $(^{40}Ar/^{36}Ar)_{m_o}=0$. The open circle denotes the paleoatmospheric composition inferred by CADOGAN (1977). Calculations based on Eqs. (9) and (15), with input parameters as in caption to Fig. 4.

An equivalent statement is that $(^{40}K/^{36}Ar)_{m_o}$ is fairly narrowly bounded (Fig. 4c). This is a general feature of first order models. The degree of restriction is a function of the parent half-life, with greater restriction corresponding to shorter life-times.

In contrast, OZIMA (1975) has pointed out the importance of $(^{40}Ar/^{36}Ar)_m$, the present composition of the source reservoir: a low value would indicate slow degassing and a high value would indicate rapid degassing (Fig. 4b). There are now several analyses available for materials of likely mantle origin (see other papers in this volume). Some samples indicate high and some low $(^{40}Ar/^{36}Ar)_m$ ratios. HAMANO and OZIMA (1978) consider $(^{40}Ar/^{36}Ar)_m \gg 296$ favoring rapid degassing. There are, however, some samples, possibly representing the deeper mantle, which give low ratios. The problem here, aside from the possibility of modern atmospheric contamination, is clearly that the mantle is heterogeneous, and that any accessible sample, by definition, is atypical. If there is a global scale reservoir which behaves reasonably closely as postulated in the first order model, we do not yet have an identified sample of its argon.

7.3 *Helium*

The situation for helium exhibits several interesting contrasts with the case for argon. The biggest contrast, of course, is that helium escapes from the atmosphere, so that an accumulation model is inapplicable: the helium inventory in the atmosphere reflects the conditions of the last 10^6 yr or so rather than an integration over geologic time.

The case for ^4He also differs from that of ^{40}Ar in that the parent abundance is more reasonably constrained (by heat flow data); ^{4}He may thus be used to constrain models through degassing rate rather than through accumulation. In the first order model we may modify Eq. (6) by ignoring any primordial ^{4}He and summing over three source terms:

$$D_m = \sum \frac{\lambda y P_{m_0}}{\lambda - \alpha}(e^{-\alpha t} - e^{-\lambda t}), \tag{10}$$

where the summation is over λ, y, and P_{m_0} of ^{238}U, ^{235}U, and ^{232}Th. If the rate at which ^{4}He is currently being degassed into the atmosphere is known it may be set equal to αD_m in Eq. (10) and this relationship may then be solved for α. Unfortunately, the juvenile flux of ^{4}He is not known. We mentioned earlier that the oceanic ^{4}He flux is more than order of magnitude lower than the current production rate of ^{4}He. If we equate the oceanic flux to αD_m we obtain $\alpha = 5.6 \times 10^{-12}$ yr^{-1} ($\alpha T = 0.025$) as a lower limit. Unless the continental/volcanic ^{4}He flux is an order of magnitude higher than the oceanic flux this approach would indicate a low α and an earth which is not extensively degassed. A good estimate of the global ^{4}He flux, via either estimating the continental flux or the lifetime in the atmosphere, would obviously be very valuable.

^{3}He may also be useful in a rate calculation; this is a particularly important instance since ^{3}He is the only case where a global juvenile flux can be computed. Craig *et al.* (1975) compute a global average ^{3}He flux of 4 atoms cm^{-2} sec^{-1}, or 2.4×10^7 (cm^{3}STP) yr^{-1}. This is in itself not immediately applicable to the first order model since neither the initial abundance nor the accumulation of ^{3}He is known. If, however, we assume that the original volatile inventory was planetary in composition (Table 2, line 2) we could infer the juvenile flux of, say, ^{36}Ar using ^{36}Ar/^{3}He$=129$, obtaining a ^{36}Ar flux of 3.1×10^9 (cm^{3}STP) yr^{-1}. For a total atmospheric ^{36}Ar inventory of 1.3×10^{20} cm^{3}STP we would then have, from Eqs. (1)–(3),

$$\frac{S_a}{dS_a/dt} = \frac{1}{\alpha}(e^{\alpha T} - 1) = 4.1 \times 10^{10} \text{ yr}, \tag{11}$$

whence $\alpha = 7.6 \times 10^{-10}$ yr^{-1}. ($\alpha T = 3.5$). This would indicate that the earth is almost completely degassed, the undegassed fraction being $e^{-\alpha t} = 3\%$. The same conclusion would follow from use of an isotope other than ^{36}Ar, e.g., for ^{84}Kr $\alpha T = 4.1$ and $e^{-\alpha T} = 1.6\%$.

While this calculation is straightforward it too, unfortunately, cannot be accepted uncritically since the presumed initial abundance of ^{3}He, relative to the other noble gases, is very uncertain and the value we used could easily be in error by an order of magnitude or more, enough to completely change the sense of the result. If the initial ^{3}He relative abundance is higher (including a possible contribution, perhaps dominant, from imported spallation), α must be larger; if ^{3}He is overestimated, perhaps because meteoritic "planetary" helium is more contaminated with solar helium than were the solids which formed the earth, α would have to be correspondingly reduced.

The ^{3}He flux corresponds to an atmospheric ^{3}He lifetime of 1.0×10^6 yr (Craig *et al.*, 1975). The oceanic ^{4}He flux extablishes a maximum ^{4}He lifetime of 10^7 yr. We may presumably assume that the ^{4}He lifetime is not less than that of ^{3}He; this sets a lower limit to the ^{4}He lifetime which in turn sets an upper limit, 2.1×10^{13} cm^{3}STP/yr, to the total rate of ^{4}He entry into the atmosphere. This flux is an order of magnitude higher than the oceanic flux and about half the calculated production rate. By the calculation described above (Eq. (10)), this corresponds to an upper limit $\alpha=8\times10^{-11}$ yr^{-1}, or $1-e^{-\alpha t}=30\%$ as the fraction to which the earth is degassed. Again, however, the uncertainty in the input parameters (^{3}He flux, uranium abundance) is sufficiently large to preclude a definitive statement that the earth is not thoroughly degassed.

7.4 *Xenon*

Since atmospheric xenon contains at least one (^{129}Xe) and quite probably two (^{136}Xe) isotopic enhancements due to radiogenic nuclear components its evolution can be described in the same fashion as argon evolution (Figs. 6 and 7). It is unfortunate that previous treatments have concentrated on argon and helium to the almost complete exclusion of xenon.

The situation for xenon differs from that for argon and helium in several qualitatively important respects. One is that the radiogenic contributions do not dominate the primordial contributions, so that the isotopic effects are relatively small and, in the case of ^{136}Xe, difficult to distinguish. In Figs. 6 and 7, diagrams for the evolution of atmospheric composition from Eq. (9), we have thus plotted $[(D/S)_a(t)-(D/S)_o]/[(D/S)_a(T)-(D/S)_o]$, the fractional enhancement over primordial composition, as well as the absolute composition $(D/S)_a(t)$, since the former quantity is not dependent on identification of the primordial composition. While the isotopic effects are relatively small, however, the resolution of significant variation is well within present experimental capabilities.

Another important difference is that both xenon parent activities are very short-lived (following the previous discussion we will now assume that ^{244}Pu rather than ^{238}U is primarily responsible for any fission component in atmospheric xenon). Because of these short lifetimes the evolutionary field accessible in first-order models (Figs. 6 and 7) is even narrower than in the case of argon (Fig. 5). For the same reason $(P/S)_{m_o}$ (Eq. (9)) is determined essentially independently of α; assuming that radiogenic ^{129}Xe is 6.7% of atmospheric ^{129}Xe and that the ^{244}Pu contribution to atmospheric ^{136}Xe is 4.7% (both from PEPIN and PHINNEY, 1976), we have $(^{129}\mathrm{I}/^{130}\mathrm{Xe})_{m_o}=0.44$ and $(^{244}\mathrm{Pu}/^{130}\mathrm{Xe})_{m_o}=1.5\times10^3$ (atomic ratios).

These data can be used to compute a "Xe-Xe" formation interval for the earth (WETHERILL, 1975). We have

$$(^{129}\mathrm{I}/^{244}\mathrm{Pu})_{m_o}=(^{129}\mathrm{I}/^{127}\mathrm{I})_{m_o}(^{127}\mathrm{I}/^{238}\mathrm{U})_{m_o}(^{238}\mathrm{U}/^{244}\mathrm{Pu})_{m_o}=2.9\times10^{-4}.$$

Taking $(^{129}\mathrm{I}/^{127}\mathrm{I}_{m_o})=8.0\times10^{-5}\times e^{-\lambda_{129}\Delta t}$ and $(^{244}\mathrm{Pu}/^{238}\mathrm{U})_{m_o}=0.015\times e^{-\lambda_{244}\Delta t}$, where the constants refer to measurements in the chondrite St. Severin (PODOSEK, 1970, 1972) and Δt is the delay from the time of formation of St. Severin, we obtain $(^{127}\mathrm{I}/^{238}\mathrm{U})_{m_o}$ $e^{-(\lambda_{129}-\lambda_{244})\Delta t}=0.055$. As noted, this result is not (in a first order model) dependent on the degree of degassing. The biggest uncertainty here is in geochemical estimates of I/U. For the crustal value used by PEPIN and PHINNEY (1976), $(^{127}\mathrm{I}/^{238}\mathrm{U})_{m_o}=0.38$, we get $\Delta t=60\times60^6$ yr. Values of similar magnitude follow from calculations based on excess ^{129}Xe and ^{136}Xe in well gases (WETHERILL, 1975; PEPIN and PHINNEY, 1976). The formation interval calculated from atmospheric xenon composition would be much longer, about 2×10^8 yr, if we used the cosmic (CAMERON, 1973) ratio $(^{127}\mathrm{I}/^{238}\mathrm{U})_{m_o}=55$.

It is of interest to attempt a comparison of the apparent radiogenic components in

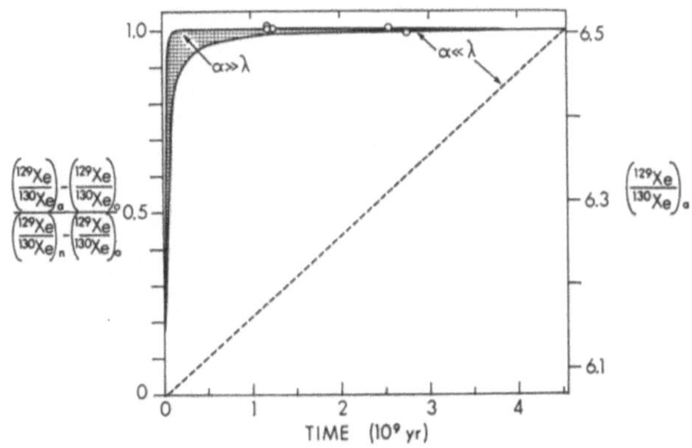

Fig. 6. Evolution of atmospheric xenon composition for continuous and catastrophic degassing models. Format and calculations are as in Fig. 5. The range accessible in continuous degassing models is narrower than in Fig. 5 because of the shorter half-life of the parent ^{129}I. In the normalized scale at left subscript *a* designates composition as a function of time, *n* the present composition, and *o* the primordial composition; this scale is independent of the value of $(^{129}Xe/^{130}Xe)_o$. The absolute scale at right assumes $(^{129}Xe/^{130}Xe)_o=6.70$, i.e., that 6.7% of modern atmospheric ^{129}Xe is radiogenic. The open circles are xenon compositions measured in ancient sedimentary rocks by PHINNEY (1972), and FRICK and CHANG (1977) and here assumed to represent ancient atmospheric composition.

the atmosphere with an estimate of the total amount available. As in the previous discussion of this problem for ^{40}Ar the principal difficulty here is the geochemistry, i.e., in estimating the amount of the "parent" element in the earth; in this case, the problem is compounded by the fact that ^{129}I and ^{244}Pu are extinct and unmeasurable today and that their abundance varied rapidly in the early stages of solar system evolution. If we accept the Xe-Xe formation interval discussed above, the earth formed with $(^{244}Pu/^{238}U)_{m_o}=0.0090$. For a present uranium abundance of 32 ppb the decay of ^{244}Pu (and ^{238}U) produces a total of 2.3×10^{16} cm^{3}STP. If 4.7% of atmospheric ^{136}Xe is fissiogenic, fission ^{136}Xe in the atmosphere is about 1.5×10^{15} cm^{3}STP, about 6% of the amount available; the corresponding degassing rate is $\alpha=1.4\times10^{-11}$ yr^{-1}. An ab initio calculation of the same sort for ^{129}Xe is essentially unmanageable, since neither the iodine abundance of the earth nor the $(^{129}I/^{127}I)_{m_o}$ ratio is even approximately constrained; for the same assumptions that are involved in computation of the Xe-Xe formation interval, however, the degree to which radiogenic ^{129}Xe is degassed is the same as the degree to which radiogenic ^{136}Xe is degassed, about 6%. It is important to note that this calculation is based on the xenon inventory in air. If 75% of the xenon degassed from the interior is now adsorbed on shales rather than in air (FANALE and CANNON, 1971), the degree to which radiogenic ^{136}Xe is degassed would be raised to 25%. In order to match the meteoritic planetary pattern (Fig. 1) we would have to assume that 90–95% of all degassed xenon is now in sediments; the corresponding conclusion is that essentially all radiogenic ^{136}Xe has been degassed.

A third major difference between the cases for xenon and argon is that xenon is more likely to be retained in samples exposed to the ancient atmosphere and is thus more

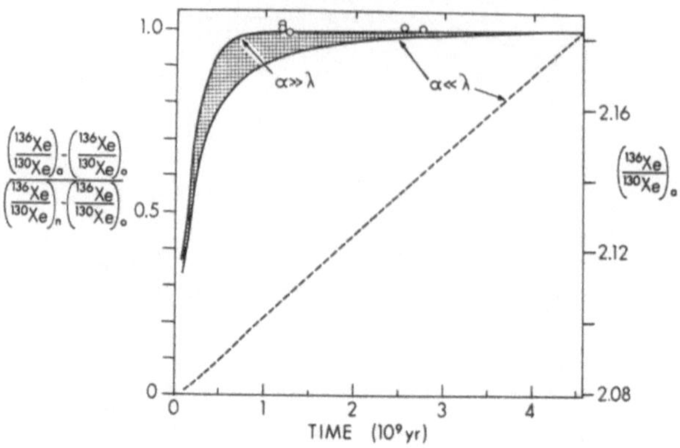

Fig. 7. Evolution of atmospheric xenon composition for continuous and catastrophic degassing models. Compare with Figs. 5 and 6. Format, normalized scale, and data as in Fig. 6. The absolute scale at right assumes that 4.7% of modern atmospheric ^{136}Xe originates in fission. The curves are calculated assuming $(^{244}\mathrm{Pu}/^{238}\mathrm{U})_{m_0}=0.015$; their basic form differs slightly from that of Figs. 5 and 6 because there are two contributions to ^{136}Xe, a dominant contribution from short-lived ^{244}Pu and a smaller one from long-lived ^{238}U.

likely to permit us an examination of its composition. In fact, while considerable but largely unsuccessful effort has been applied to measuring ancient atmospheric argon, comparable measurements for xenon already exist (Figs. 6 and 7), a circumstance which has not been well appreciated. The available data, in principle measuring atmospheric xenon as far back as 3.3×10^9 yr ago, show no change in xenon composition over this time. While any measurement of something that looks like modern air is immediately suspect of being just that, a caution that must not be forgotten, still there is no prima facie reason not to accept these results at face value. The available data are not quite old enough to clearly distinguish between rapid and slow first-order degassing for ^{129}Xe (Fig. 6). If, however, the data do truly measure ancient atmospheric xenon and the assessment of the fission ^{136}Xe content in air (PEPIN and PHINNEY, 1976) is even approximately correct, the ^{136}Xe data (Fig. 7) clearly favor rapid over slow first-order degassing.

7.5 Catastrophic degassing

Often the question of whether or not the earth is fully degassed is framed in terms of whether or not it has undergone a "catastrophic" degassing, an extensive degassing of all volatiles very early in the history of the earth. A limiting case of this phenomenon can be included in our discussion with only a small modification of the first-order model, and one which involves no new parameters. If it is assumed that the earth initially released all its volatiles, and continues to release them as quickly as any new (radiogenic) volatiles are formed, this is just a special case of first order degassing with $\alpha T\gg1$. We will, therefore, examine the case for immediate ($t=0$) release of all primordial volatiles, followed by a regime in which subsequently produced volatiles are released according to the first order model, not necessarily with $\alpha T\gg1$.

With this modification Eqs. (2) and (3) become:

$$S_m = 0, \tag{10}$$

$$S_a = S_{m_o}. \tag{11}$$

Eqs. (6) and (7) become:

$$D_m = \frac{\lambda_y P_{m_o}}{\lambda - \alpha} (\mathrm{e}^{-\alpha t} - \mathrm{e}^{-\lambda t}), \tag{12}$$

$$D_a = D_{m_o} + y P_{m_o} \left(1 - \frac{\lambda \mathrm{e}^{-\alpha t} - \alpha \mathrm{e}^{-\lambda t}}{\lambda - \alpha}\right). \tag{13}$$

And Eqs. (8) and (9) become:

$$\left(\frac{D}{S}\right)_m = \infty, \tag{14}$$

$$\left(\frac{D}{S}\right)_a = \left(\frac{D}{S}\right)_{m_o} + y\left(\frac{P}{S}\right)_{m_o} \left(1 - \frac{\lambda \mathrm{e}^{-\alpha t} - \alpha \mathrm{e}^{-\lambda t}}{\lambda - \alpha}\right). \tag{15}$$

The evolution of atmospheric composition described by Eq. (15) for various values of α is shown in Figs. 5, 6, and 7. The principal effect of early catastrophic degassing is that it decouples the primordial from the radiogenic gases and for low rates of subsequent continuous degassing the accessible field is broadened over that accessible to simple first order models, and even for short-lived parents (Figs. 6 and 7) a closer approach to linear growth is allowed.

In the strict sense of instantaneous and complete degassing the catastrophic degassing model cannot be and has not been seriously advocated, since it is clear that even now the earth is not totally degassed. In the broader sense of nearly complete degassing, however, it is a more viable model and whether or not it is a valid model is an important consideration not only for the evolution of the atmosphere but also in study of the kinetics of the evolution of the earth as a whole. In a similarly broad sense it is unrealistic to suppose a truly instantaneous degassing which was finished at $t=0$; the important question is whether or not there was a qualitative change from rapid to slow degassing early in earth history. If a period of early rapid degassing lasted for 10^8 yr, say, use of a model of catastrophic degassing only at $t=0$ would make no difference for ^{4}He and ^{40}Ar but would make a major qualitative difference for ^{129}Xe and ^{136}Xe, since the major production of these species occurs in the first 10^8 yr of earth history.

The atmospheric development expressed in Eqs. (10)–(15) takes no account of less than complete and less than instantaneous early catastrophic degassing. To do so would require two additional free parameters (for example, see HAMANO and OZIMA (1978)) and, as already discussed, we consider that available data are not sufficient to constrain even the one free parameter inherent in the simple first order model. The consideration of even this extreme form of catastrophic degassing is nevertheless important because it does allow qualitative discrimination among different models. As seen in Figs. 6 and 7, for example, if the existing data are accepted at face value then one model history which is not allowed is catastrophic degassing followed by slow ($\alpha T \ll 1$) continuous degassing. It must be recognized, however, that the qualitative conclusion is that the degassing of primordial and of radiogenic xenon have not been decoupled, i.e., if primordial xenon is extensively degassed, so too is radiogenic xenon. Nearly complete instantaneous catastrophic degassing followed by slow continuous degassing is also not allowed; if there was an early catastrophic degassing followed by slow continuous degassing the early episode must have occurred or persisted at least late enough to allow for essentially complete decay of ^{129}I and ^{244}Pu.

8. Summary

We have discussed the probable or possible presence of nuclear components in atmospheric gases, as summarized in Table 3, and have illustrated their potential utility in addressing problems concerning the formation of the earth and its subsequent development of an atmosphere.

Table 3. Nuclear components[1] in atmospheric gases.

Isotope	Fractional abundance	Status	Source
^{40}Ar	1	Uncontested	^{40}K
^{4}He	1	Uncontested	^{238}U, ^{235}U, ^{232}Th
^{129}Xe	7%	Almost surely	^{129}I
^{136}Xe[2]	≤ 5% (?)	Probable	Fission (^{244}Pu, ^{238}U)
^{21}Ne	≤20%	Conjectural	Imported spallation, ^{18}O (α, n)^{21}Ne
^{3}He[3]	?	Conjectural	Imported spallation
^{14}N	?	Conjectural	^{14}C
^{86}Kr	1%	Unlikely	Fission?

1) This table does not include components which may be major constituents of the total solar system inventory of their respective isotopes and which, because of large scale inhomogeneities in the early solar system, may be present in different proportions in various solar system materials; possible examples are oxygen and neon (see text).

2) Along with corresponding amounts of ^{134}Xe, ^{132}Xe, ^{131}Xe, and ^{129}Xe.

3) This entry refers to the juvenile ^{3}He observed in seawater (CLARKE *et al.*, 1969; CRAIG *et al.*, 1975); production of ^{3}He by spallation in the atmosphere is well documented and probably accounts for about 25% of the atmospheric inventory (CRAIG *et al.*, 1975).

8.1 Planetary gases

Although in general there seems to be no lack of sources for identified or hypothesized nuclear components, the primary identification comes principally from comparison of terrestrial atmospheric isotopic compositions with extraterrestrial isotopic compositions. To this end, the comparison with planetary gases observed in carbonaceous chondrites (and with lesser clarity in other classes of meteorites) has been especially valuable. The comparison with meteorites is very valuable in another sense also. The meteorites incorporate volatiles—the major volatiles as well as the rare gases—in proportions which can be sensibly reckoned as similar to those which we infer to have been present in the materials of which the earth was formed. Even though we do not yet understand how these proportions were generated we need not postulate that this phenomenon is unique to the earth.

At the same time, it would be unwise to allow the similarities of terrestrial gases to meteoritic gases to lead to extension of the analogy beyond its applicability. It is plausible that terrestrial helium and argon differ from primordial planetary gases in meteorites principally or solely by addition of nuclear components: ^{4}He, ^{40}Ar, possibly ^{3}He. The comparisons involving neon and xenon, however, are more subtle and perhaps more revealing. In meteorites both gases have a complicated and ill-understood structure. In neither case is it clear how the terrestrial gases are related to the meteoritic gases, but it is clear that terrestrial neon and xenon cannot be derived from meteoritic neon and xenon simply by addition of nuclear components generated in the earth or in material destined to become the earth. The differences must go farther back than that, either to large scale heterogeneities in the solar nebula or to the mechanism by which nebular gases were incorporated into solids.

A possibly misplaced emphasis on the attempt to "derive" the atmosphere from meteoritic planetary gas could be particularly important in the case of xenon. Both the large fractionation and the underabundances of atmospheric xenon, relative to the meteoritic gases, have usually been considered peculiar to the earth and their explanation sought in its particular evolution and circumstances. It may not be so: since terrestrial xenon is clearly not simply derived from meteoritic xenon, it may be that either or both characteristics are primordial features of the earth and that their explanation should be sought in the solar nebula. This shift in viewpoint would present no less interesting a problem but would likely lead to different interpretations of atmospheric evolution.

8.2 Degassing of the earth

The problem of the extent to which the earth has been degassed is one which has commanded interest since the recognition that the earth's atmosphere was secondary. The use of radiogenic components in modeling atmospheric evolution has played a major part in these studies, in a role similar to the use of Rb-Sr, U-Th-Pb, and Sm-Nd systematics in the study of mantle/crust reservoir evolution. In principle at least, all such studies are coupled in that they address different aspects of the broader problem of the thermal and chemical evolution of the earth, and advances in one such area are likely to lead to different perspectives in others.

While opinion has shifted from time to time, there is considerable present opinion favoring the view that the earth is nearly completely outgassed. Arguments favoring this position have been reviewed by FANALE (1971) who advocates a thorough catastrophic outgassing early in earth history. Many of these arguments, however, are negative, denying the applicability of contrary arguments and so leaving the question unsettled. Perhaps the major positive argument is indirect: the supposition that at some time in earth history the thermal regime was such as to lead to quantitative outgassing. The usual association is with formation of the core. The kinctics of the situation are not clear, however, and while it is intuitively plausible it is not necessarily true that even core formation would bring about extensive degassing. The degree of degassing would probably be coupled to the degree to which convection could bring material close to the surface, since even in a very hot earth the efficiency of diffusion over planetary distances is questionable. Our position in this paper is that the evolution of the atmosphere is a problem which can best be approached by study of the atmosphere itself. The close coupling of atmospheric evolution and thermal and chemical evolution may then be used to shed light on the latter problems.

We have illustrated a variety of ways in which the features of atmospheric structure, particularly the presence of radiogenic components, have been applied to the problem of the degree of outgassing. In all cases the approach to a definitive calculation is thwarted by uncertainty in a necessary parameter so that the situation is characterized by underconstraint of theoretical models, even so simple a model as first order degassing. We will argue below in favor of a limited degree of degassing, but our basic conclusion is that a definitive assessment of the degree of degassing cannot yet be made.

We have described a number of calculations, to varying degrees within the format of formal first-order degassing models outlined in the previous section, which are relevant to the question of the extent to which the earth is degassed. As already indicated, we do not consider any of them definitive. It is perhaps significant, however, that calculations which involve the total inventory of radiogenic gases suggest a low degree of degassing, while those which involve the ratios of radiogenic to stable isotopes suggest, within the confines of first-order models, a high degree of degassing.

Inventory calculations of the extent of degassing can be made for the radiogenic species ^{4}He, ^{40}Ar, and ^{129}Xe and ^{136}Xe. For ^{4}He we calculate an upper limit $1-e^{-\alpha t} \lesssim 30\%$. The principal ambiguity here is in the ^{3}He lifetime (which serves as an upper limit to the ^{4}He lifetime) and the uranium abundance of the earth; uncertainties in these two parameters could easily account for a factor of two in the ^{4}He flux rate, enough to completely change the sense of the calculation. For ^{40}Ar, the atmospheric inventory accounts for only 10% of the ^{40}Ar which would be generated in an earth with chondritic potassium; however, geochemical arguments can be mustered in favor of the proposition that the earth's potassium content is an order of magnitude below chondritic level. We have also calculated that only about 6% of radiogenic ^{129}Xe and ^{136}Xe is now in the atmosphere, although this calculation too rests on relatively unsubstantiated assumptions about the initial abundances of the parents ^{129}I and ^{244}Pu and its sense might also be reversed if the major portion of degassed xenon resides in sediments.

The converse arguments, for a high degree of degassing, follow from the inferred compositions of ancient atmospheric argon (Fig. 5) and xenon (Figs. 6 and 7), and from the inferred high $^{40}Ar/^{36}Ar$ ratios (HAMANO and OZIMA, 1978) in the present mantle (Fig. 4). Here the suggestion is clearly that $\alpha T \gg 1$, i.e., that the earth is extensively degassed. The problems with this interpretation, of course, are in whether the data actually measure ancient atmosphere or whether the high $^{40}Ar/^{36}Ar$ ratios are in any sense representative of the mantle. Equally important, this suggestion is much more model-dependent than in the case of the inventory calculations, so that even if the data are correct the conclusion follows only if the model is also valid.

Because of the possibility of an extensive early catastrophic degassing, the input of primordial volatiles into the atmosphere may be decoupled from that of the radiogenic components and the problem should be evaluated separately. A possible approach is to predict an initial whole earth volatile content by comparison with meteorites; this approach is unfruitful, however, since meteorite volatile concentrations vary by several orders of magnitude, among which the earth is intermediate. Another approach is to evaluate the present juvenile flux. The only species for which this can now be done is ^{3}He; a calculation based on this rate in a first order model indicates that all but a few percent of the earth's volatiles have been degassed. This calculation depends on the assumed primordial abundance of ^3He (relative to the heavier noble gases); if the abundance assumed for planetary gas (Table 2) is too high by an order of magnitude, a not impossible overestimate, the sense of the calculation would be reversed and indicate slow uniform degassing. We note that this would about correspond to the ^3He level conjectured to originate in spallation.

The most direct way to estimate the degree of primordial volatile degassing is, of course, to compare volatiles in the atmosphere with volatiles in the solid earth. The lack of a representative sample prohibits this comparison, with the possible exception of water (or perhaps also CO_2). It has been argued that the H_2O content of the mantle is of the order of 0.1%, as evidenced by petrogenetic models and apparent partial melting in the asthenosphere. Water in the mantle is thus comparable to or in excess of the amount in the oceans, indicating that the earth is not extensively degassed. Even this approach is inconclusive, however, as FANALE (1971) has argued that this much water could be dissolved in equilibrium (with the present oceanic water levels) in a hot and otherwise thoroughly degassed earth.

Insight into the problem may also be gained by considering the qualitative features of formal degassing models, such as the first order model described here. For example, the high $^{40}Ar/^{36}Ar$ observed in some presumably mantle derived samples is, in essentially any model, decisively indicative of nearly complete degassing of at least the primordial component and, in first order models, of the radiogenic component as well (Fig. 4). This is probably the strongest argument favoring the view that the earth is nearly completely degassed (OZIMA, 1975, 1977). Still, this interpretation cannot be accepted uncritically, and does not necessarily conflict with the inventory calculations which indicate that at least the radiogenic components are not extensively outgassed. The problem, in this case, is probably just a limitation of the model. Certainly some parts of the earth *are* degassed, specifically crustal and upper mantle materials, in that they have lower gas concentrations than the minimum average earth (atmospheric gases divided by the mass of the earth). This does not necessarily imply that the rest of the earth, the lower mantle, say, is also degassed, and in fact there are also mantle derived materials with low $^{40}Ar/^{36}Ar$ which, in essentially any model, indicate low degrees of degassing.

A perhaps more realistic conception, at least for describing the radiogenic gases, is that the earth cannot be adequately treated as a single uniform reservoir, but that one part, the upper mantle and crust, has been extensively degassed and another, the lower mantle, has not. An identification with specific parts of an inhomogeneous mantle is an unfounded hypothesis of course, but the important feature is maintenance of separate reservoirs. This sort of compartmentalization has evidently also been observed in studies of nonvolatile radiochronometers. DE PAOLO and WASSERBURG (1976), for example, infer the existence of separate reservoirs, a "primitive" one (with chondritic Sm/Nd) associated with continental igneous rocks and an "evolved" one associated with oceanic volcanism, with higher Sm/Nd (depleted in light rare earths). To the extent than an ancient undifferentiated reservoir can be identified it is attractive (although probably an oversimplification) to suppose that such a reservoir is a repository of undegassed volatiles. If a more evolved reservoir also has been principally responsible for generation of the radiogenic components of the atmosphere, the conclusions reached from degassing models (e.g., HAMANO and OZIMA, 1978) might with some fidelity apply to the evolution of this reservoir if not to the earth as a whole.

Even if the radiogenic gases are not extensively outgassed, the same need not be true for the primordial gases, since they could have been outgassed before the generation of these nuclear components. A thorough catastrophic degassing would have to occur, and to terminate, very early in earth history, however, to avoid degassing of the daughters of short-lived ^{129}I and ^{244}Pu. Independently of inventory and model calculations, the same conclusion follows from the observation of ^{129}Xe excesses, relative to air composition, in well gases and rocks (BOULOS and MANUEL, 1971; HENNECKE and MANUEL, 1975). A period of major degassing would presumably also homogenize any residual gas left in the earth, so this appears not to have happened at any time after decay of ^{129}I. The model of very early thorough catastrophic degassing followed by slow continuous degassing of subsequently produced radiogenic components, however, is inconsistent with the observation of a constant ^{129}Xe/^{130}Xe atmospheric ratio over more than half the age of the earth (Fig. 6). The case of ^{129}Xe thus leads us to conclude that the degassing of primordial and radiogenic gases have not been decoupled. We therefore feel that primordial gases, like radiogenic gases, have not been extensively outgassed, so that the major part of the earth's volatiles are still within the earth.

We wish to express our appreciation to the organizers of the U.S.-Japan Seminar on Rare Gas Abundance and Isotopic Constraints on the Origin and Evolution of the Earth's Atmosphere for the successful engineering of a meeting which materially advanced the state of the art. We also are grateful for the good humor with which Ann Tolin perservered through the preparation of a long manuscript. This work was supported by the National Science Foundation through grant DES74-22752.

REFERENCES

ALEXANDER, E. C., Jr., ^{40}Ar-^{39}Ar studies of Precambrian cherts: An unsuccessful attempt to measure the time evolution of the atmospheric ^{40}Ar/^{36}Ar ratio, *Precambr. Res.*, **2**, 329–344, 1975.

ANDERS, E. and D. HEYMANN, Elements 112 to 119: Were they present in meteorites? *Science*, **164**, 821–823, 1969.

ANDERS, E., D. HEYMANN, and E. MAZOR, Isotopic composition of primordial helium in carbonaceous chondrites, *Geochim. Cosmochim. Acta*, **34**, 127–132, 1970.

AXFORD, W. I., The polar wind and the terrestrial helium budget, *J. Geophys. Res.*, **73**, 6855–6859, 1968.

AUDOUZE, J., J. P. BIBRING, J. C. DRAN, M. MAURETTE, and R. M. WALKER, Heavily irradiated grains and neon isotope anomalies in carbonaceous chondrites, *Astrophys. J.*, **206**, L185–L189, 1976.

BASFORD, J. R., J. G. BRADLEY, J. C. DRAGON, and R. O. PEPIN, Krypton and xenon in lunar fines, *Lunar Sci.*, **IV**, 53–53b, 1972.

BECKER, R. H. and R. N. CLAYTON, Nitrogen abundances and isotopic compositions in lunar samples, *Proc. Lunar Sci. Conf. 6th*, 2131–2149, 1975.

BEGEMANN, F. H., W. WEBER, and H. HINTENBERGER, On the primordial abundance of argon-40, *Astrophys. J.*, **203**, L155–L157, 1976.

BLACK, D. C., Trapped helium-neon isotopic correlations in gas-rich meteorites and carbonaceous chondrites, *Geochim. Cosmochim. Acta*, **34**, 132–140, 1970.

BLACK, D. C., Trapped neon-argon isotopic correlations in gas-rich meteorites and carbonaceous chondrites, *Geochim. Cosmochim. Acta*, **35**, 230–235, 1971.

BLACK, D. C., On the origins of trapped helium, neon and argon isotopic variations in meteorites—I. Gas-rich meteorites, lunar soil, and breccia, *Geochim. Cosmochim. Acta*, **36**, 347–375, 1972a.

BLACK, D. C., On the origins of trapped helium, neon and argon isotopic variations in meteorites—II. Carbonaceous meteorites, *Geochim. Cosmochim Acta*, **36**, 377–394, 1972b.

BLACK, D. C. and R. O. PEPIN, Trapped neon in meteorites II, *Earth Planet. Sci. Lett.*, **6**, 395–405, 1969.

BOULOS, M. S. and O. K. MANUEL, The xenon record of extinct radioactivities in the earth, *Science*, **174**, 1334–1336, 1971.

CADOGAN, P. H., Paleoatmospheric argon in Rynie cherts, *Nature*, **268**, 38–40, 1977.

CAMERON, A. G. W., Abundances of the elements in the solar system, *Space Sci. Rev.*, **15**, 121–146, 1973.

CLARKE, W. B., M. A. BEG, and H. CRAIG, Excess ^{3}He in the sea: Evidence for terrestrial primordial helium, *Earth Planet. Sci. Lett.*, **6**, 213–220, 1969.

CLAYTON, D. D., Extinct radioactivities: Trapped residuals of presolar grains, *Astrophys. J.*, **199**, 765–769, 1975.

CLAYTON, R. N., L. GROSSMAN, and T. K. MAYEDA, A component of primitive nuclear composition in carbonaceous meteorites, *Science*, **182**, 485–488, 1973.

CLAYTON, R. N., N. ONUMA, and T. K. MAYEDA, A classification of meteorites based on oxygen isotopes, *Earth Planet. Sci. Lett.*, **30**, 10–18, 1976.

CRAIG, H., W. B. CLARKE, and M. A. BEG, Excess ^{3}He in deep water on the East Pacific Rise, *Earth Planet. Sci. Lett.*, **26**, 125–132, 1975.

COWAN, G. A., A natural fission reactor, *Sci. Am.*, **235**, 36–47, 1976.

DAKOWSKI, M., The possibility of extinct super-heavy elements occurring in meteorites, *Earth Planet. Sci. Lett.*, **6**, 152–154, 1969.

DAMON, P. E. and J. L. KULP, Excess helium and argon in beryl and other minerals, *Am. Min.*, **43**, 433–459, 1958.

DE PAOLO, D. J. and G. J. WASSERBURG, Inferences about magma sources and mantle structure from variations of ^{143}Nd/^{144}Nd, *Geophys. Res. Lett.*, **3**, 743–746, 1976.

DROZD, R. J., Krypton and xenon in lunar and terrestrial samples, Ph.D. thesis, Washington University, 1974.

DROZD, R. J. and F. A. PODOSEK, Primordial ^{129}Xe in meteorites, *Earth Planet. Sci. Lett.*, **31**, 15–30, 1976.

EBERHARDT, P., A Neon-E-rich phase in the Orgueil carbonaceous chondrite, *Earth Planet. Sci. Lett.*, **24**, 182–187, 1974.

EBERHARDT, P., Ne-E rich phase in Orgueil, *Meteoritics*, **10**, 401, 1975.

EBERHARDT, P., J. GEISS, H. GRAF, N. GRÖGLER, U. KRÄHENBUHL, H. SCHWALLER, J. SWARTZMÜLLER, and A. STETTLER, Trapped solar wind noble gases, exposure age and K/Ar-age in Apollo 11 lunar fine material, *Proc. Apollo 11 Lunar Sci. Conf.*, 1037–1070, 1970.

EBERHARDT, P., J. GEISS, H. GRAF, N. GRÖGLER, M. D. MENDIA, M. MÖRGELI, H. SCHWALLER, and A. STETTLER, Trapped solar wind noble gases in Apollo 12 lunar fines 12001 and Apollo 11 breccia 10046, *Proc. Lunar Sci. Conf. 3rd*, 1821–1856, 1972.

EUGSTER, O., P. EBERHARDT, and J. GEISS, Isotopic analyses of krypton and xenon in fourteen stone meteorites, *J. Geophys. Res.*, **74**, 3874–3896, 1969.

FANALE, F. P., A case for catastrophic early degassing of the earth, *Chem. Geol.*, **8**, 79–105, 1971.

FANALE, F. P. and W. A CANNON, Physical adsorption of rare gas on terrigenous sediments, *Earth Planet. Sci. Lett.*, **11**, 362–368, 1971.

FANALE, F. P. and W. A. CANNON, Surface properties of the Orgueil meteorite: Implications for the early history of solar system volatiles, *Geochim. Cosmochim. Acta*, **38**, 453–470, 1974.

FRICK, U. and S. CHANG, Ancient carbon and noble gas fractionation, Submitted to *Proc. Lunar Sci. Conf. 8th*, 1977.

FOWLER, W. A., J. L. GREENSTEIN, and F. HOYLE, Nucleosynthesis during the early history of the solar system, *Geophys. J.*, **6**, 148–220, 1962.

GEISS, J., F. BUEHLER, H. CERUTTI, and P. EBERHARDT, Solar wind composition experiment, *Apollo 15 Prelim. Sci. Rep.*, Ch. 15, 1972.

GOETTEL, K. A., Models for the origin and composition of the earth, and the hypothesis of potassium in the earth's core, *Geophys. Surv.*, **2**, 369–397, 1976.

HAMANO, Y. and M. OZIMA, Earth-atmosphere evolution model based on Ar isotopic data, This volume, 1978.

HENNECKE, E. W. and O. K. MANUEL, Noble gases in a Hawaiian xenolith, *Nature*, **257**, 778–780, 1975.

HEYMANN, D. and M. DZICZKANIEC, Isotopic effects in magnesium and neon caused by proton bombardment of a gas of solar composition, *Meteoritics*, **11**, 300, 1976.

HEYMANN, D., M. DZICZKANIEC, and R. PALMA, Limits for the accretion time of the earth from cosmogenic ^{21}Ne produced in planetesimals, *Proc. Lunar Sci. Conf. 7th*, 3411–3419, 1976.

HOHENBERG, C. M., F. A. PODOSEK, and J. H. REYNOLDS, Xenon-iodine dating: Sharp isochronism in chondrites, *Science*, **156**, 202–206, 1967.

JEFFERY, P. M. and E. ANDERS, Primordial noble gases in separated meteoritic minerals—I., *Geochim. Cosmochim. Acta*, **34**, 1175–1198, 1970.

JEFFERY, P. M. and J. H. REYNOLDS, Origin of excess ^{129}Xe in: Stone meteorites, *J. Geophys. Res.*, **66**, 3582–3583, 1961.

JOKIPII, J. R., The distribution of gases in the protoplanetary nebula, *Icarus*, **3**, 248–252, 1964.

KERRIDGE, J. F., Solar nitrogen: Evidence for a secular increase in the ratio of nitrogen-15 to nitrogen-14, *Science*, **188**, 162–164, 1975.

KRUMMENACHER, D., C. M. MERRIHUE, R. O. PEPIN, and J. H. REYNOLDS, Meteoritic krypton and barium versus the general isotopic anomalies in meteoritic xenon, *Geochim. Cosmochim. Acta*, **26**, 231–251, 1962.

LAL, D., Study of long and short term geophysical processes using natural radioactivity, in *Radioactive Dating*, pp. 149–157, International Atomic Energy Agency, 1963.

LANCET, M. S. and E. ANDERS, Solubilities of noble gases in magnetite: Implications for planetary gases in meteorites, *Geochim. Cosmochim. Acta*, **37**, 1371–1388, 1973.

LANGSETH, M. G., S. J. KEIHM, and K. PETERS, Revised lunar heat-flow values, *Proc. Lunar Sci. Conf. 7th*, 3143–3171, 1976.

LEICH, D. A. and S. NIEMEYER, Trapped xenon in lunar anorthositic breccia 60015, *Proc. Lunar Sci. Conf. 6th*, 1953–1965, 1975.

LEWIS, R. S., Rare gases in separated whitlockite from the St. Severin chondrite: Xenon and krypton from fission of extinct ^{244}Pu, *Geochim. Cosmochim. Acta*, **39**, 417–432, 1975.

LEWIS, R. S., B. SRINIVASAN, and E. ANDERS, Host phase of a strange xenon component in Allende, *Science*, **190**, 1251–1262, 1975.

LIGHTNER, B. D. and K. MARTI, Lunar trapped xenon, *Proc. Lunar Sci. Conf. 5th*, 2023–2031, 1974.

LUPTON, J. E. and H. CRAIG, Excess ^{3}He in oceanic basalts: Evidence for terrestrial primordial helium, *Earth Planet. Sci. Lett.*, **26**, 133–139, 1975.

MARTI, K., Solar-type xenon: A new isotopic composition of xenon in the Pesyanoe meteorite, *Science*, **166**, 1262–1265, 1969.

MAZOR, E., D. HEYMANN, and E. ANDERS, Noble gases in carbonaceous chondrites, *Geochim. Cosmochim. Acta*, **34**, 781–824, 1970.

NIEMEYER, S. and D. A. LEICH, Atmospheric rare gases in lunar rock 60015, *Proc. Lunar Science Conf. 7th*, 587–597, 1976.

OWEN, T., K. BIEMANN, D. R. RUSHNECK, J. E. BILLER, D. W. HOWARTH, and A. L. LAFLEUR, The composition of the atmosphere at the surface of Mars, *J. Geophys. Res.*, **82**, 4635–4839, 1977.

OZIMA, M., Ar isotopes and earth-atmosphere evolution models, *Geochim. Cosmochim. Acta*, **39**, 1127–1134, 1975.

OZIMA, M. and K. KUDO, Excess argon in submarine basalts and an earth-atmosphere evolution model, *Nature Phys. Sci.*, **239**, 23–24, 1972.

PEPIN, R. O., Trapped neon in meteorites, *Earth Planet. Sci. Lett* , **2**, 13–18, 1967.

PEPIN, R. O., Neon and xenon in carbonaceous chondrites, in *Origin and Distribution of the Elements*, edited by L. H. Ahrens, pp. 379–386, Pergamon Press, 1968.

PEPIN, R. O., Oral presentation at American Geophysical Union meeting, 1976.

PEPIN, R. O. and D. PHINNEY, The formation interval of the earth, *Lunar Sci.*, **VII**, 682–684, 1976.

PHINNEY, D., ^{36}Ar, Kr, and Xe in terrestrial materials, *Earth Planet. Sci. Lett.*, **16**, 413–420, 1972.

PODOSEK, F. A., Dating of meteorites by the high-temperature release of iodine-correlated Xe^{129}, *Geochim. Cosmochim. Acta*, **34**, 341–365, 1970.

PODOSEK, F. A., Gas retention ages of Petersburg and other meteorites, *Geochim. Cosmochim. Acta*, **36**, 755–772, 1972.

PODOSEK, F. A., J. C. HUNEKE, D. S. BURNETT, and G. J. WASSERBURG, Isotopic composition of xenon and krypton in the lunar soil and in the solar wind, *Earth Planet. Sci. Lett.*, **10**, 199–216, 1971.

REYNOLDS, J. H., Isotopic composition of primordial xenon, *Phys. Rev. Lett.*, **4**, 351–354, 1960.

REYNOLDS, J. H. and G. TURNER, Rare gases in the chondrite Renazzo, *J. Geophys. Res.*, **69**, 3263–3281, 1964.

RUBEY, W. W., Geologic history of sea water, *Bull. Geol. Soc. Am.*, **62**, 1111–1148, 1951.

SABU, D. D. and O. K. MANUEL, Xenon record of the early solar system, *Nature*, **262**, 28–32, 1976.

SCHWARTZMAN, D. W., Ar degassing and the origin of the sialic crust, *Geochim. Cosmochim. Acta*, **37**, 2479–2495, 1973.

SIGNER, P. and H. E. SUESS, Rare gases in the sun, in the atmosphere, and in meteorites, in *Earth Science and Meteoritics*, edited by J. Geiss and E. D. Goldberg, Ch. 13, pp. 241–272, 1963.

SMITH, S. P., J. C. HUNECKE, and G. J. WASSERBURG, Neon in gas-rich samples of the carbonaceous chondrites Mokoia, Murchison, and Cold Bokkeveld, Submitted to *Earth Planet. Sci. Lett.*, 1977.

SRINIVASAN, B., E. C. ALEXANDER, and O. K. MANUEL, Iodine-129 in terrestrial ores, *Science*, **173**, 327–328, 1971.

TUREKIAN, K. K., The terrestrial economy of helium and argon, *Geochim. Cosmochim, Acta*, **17**, 37–43, 1959.

VON WEIZSÄCKER, C. F., Über die Möglichkeit eines dualen β-Zerfalls von Kalium, *Physik. Z.*, **38**, 623–624, 1937.

WALTON, J .R., D. HEYMANN, J. L. JORDAN, and A. YANIV, Evidence for solar cosmic ray proton-induced neon in fines 67701 from the rim of North Ray Crater, *Proc. Lunar Sci. Conf. 5th*, 2045–2060, 1974.

WASSERBURG, G. J., G. J. MACDONALD, F. HOYLE, and W. A. FOWLER, Relative contributions of uranium, thorium, and potassium to heat production in the earth, *Science*, **143**, 465–467, 1964.

WETHERILL, G. W., Variations in the isotopic abundances of neon and argon extracted from radioactive minerals, *Phys. Rev.*, **96**, 679–683, 1954.

WETHERILL, G. W., Radiometric chronology of the early solar system, *Annu. Rev. Nucl. Sci.*, **25**, 283–328, 1975.

WILLIAMS, D. L. and R. P. VON HERZEN, Heat loss from the earth: New estimate, *Geology*, **2**, 305–368, 1974.

WYLLIE, P. J., *The Dynamic Earth: Textbook in Geosciences*, John Wiley and Sons, Inc., 1971.

Trapped Xenon and Cosmic-Ray Effects in Meteorites, in Lunar Samples, and in the Earth's Materials

Koh Sakamoto

Department of Chemistry, Kanazawa University, Kanazawa-shi 920, Japan

A trapped composition of ^{124}Xe: ^{126}Xe: ^{128}Xe: ^{130}Xe: ^{131}Xe: ^{132}Xe: ^{134}Xe: ^{136}Xe= (0.027±0.001): (0.024±0.004): (0.49±0.10): 1: (5.00±0.08): (6.1±0.6): (2.42±0.01): (2.07±0.07) was tentatively obtained for carbonaceous chondrites (C.C.) with a new method of component analysis. There are indications in the light xenon isotopes of non-carbonaceous meteorites and lunar samples of a different trapped composition from that of C.C. Atmospheric xenon seems to be the trapped component for the earth's samples derived from deep-underground. Cosmogenic xenon, $(^{124}\mathrm{Xe}/^{126}\mathrm{Xe})_c$, was found to vary widely: $-0.3^{+0.6}_{-0.2}$ in lunar breccias, a granite sample from Carney Lake, Michigan, and others; 0.4~0.7 in lunar rocks, lunar fines, metamorphosed meteorites and a thucholite; ≈1 in chondrules and well gases; 1.6~2.6 or more in C.C., some low-temperature fractions of lunar and meteorite samples and the Red Rock granite. Most of the samples are heterogeneous with respect to $(^{124}\mathrm{Xe}/^{126}\mathrm{Xe})_c$.

1. Introduction

The present paper presents a new method of xenon component analysis which is based on linear correlations of the observed isotopic ratios in meteorites, lunar samples and the earth's materials. We will search for the trapped components in these materials first. Existence of peculiar cosmogenic components in the lighter xenon isotopes in all three objects will be derived next. Fissiogenic components in heavier xenon isotopes can be determined by extending the present analytical method, for which results will be presented in a separate publication.

2. Experimental Data and Their Treatments

The present work examines the reported data on meteoritic and lunar xenon isotopes from stepwise heating experiments of bulk samples and of mineral separates of Allende (Manuel *et al.*, 1972b; Lewis *et al.*, 1975, 1977), Leoville (Manuel *et al.*, 1970), Mokoia (Manuel *et al.*, 1972b), Murchison (Kuroda *et al.*, 1975; Srinivasan *et al.*, 1977), Murray (Kuroda *et al.*, 1974), Renazzo (Reynolds and Turner, 1964), Bruderheim (Merrihue, 1966), Pasamonte (Hohenberg *et al.*, 1967), Angra Dos Reis (Hohenberg, 1970; Wasserburg *et al.*, 1977; Lugmair and Marti, 1977), lunar fines 10084-59 (Hohenberg *et al.*, 1970), lunar breccias 14318 (Behrmann *et al.*, 1973; Reynolds *et al.*, 1974), 14313 (Behrmann *et al.*, 1973), 14301 (Drozd *et al.*, 1972), 60025, 62255, 67075 (Lichtner and Marti, 1974) and 60015 (Leich and Niemeyer, 1975), and lunar rocks 10057–20 and 10044–20 (Hohenberg *et al.*, 1970). This choice of data is arbitrary. Many other data from temperature releases of bulk samples and grain size fractions were also examined but excluded from this paper for simplicity of presentation. The terrestrial xenon data are of temperature-release experiments on granites (KURODA *et al.*, 1977), a thucholite (KURODA and SHERRILL, 1977) and a lava rock (HENNECKE and MANUEL, 1975b). The data on barites (SRINIVASAN, 1976) and well gases (BUTLER *et al.*, 1963; BENNETT and MANUEL, 1970; BOULOS and MANUEL, 1971, 1972; HENNECKE and MANUEL, 1975a) were also included.

The isotopic ratios of the temperature fractions from the individual samples were examined to look for any linear correlations which will be described in the following sections. A computer program was constructed for this search on the basis of the least squares fitting method described by YORK (1966). The calculation was repeated by rejecting data points whose x- and y-residuals were larger than the experimental errors (1σ) and confirming the validity of the rejections by a visual investigation of the plots on a graph. In some cases where there were wide scatters and therefore many rejections, the input data were visually divided into groups. The well-gas data were treated as fractions of one sample.

3. Nature of Linear Correlations and Trapped Components

Xenon isotopes with the fewest components may be those with mass numbers of 124, 126 and 130; they are assumed to be composed of trapped and cosmogenic (spallation-produced) components. A two-component correlation for these isotopes has been demonstrated most clearly and interpreted by the equation (PODOSEK, 1970),

$$\left(\frac{^{124}\mathrm{Xe}}{^{130}\mathrm{Xe}}\right)_m = S_{26}^{24}\cdot\left(\frac{^{126}\mathrm{Xe}}{^{130}\mathrm{Xe}}\right)_m + \left(\frac{^{124}\mathrm{Xe}}{^{130}\mathrm{Xe}}\right)_t - S_{26}^{24}\cdot\left(\frac{^{126}\mathrm{Xe}}{^{130}\mathrm{Xe}}\right)_t, \tag{1}$$

where

$$S_{26}^{24} = \frac{(^{124}\mathrm{Xe}/^{130}\mathrm{Xe})_c - (^{124}\mathrm{Xe}/^{130}\mathrm{Xe})_t}{(^{126}\mathrm{Xe}/^{130}\mathrm{Xe})_c - (^{126}\mathrm{Xe}/^{130}\mathrm{Xe})_t}.$$

The subscripts m, c, and t denote the measured, cosmogenic and trapped components, respectively. The relative contributions to the m value of the t and c components which are the end points of the straight line expressed by Eq. (1) can not be evaluated without assumptions of the t and c compositions (PODOSEK, 1970).

However, one can determine the trapped components in the following way; the intercept (I_{26}^{24}) of Eq. (1) is expressed as

$$I_{26}^{24} = -S_{26}^{24}\cdot(^{126}\mathrm{Xe}/^{130}\mathrm{Xe})_t + (^{124}\mathrm{Xe}/^{130}\mathrm{Xe})_t. \tag{2}$$

If the trapped components are common to a group of samples, the plot of I_{26}^{24} versus S_{26}^{24} would define a straight line and the values of the $(^{124}\mathrm{Xe}/^{130}\mathrm{Xe})_t$ and $(^{126}\mathrm{Xe}/^{130}\mathrm{Xe})_t$ ratios are determined from the intercept and slope of this straight line with the currently accepted relation of $(^{i}\mathrm{Xe}/^{130}\mathrm{Xe})_c \geq (^{i}\mathrm{Xe}/^{130}\mathrm{Xe})_m \geq (^{i}\mathrm{Xe}/^{130}\mathrm{Xe})_t$. The superscript i denotes 124 and 126. Figure 1a is an example. The $I_{26}^{24}-S_{26}^{24}$ points determined by the least squares fit as described in section 2 from the carbonaceous chondrites (C.C., open circles), fall beautifully on a straight line whose intercept, $(^{124}\mathrm{Xe}/^{130}\mathrm{Xe})_t$, and slope, $(^{126}\mathrm{Xe}/^{130}\mathrm{Xe})_t$, are 0.0265 ± 0.0007 and 0.0242 ± 0.0038 respectively. The terrestrial samples (open squares) seem to constitute a different line which yields the values of 0.0232 ± 0.0029 and 0.0222 ± 0.0113 for $(^{124}\mathrm{Xe}/^{130}\mathrm{Xe})_t$ and $(^{126}\mathrm{Xe}/^{130}\mathrm{Xe})_t$ respectively, comparable to the corresponding values

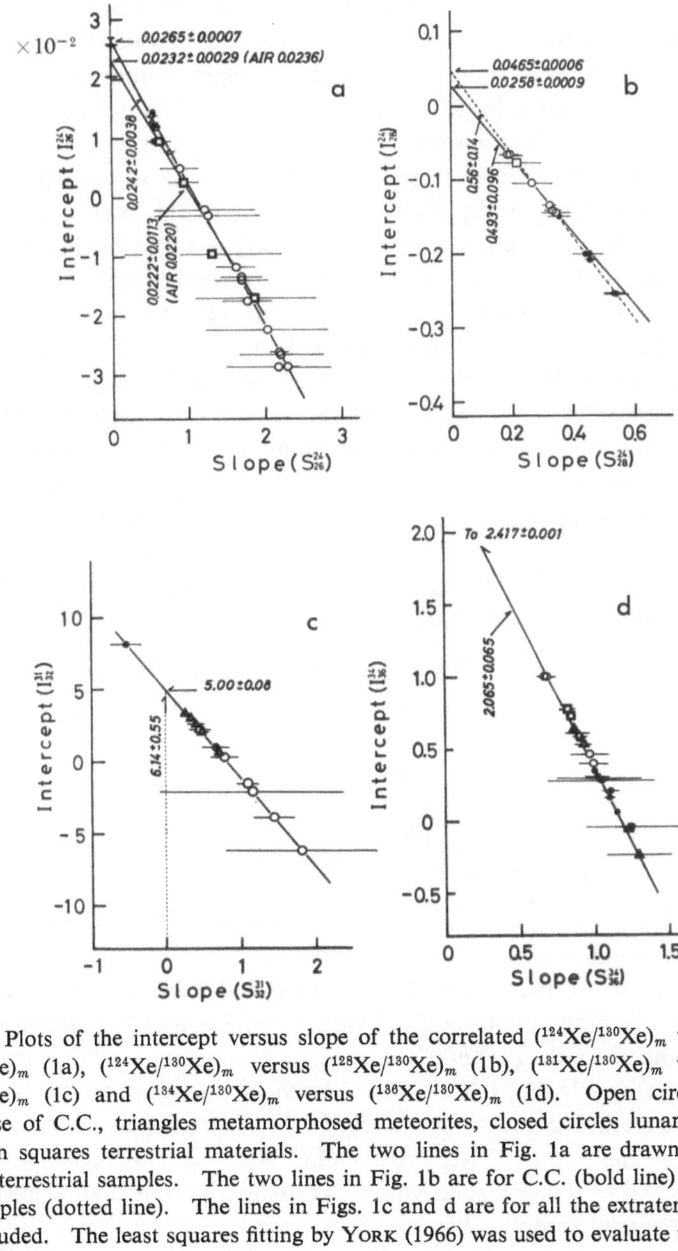

Fig. 1. Plots of the intercept versus slope of the correlated $(^{124}Xe/^{130}Xe)_m$ versus $(^{126}Xe/^{130}Xe)_m$ (1a), $(^{124}Xe/^{130}Xe)_m$ versus $(^{128}Xe/^{130}Xe)_m$ (1b), $(^{131}Xe/^{130}Xe)_m$ versus $(^{132}Xe/^{130}Xe)_m$ (1c) and $(^{134}Xe/^{130}Xe)_m$ versus $(^{136}Xe/^{130}Xe)_m$ (1d). Open circles represent those of C.C., triangles metamorphosed meteorites, closed circles lunar samples and open squares terrestrial materials. The two lines in Fig. 1a are drawn for C.C. and for terrestrial samples. The two lines in Fig. 1b are for C.C. (bold line) and for lunar samples (dotted line). The lines in Figs. 1c and d are for all the extraterrestrial points included. The least squares fitting by York (1966) was used to evaluate the slopes and intercepts.

of the earth's atmosphere of 0.0236 and 0.0220 (Nier, 1950). The S^{24}_{26} values of lunar samples (closed circles) and non-carbonaceous meteorites (triangles) are very close to each other, but the I^{24}_{26}, except for those of the lunar rocks and fines and the high temperature fractions of lunar breccia 14318 and Angra Dos Reis, tend to show smaller values than the corresponding C.C. values. It is noted that the $I^{24}_{26}-S^{24}_{26}$ points of some of Apollo 14 breccias are located very close to the terrestrial line. This supports the recent finding of a terrestrial-like trapped xenon in the Apollo 16 breccias (LICHTNER and MARTI, 1974; LEICH and NIEMEYER, 1975; NIEMEYER and LEICH, 1976).

The 128 isotope may have three components: trapped, cosmogenic and $^{127}I(n, \gamma\beta^-)$ produced. We, however, see a linear correlation of $(^{124}Xe/^{130}Xe)_m$ and $(^{128}Xe/^{130}Xe)_m$ in some samples for which the two-component (t and c) assumption may be valid. In order to evaluate the two components from graphical presentation, the slopes, S_{28}^{24}, and intercepts, I_{28}^{24}, defined in the same way as in Eq. (2), were calculated by the least squares method and plotted in Fig. 1b. The C. C. points (open circles) define a line yielding 0.0258 ± 0.0009 and 0.493 ± 0.096 for $(^{124}Xe/^{130}Xe)_t$ and $(^{128}Xe/^{130}Xe)_t$ respectively. The former value agrees expectedly with the one determined above from the I_{26}^{24}—S_{26}^{24} plots. The least squares determination of the lunar points (closed circles) yields the dotted line in Fig. 1 b whose intercept and slope are 0.0465 ± 0.0006 and 0.56 ± 0.14 respectively. The significance of these values can not be determined at the present stage because one can not tell whether the separate treatment of the lunar points is meaningful or not in view of the size of the attached errors. Only one point (the open square from the Red Rock granite) was available for the terrestrial $I_{28}^{24} - S_{28}^{24}$ plot and it is indistinguishable from the array of extraterrestrial points.

Linear correlations among heavier isotopes are currently known to hold for a suite of temperature fractions from individual samples:

$$\left(\frac{^{i}Xe}{^{130}Xe}\right)_m = S_j^i \cdot \left(\frac{^{j}Xe}{^{130}Xe}\right)_m + \left(\frac{^{i}Xe}{^{130}Xe}\right)_t - S_j^i \cdot \left(\frac{^{j}Xe}{^{130}Xe}\right)_t, \tag{3}$$

where

$$S_j^i = \frac{\dfrac{(^{i}Xe)_f + (^{i}Xe)_c}{(^{130}Xe)_c} - (^{i}Xe/^{130}Xe)_t}{\dfrac{(^{j}Xe)_f + (^{j}Xe)_c}{(^{130}Xe)_c} - (^{j}Xe/^{130}Xe)_t}.$$

The subscript f denotes fissiogenic xenon and the i and j are 134 and 136, or 131 and 132 respectively. It is assumed in Eq. (3) that $(^{i,j}Xe)_m = (^{i,j}Xe)_t + (^{i,j}Xe)_f + (^{i,j}Xe)_c$. A linear array of the data points in the $(^{i}Xe/^{130}Xe)_m$ versus $(^{j}Xe/^{130}Xe)_m$ diagram means that S_j^i is constant among the fractions involved of individual samples, and that the ratios of $(^{i,j}Xe)_f/(^{130}Xe)_c$ are constant, since the $(^{i,j}Xe/^{130}Xe)_t$ ratios are constant as shown in Figs. 1c and d, where $I_j^i = -S_j^i \cdot (^{j}Xe/^{130}Xe)_t + (^{i}Xe/^{130}Xe)_t$. Each of the $I_j^i - S_j^i$ points was determined by the least squares method as mentioned in section 2. The trapped compositions obtained for $(^{131}Xe/^{130}Xe)_t$, $(^{132}Xe/^{130}Xe)_t$, $(^{134}Xe/^{130}Xe)_t$ and $(^{136}Xe/^{130}Xe)_t$ are 5.00 ± 0.08, 6.14 ± 0.55, 2.417 ± 0.001, and 2.065 ± 0.065, respectively. No distinction was made between C.C. and the other extraterrestrial samples, because they were indistinguishable. The terrestrial points (open squares for the granites and thucholite) deduced for the 134–136 pair seem to be indistinguishable from the extraterrestrial line in the figure, but one point from the thucholite actually lies off the line, suggesting, with the granite point, the direction to the atmospheric values.

It is interesting to note that the f/c ratios are constant through some temperature fractions. Constancies of the f/c ratios are also demonstable by plotting $(^{i,j}Xe/^{130}Xe)_m$ against $(^{124,126}Xe/^{130}Xe)_m$ and are more distinct for C. C. than for other materials. This provides the basis for the theory of the X component by MANUEL *et al.* (1972a), and SABU and MANUEL (1976). LEWIS *et al.* (1975, 1977) ascribed the heavy xenon enrichment to the fission of unidentified superheavy elements and the lighter xenon enrichment to mass dependent fractionation during trapping. This point is still open and deserves further study. It should be noted that materials other than C.C. also exhibit an approximate correlation of *f* and *c* components.

For the xenon isotope of mass 129, the above correlation method may not be applied, because it is possible that the trapped component of ^{129}Xe has been altered by the addition of the decay product of the extinct nuclide ^{129}I at least to the extent corresponding to the sample ages (LEWIS and ANDERS, 1975; DROZD and PODOSEK, 1976).

4. Cosmic-Ray Production Ratio, $(^{124}Xe/^{126}Xe)_c$

Rewriting Eq. (1), one obtains

$$\left(\frac{^{130}Xe}{^{126}Xe}\right)_m = S_{24}^{30}\cdot\left(\frac{^{124}Xe}{^{126}Xe}\right)_m + \left(\frac{^{130}Xe}{^{126}Xe}\right)_t - S_{24}^{30}\cdot\left(\frac{^{124}Xe}{^{126}Xe}\right)_t, \tag{4}$$

where

$$S_{24}^{30} = \frac{(^{130}Xe/^{126}Xe)_c - (^{130}Xe/^{126}Xe)_t}{(^{124}Xe/^{126}Xe)_c - (^{124}Xe/^{126}Xe)_t},$$

from which

$$\left(\frac{^{124}Xe}{^{126}Xe}\right)_c = \frac{[(^{130}Xe/^{126}Xe)_c - (^{130}Xe/^{126}Xe)_t]}{S_{24}^{30}} + (^{124}Xe/^{126}Xe)_t \tag{5}$$

$$= \frac{[(^{130}Xe/^{126}Xe)_c - 41.3]}{S_{24}^{30}} + 1.10 \quad \text{(for C.C.)} \tag{6}$$

$$= \frac{[(^{130}Xe/(^{126}Xe)_c - 45.5]}{S_{24}^{30}} + 1.07. \quad \text{(for terrestrial samples)} \tag{7}$$

The variation of the quantity $(^{130}Xe/^{126}Xe)_c$ does not affect the $(^{124}Xe/^{126}Xe)_c$ for any S_{24}^{30} unless $(^{130}Xe/^{126}Xe)_c \gg 1$(see also Eq. (3) of PODOSEK and HUNEKE (1971)) and can be neglected in Eqs. (5)–(7). The values of $(^{124}Xe/^{126}Xe)_c$ are then estimated from the slope S_{24}^{30}.

The cosmic-ray production ratio $(^{124}Xe/^{126}Xe)_c$ has been reported to be in the range of 0.6 ± 0.2 for meteorites and lunar samples (see, for example, PODOSEK and HUNEKE, 1971; BOGARD *et al.*, 1971), which is consistent with the results of high-energy proton bombardment of Ba and rare-earth targets and spallation systematics. The same value were obtained in the present work for Pasamonte, Bruderheim, Angra Dos Reis, and lunar rocks and fines by assuming the trapped composition to be the same as that of C.C. The ratios found from the Besner mine thucholite and the barites are also in this range, as shown in Fig. 2. The lunar breccias and the Carney Lake granite indicate a ratio as low as $-0.3^{+0.6}_{-0.2}$, for which no known processes can be assigned. On the other hand, values ranging from ≈ 1 to more than 2.3 were obtained from well gases, Red Rock granite, C.C. and low temperature fractions of metamorphosed meteorites. KURODA *et al.* (1974) have explained the high values of $(^{124}Xe/^{126}Xe)_c$ for C.C. as due to the addition of spallation produced ^{124}Xe and ^{126}Xe, to the original xenon which has a

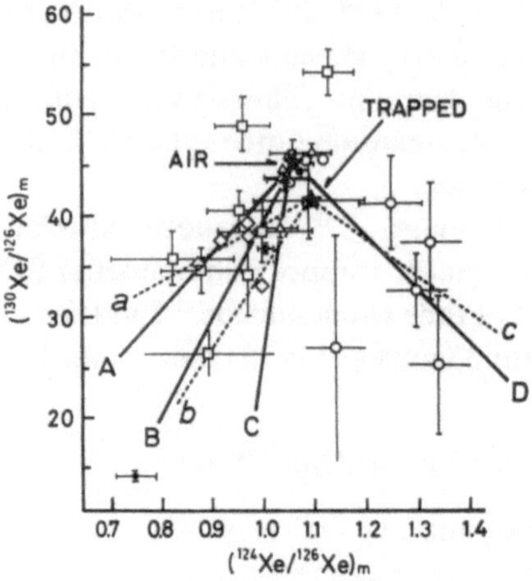

Fig. 2. Terrestrial (^{130}Xe/^{126}Xe)$_m$ versus (^{124}Xe/^{126}Xe)$_m$ plots. Data points are of stepwise heatings of Red Rock granite (circles), Carney Lake granite (diamonds), Besner mine thucholite (open squares), Mt. Capulin lava rocks and barites (closed circles). The error bars without limits indicate to be extended more than attached. The trapped point is that of C.C. inferred in this work. The bold lines are the least squares fittings for the Carney Lake granite (line A, (^{124}Xe/^{126}Xe)$_c$=0.3±0.1), the thucholite (B, 0.64±0.15), well gases (C, 0.95±0.22) and the Red Rock granite (D, 1.8±0.8). The dotted lines are shown for comparison; lunar breccias (line a, (^{124}Xe/^{126}Xe)$_c$ = $-0.3^{+0.6}_{-0.2}$), Pasamonte (b, 0.60±0.04) and a typical group of Allende and Murchison (c, 2.28±0.16), where the trapped compositions are assumed to be those of C.C.

(^{124}Xe/^{126}Xe) ratio greater than 1.274, deduced from Allende by KURODA *et al.* (1974). This may have resulted in a decrease of the ^{124}Xe/^{126}Xe ratio within the solar system in its early stages. This might be shown in Eq. (1) or (4) by exchanging the *c* and *t* terms. The assumption that (^{i}Xe/^{130}Xe)$_c$≥(^{i}Xe/^{130}Xe)$_m$≥(^{i}Xe/^{130}Xe)$_t$ for i=124 and 126 would be reversed. Hence, it should be concluded that the trapped values of 0.0265 and 0.0242 inferred above for the C.C. The values of (^{124}Xe/^{130}Xe)$_t$ and (^{126}Xe/^{130}Xe)$_t$ ratio respectively, are regarded to be the cosmic-ray production ratios and the trapped compositions are variable as shown for the (^{124}Xe/^{126}Xe)$_c$. This conclusion is not consistent, however, with experimental determinations of spallation yields. Manuel and his colleagues (MANUEL *et al.*, 1972a; SABU *et al.*, 1974; SABU and MANUEL, 1976) believe a component *X* to have originated in the p-process in a near-by supernova. However, we know of no nuclear processes which produce (^{124}Xe/^{126}Xe)$_c$ as high as 2 to 3 even in the supernova.

5. Concluding Remarks

Rare-gas isotope data on terrestrial samples are not abundant. Before proceeding into details for the constraints, precise measurements of isotopic abundances are needed for as many terrestrial samples as possible. Many of the meteorite and lunar data, especially for the lighter xenon isotopes, require more precise measurement with fractional gas release methods. The values derived in this paper are regarded to be tentative and are to be revised by including more precise data. Extensive laboratory experiments on nuclear reactions producing xenon isotopes are also necessary.

The author thanks Dr. K. Komura for his valuable discussion and for his help in computer calculations and Miss N. Fukushima for her renormalization calculation of the isotopic ratios and for typing.

REFERENCES

BEHRMANN, C. J., R. J. DROZD, and C. M. HOHENBERG, Extinct lunar radioactivities: Xenon from ^{244}Pu and ^{129}I in Apollo 14 breccias, *Earth Planet. Sci. Lett.*, **17**, 446–455, 1973.

BENNETT, G. A. and O. K. MANUEL, Xenon in natural gases, *Geochim. Cosmochim. Acta*, **34**, 593–610, 1970.

BOGARD, D. D., J. C. HUNEKE, D. S. BURNETT, and G. J. WASSERBURG, Xe and Kr analyses of silicate inclusions from iron meteorites, *Geochim. Cosmochim. Acta*, **35**, 1231–1254, 1971.

BOULOS, M. S. and O. K. MANUEL, The xenon record of extinct radioactivities in the earth, *Science*, **174**, 1334–1336, 1971.

BOULOS, M. S. and O. K. MANUEL, Extinct radioactive nuclides and production of xenon isotopes in natural gas, *Nature Phys. Sci.*, **235**, 150–152, 1972.

BUTLER, W. A., P. M. JEFFERY, J. H. REYNOLDS, and G. J. WASSERBURG, Isotopic variations in terrestrial xenon, *J. Geophys. Res.*, **68**, 3283–3291, 1963.

DROZD, R., C. M. HOHENBERG, and D. RAGAN, Fission xenon from extinct ^{244}Pu in 14301, *Earth Planet. Sci. Lett.*, **15**, 338–346, 1972.

DROZD, R. J. and F. A. PODOSEK, Primordial ^{129}Xe in meteorites, *Earth Planet. Sci. Lett.*, **31**, 15–30, 1976.

HENNECKE, E. W. and O. K. MANUEL, Noble gases in CO_2 well gas, Harding County, New Mexico, *Earth Planet. Sci. Lett.*, **27**, 346–355, 1975a.

HENNECKE, E. W. and O. K. MANUEL, Noble gases in lava rock from Mount Capulin, New Mexico, *Nature*, **256**, 284–287, 1975b.

HOHENBERG, C. M., Xenon from the Angra Dos Reis meteorite, *Geochim. Cosmochim. Acta*, **34**, 185–191, 1970.

HOHENBERG, C. M., M. N. MUNK, and J. H. REYNOLDS, Spallation and fissiogenic xenon and krypton from stepwise heating of the Pasamonte achondrite; The case for extinct plutonium 244 in meteorites; Relative ages of chondrites and achondrites, *J. Geophys. Res.*, **72**, 3139–3177, 1967.

HOHENBERG, C. M., P. K. DAVIS, W. A. KAISER, R. S. LEWIS, and J. H. REYNOLDS, Trapped and cosmogenic rare gases from stepwise heating of Apollo 11 samples, *Proc. Apollo 11 Lunar Sci. Conf.*, **2**, 1283–1309, 1970.

KURODA, P. K. and R. D. SHERRILL, Xenon and krypton isotope anomalies in the Besner Mine, Ontario, thucholite, *Geochem. J.*, **11**, 9–19, 1977.

KURODA, P. K., R. D. SHERRILL, and K. C. JACKSON, Abundances and isotopic compositions of rare gases in granites, *Geochem. J.*, **11**, 75–90, 1977.

KURODA, P. K., J. N. BECK, D. W. EFURD, and D. K. MILLER, Xenon isotope anomalies in the carbonaceous chondrite Murray, *J. Geophys. Res.*, **79**, 3981–3992, 1974.

KURODA, P. K., R. D. SHERRILL, D. W. EFURD, and J. N. BECK, Xenon isotope anomalies in the carbonaceous chondrite Murchison, *J. Geophys. Res.*, **80**, 1558–1570, 1975.

LEICH D. A. and NIEMEYER, Trapped xenon in lunar anorthositic breccia 60015, *Proc. 6th Lunar Sci. Conf., Suppl. 6, Geochim. Cosmochim. Acta*, **2**, 1953–1965, 1975.

LEWIS, R. S. and E. ANDERS, Condensation time of the solar nebula from extinct ^{129}I in primitive meteorites, *Proc. Natl. Acad. Sci. USA*, **72**, 268–273, 1975.

LEWIS, R. S., J. GROS, and E. ANDERS, Isotopic anomalies of noble gases in meteorites and their origins. 2. Separated minerals from Allende, *J. Geophys. Res.*, **82**, 779–792, 1977.

LEWIS, R. S., B. SRINIVASAN, and E. ANDERS, Host phase of a strange xenon component in Allende, *Science*, **190**, 1251–1262, 1975.

LICHTNER, B. D. and K. MARTI, Lunar trapped xenon, *Proc. Fifth Lunar Conf. Suppl. 5, Geochim. Cosmochim. Acta*, **2**, 2023–2031, 1974.

LUGMAIR, G. W. and K. MARTI, Sm-Nd-Pu time pieces in the Angra Dos Reis meteorite, *Earth Planet. Sci. Lett.*, **35**, 273–284, 1977.

MANUEL, O. K., E. W. HENNECKE, and D. D. SABU, Xenon in carbonaceous chondrites, *Nature*, **240**, 99–101, 1972a.

MANUEL, O. K., R. J. WRIGHT, D. K. MILLER, and P. K. KURODA, Heavy noble gases in Leoville: The case for mass fractionated xenon in carbonaceous chondrites, *J. Geophys. Res.*, **75**, 5693–5701, 1970.

MANUEL, O. K., R. J. WRIGHT, D. K. MILLER, and P. K. KURODA, Isotopic compositions of rare gases in the carbonaceous chondrites Mokoia and Allende, *Geochim. Cosmochim. Acta*, **36**, 961–983, 1972b.

MERRIHUE, C., Xenon and krypton in the Bruderheim meteorite, *J. Geophys. Res.*, **71**, 263–312, 1966.

NIEMEYER, S. and D. A. LEICH, Atmospheric rare gases in lunar rock 60015, *Proc. 7th Lunar Sci. Conf., Suppl. 7, Geochim. Cosmochim. Acta*, **1**, 587–597, 1976.

NIER, A. O., A redetermination of the relative abundances of the isotopes of neon, krypton, rubidium, xenon and mercury, *Phys. Rev.*, **79**, 450–454, 1950.

PODOSEK, F. A., Dating of meteorites by the high-temperature release of iodine-correlated Xe^{129}, *Geochim. Cosmochim. Acta*, **34**, 341–365, 1970.

PODOSEK, F. A. and J. C. HUNEKE, Isotopic composition of ^{244}Pu fission xenon in meteorites: Reevaluation using lunar spallation xenon systematics, *Earth Planet. Sci. Lett.*, **12**, 73–82, 1971.

REYNOLDS, J. H. and G. TURNER, Rare gases in the chondrite Renazzo, *J. Geophys. Res.*, **69**, 3263–3281, 1964.

REYNOLDS, J. H., E. C. ALEXANDER, Jr., P. K. DAVIS, and B. SRINIVASAN, Studies of K-Ar dating and xenon from extinct radioactivities in breccia 14318; implications for early lunar history, *Geochim. Cosmochim. Acta*, **38**, 401–417, 1974.

SABU, D. D. and O. K. MANUEL, Xenon record of the early solar system, *Nature*, **262**, 28–32, 1976.

SABU, D. D., E. W. HENNECKE, and O. K. MANUEL, Trapped xenon in meteorites, *Nature*, **251**, 21–24, 1974.

SRINIVASAN, B., Barites: Anomalous xenon from spallation and neutron-induced reactions, *Earth Planet. Sci. Lett.*, **31**, 129–141, 1976.

SRINIVASAN, B., J. GROS, and E. ANDERS, Noble gases in separated meteoritic minerals: Murchison (C2), Ornans (C3), Karoonda (C5), and Abee (E4), *J. Geophys. Res.*, **82**, 762–778, 1977.

WASSERBURG, G. J., F. TERA, D. A. PAPANASTASSIOU, and J. C. HUNEKE, Isotopic and chemical investigations on Angra Dos Reis, *Earth Planet. Sci. Lett.*, **35**, 294–316, 1977.

YORK, D., Least-squares fitting of a straight line, *Can. J. Phys.*, **44**, 1079–1086, 1966.

Classification and Generation of Terrestrial Rare Gases

Kazuo Saito

Department of Geology and Geophysics, University of Minnesota, Minneapolis, Minnesota 55455, U. S. A.

A $^{84}Kr/^{130}Xe$ versus $^{20}Ne/^{36}Ar$ diagram is a very useful format with which to study the elemental ratios of rare gases from terrestrial materials. It can separate not only the three types of rare gases which Ozima and Alexander (1976) classified but also the 'planetary' type rare gases from the other three types of rare gases.

When all the available terrestrial rare gas data are plotted in a $^{84}Kr/^{130}Xe$ versus $^{20}Ne/^{36}Ar$ diagram, several observations can be made. First, most of the analyses of rare gases from shales yield $^{84}Kr/^{130}Xe$ ratios between the 'planetary' and atmospheric values. If, however, the atmosphere's high $^{84}Kr/^{130}Xe$ ratio was produced by the selective adsorption of xenon onto shales from an initially 'planetary' atmosphere, as is widely accepted, then the $^{84}Kr/^{130}Xe$ ratio in shales should be even lower than the 'planetary' value. Second, the rare gas pattern in the quenched rims of submarine basalts may be explained as fractionated samples of the rare gases in sea water.

1. Introduction

A unifying theme behind much of the recent terrestrial rare gas work has been the search for primordial rare gases in the solid earth, whose isotopic or elemental ratios differ from atmospheric rare gases. There is reasonably clear isotopic evidence that such primordial rare gases have been retained in the earth's deep interior. $^{3}He/^{4}He$ ratios higher than the atmospheric value have been reported for deep sea water (Clarke *et al.*, 1969; Craig *et al.*, 1975), in volcanic gases and hot springs (Mamyrin *et al.*, 1969, 1972; Kononov *et al.*, 1975), in submarine rocks (Lupton and Craig, 1975; Craig and Lupton, 1976), and in mantle-derived xenoliths and xenocrysts (Tolstikhin *et al.*, 1974; Kaneoka *et al.*, 1978; Saito *et al.*, 1978). Anomalous abundances of ^{20}Ne have been measured in Kilauea volcanic gases (Craig and Lupton, 1976) and in mantle-derived minerals (Rison and Kyser, 1977). Excess ^{21}Ne has been detected in mantle-derived minerals (Rison and Kyser, 1977; Saito *et al.*, 1977) and excess ^{129}Xe has been detected in CO_2 well gas (Butler *et al.*, 1963; Boulos and Manuel, 1971; Hennecke and Manuel, 1975a) and in some mantle-derived xenoliths (Hennecke and Manuel, 1975b; Kaneoka *et al.*, 1978).

The elemental ratio evidence for primordial rare gases in mantle is, however, the subject of considerable scientific debate. For example, Dymond and Hogan (1973) reported that the quenched rims of submarine basalts contain significant quantities of rare gases whose elemental ratios are different from the atmospheric values. They interpreted their data in terms of a 'solar' primordial rare gas component in the mantle. Fisher (1974) interpreted the same data set in terms of a 'planetary' primordial rare gas component and Ozima and Alexander (1976) have argued that the quenched rim data represent a fractionated sample of 'planetary' primordial rare gases. Part of this debate can be traced to a lack of consensus of if and how the various observed rare gas patterns can be derived from one another.

OZIMA and ALEXANDER (1976) classified the rare gases contained in terrestrial samples into three types by using the apparent fractionation of elemental ratios relative to the atmospheric values. This classification becomes much clearer when the elemental ratios are plotted in a $^{84}Kr/^{130}Xe$ versus $^{20}Ne/^{36}Ar$ diagram shown in Fig. 1.

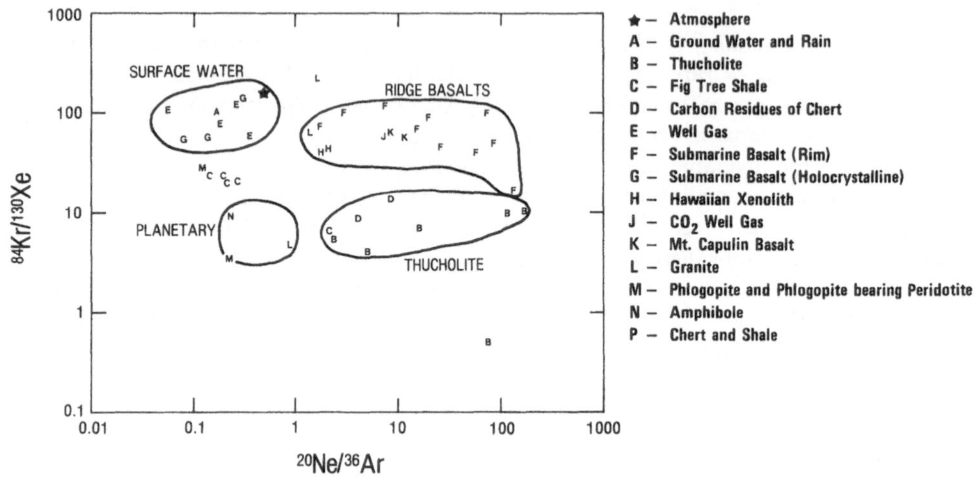

Fig. 1. Rare gas data are plotted in a $^{84}Kr/^{130}Xe$ versus $^{20}Ne/^{36}Ar$ diagram. The 'surface water' area corresponds to the 'type 1' pattern of Ozima and Alexander's classification (OZIMA and ALEXANDER, 1976), the 'ridge basalts' to the 'type 2' pattern and the 'thucholite' to the 'type 3.' When the isotopic composition were not reported in the original paper, the atmospheric composition were assumed to calculate the elemental ratios. The data were obtained from the following references: A, MAZOR (1972); B, BOGARD *et al.* (1965), SHERRILL (1976); C, CANALAS *et al.* (1968); D, FRICK and CHANG (1977); E, BENNETT and MANUEL (1970); F, DYMOND and HOGAN (1973) and unpublished data from this laboratory; G, DYMOND and HOGAN (1973); H, HENNECKE and MANUEL (1975b); J, HENNECKE and MANUEL (1975a); K, HENNECKE and MANUEL (1975c); L, KURODA *et al.* (1977); M, KANEOKA *et al.* (1977); N, SAITO *et al.* (1978); P, PHINNEY (1972).

In this paper I will first discuss the advantage of $^{84}Kr/^{130}Xe$ versus $^{20}Ne/^{36}Ar$ diagrams as a format for displaying rare gas elemental ratios and, second, will comment on possible schemes by which the patterns can be derived from each other.

2. $^{84}Kr/^{130}Xe$ versus $^{20}Ne/^{36}Ar$ Diagrams

The major advantage of a $^{84}Kr/^{130}Xe$ versus $^{20}Ne/^{36}Ar$ diagram, compared to the fractionation diagrams of OZIMA and ALEXANDER (1976), is that it clearly distinguishes not only the three patterns recognized by OZIMA and ALEXANDER but also the 'planetary' pattern. The 'planetary' primordial rare gases and the atmospheric like rare gases (the type 1 pattern in Ozima and Alexander's classification) differ mainly in the $^{84}Kr/^{130}Xe$ ratio. Rare gases in terrestrial samples show a wide range in $^{20}Ne/^{36}Ar$ and $^{84}Kr/^{130}Xe$ ratios but a rather narrow range in $^{84}Kr/^{36}Ar$ ratios. A fractionation pattern, such as that used by Ozima and Alexander which is normalized to ^{36}Ar does not emphasize the atmosphere-'planetary' distinction since the abundances of Kr and Xe are not compared directly.

Figure 1 is a $^{84}Kr/^{130}Xe$ versus $^{20}Ne/^{36}Ar$ plot of all the data used by Ozima and Alexander plus the more recent data from acid insoluble, carbonaceous residues from Precambrian cherts (FRICK and CHANG, 1977), from four new analyses of the quenched rims of submarine basalts (unpublished data from this laboratory), form granites (KURODA *et al.*, 1977) and from a peridotite and a phlogopite (KANEOKA *et al.*, 1977). In Fig. 1 the 'surface water' type rare gases correpond to the 'type 1' pattern of Ozima and Alexander's classification, the 'ridge basalts' type rare gases to the 'type 2' pattern and the 'thucholite' type rare gases to the 'type 3' pattern.

The 'planetary' area in Fig. 1 was determined as follows. All the rare gas data of carbonaceous chondrites reported in Table 1 of MAZOR *et al.* (1970), except those including italicized numbers, were plotted in a $^{84}Kr/^{130}Xe$ versus $^{20}Ne/^{36}Ar$ diagram. The region which contained the highest density of data points is shown as the 'planetary' region in Fig. 1. As can be seen in Fig. 1 the 'planetary' region is clearly separated from the other three regions.

An additional advantage of a $^{84}Kr/^{130}Xe$ versus $^{20}Ne/^{36}Ar$ diagram to a fractionation pattern diagram is that the differences between rare gas compositions can be discussed more quantitatively. This is important when one wants to discuss possible genetic relationships between rare gases in terrestrial samples.

$^{84}Kr/^{130}Xe$ versus $^{20}Ne/^{36}Ar$ plots, however, have their own disadvantages. The abundances of ^{36}Ar and ^{84}Kr are not compared directly. Since $^{36}Ar/^{84}Kr$ ratios are rather constant in terrestrial samples, there is usually no problem, but if solar wind type rare gases are present in some terrestrial samples, the data points will fall in the 'ridge basalts' or 'thucholite' area of Fig. 1. Hence, it is often useful to use a $^{84}Kr/^{130}Xe$ versus $^{36}Ar/^{84}Kr$ diagram, introduced by FISHER (1970), or a $^{36}Ar/^{84}Kr$ versus $^{20}Ne/^{36}Ar$ diagram as secondary

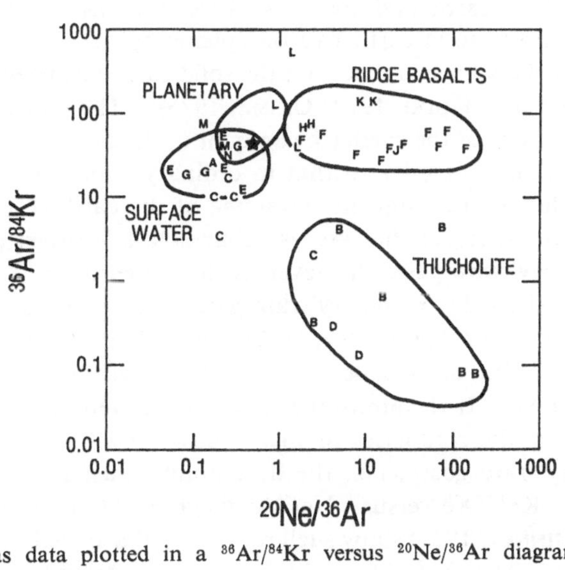

Fig. 2. Rare gas data plotted in a $^{36}Ar/^{84}Kr$ versus $^{20}Ne/^{36}Ar$ diagram. Symbols are the same as those in Fig. 1.

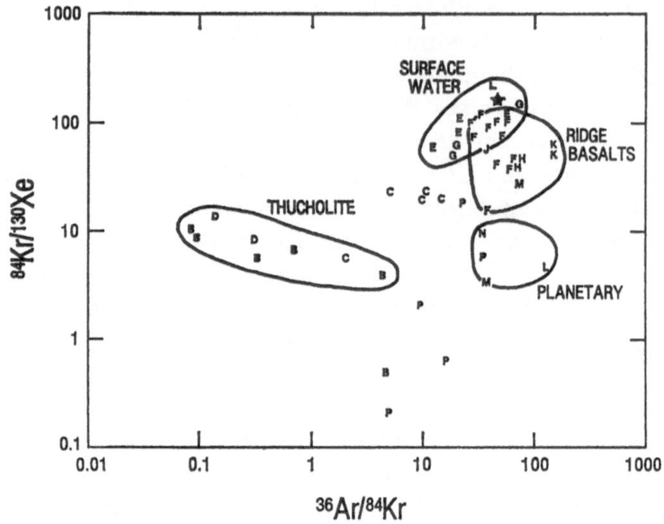

Fig. 3. Rare gas plotted in a $^{84}Kr/^{130}Xe$ versus $^{36}Ar/^{84}Kr$ diagram. Symbols are the same as those in Fig. 1.

aids. In such plots, however, the regions occupied by the 'surface water,' 'ridge basalts' and 'planetary' type rare gases overlap to varying degrees—as can be seen in Figs. 2 and 3.

3. Development of Terrestrial Rare Gas Patterns

3.1 The atmospheric rare gases

It is commonly assumed that the terrestrial atmosphere is a secondary product from the interior of the solid earth and many workers have assumed that the rare gases originally contained in the solid earth had the 'planetary' rare gas pattern.

Several recent workers have argued that the solid earth has been extensively degassed (FANALE, 1971; OZIMA and KUDO, 1972; OZIMA, 1975). The fact that the effective concentration of the rare gases of the earth (the amount of the atmospheric rare gases divided by the mass of the earth) is similar to that in ordinary chondrites has been taken as a strong evidence for the almost complete degassing of the earth (FANALE, 1971).

A conspicuous difference of the rare gas abundances between the atmospheric rare gases and the 'planetary' rare gases, however, is the depletion of Xe in the atmospheric rare gases as compared to the 'planetary' rare gases. Several workers (CANALAS *et al.*, 1968; FANALE and CANON, 1971; PHINNEY, 1972) argued that this Xe depletion is due to the secondary adsorption of Xe onto shales. If this is the case, shales must have adsorbed 90% of the terrestrial Xe. If the atmospheric rare gases were derived from a 'planetary' pattern by this process, the rare gases in shales must show $^{84}Kr/^{130}Xe$ ratios lower than that of the 'planetary' rare gases; i.e., the data points must fall in the area below the 'planetary' area in a $^{84}Kr/^{130}Xe$ versus $^{20}Ne/^{36}Ar$ diagram. Only five analyses of the rare gases from shales (PHINNEY, 1972) show such ratios and those analyses, unfortunately, did not include Ne data. The fact that only five analyses from a total of twelve shale measurements show $^{84}Kr/^{130}Xe$ ratios lower than that of the 'planetary' pattern appears to be a small fraction, if shales trapped 90% of the terrestrial Xe. This may, however, only mean that we have measured unrepresentative samples since the number of the analyses is still quite limited; or as KURODA and SHERRILL (1977) found in thucholite, Xe may have been trapped on shales very loosely and a considerable part of trapped Xe may have been pumped out before measurement, for example, in a preheating stage.

There is, however, another way to explain the Xe depletion in the earth's atmosphere. It is possible that a considerable portion of the terrestrial inventory of Xe still remains in the earth's interior (KANEOKA *et al.*, 1977; SAITO *et al.*, 1978). Also the striking resemblance of the rare gas abundances in the martian atmosphere and the terrestrial atmosphere (OWEN *et al.*, 1977) may be more compatible with the preferential retention of Xe in the interior of the planet than to adsorption onto shales.

3.2 The 'surface water' type rare gas pattern

OZIMA and ALEXANDER (1976) explained their type 1 pattern as atmospheric rare gases fractionated by the difference in the solubility of rare gases at low temperature into water. Basically I have no objection to their explanation. My only comment is that shale gases should be excluded from the type 1 pattern because shales preferentially adsorb Xe. This is the reason that the Fig Tree shale data were not used to help define the 'surface water' area in the figures.

3.3 The 'ridge basalts' type rare gases

OZIMA and ALEXANDER (1976) proposed that "the type 2 pattern represents a fractionated sample of 'planetary' primordial rare gas from the mantle." They based their model on the high temperature solubility data (experimental for He, Ne, and Ar, but extrapolated for Kr and Xe) of rare gases in molten enstatite (KIRSTEN, 1968). Their model is rigorously valid only under the following assumptions: either 1) the laboratory gas/liquid equilibrium must be the same as the liquid/solid equilibrium in the mantle, or 2) there must be a gas phase present in the mantle. In either case the laboratory data at <1 atmosphere pressure must be assumed to be valid at mantle pressures. All of these assumptions appear to be unrealistic.

One could explain the rare gases in the quenched rims of submarine basalts as a mixture of the rare gas in sea water and with another type of rare gas which has a high $^{20}Ne/^{36}Ar$ ratio and a low $^{84}Kr/^{130}Xe$ ratio. Indeed, if we assume $^{20}Ne/^{36}Ar$ ratio of 100, $^{84}Kr/^{130}Xe$ ratio of 8 and $^{36}Ar/^{84}Kr$ ratio of 50 for this unknown gas, data points of the quenched rims fall along the mixing curve of these two gases. Also, it is worth noting that samples which show high $^{20}Ne/^{36}Ar$ ratios contain small amounts of gas. While the elemental rare gas ratios in the quenched rims could be explained as mixtures of the rare gases dissolved in sea water with a hypothetical mantle composition, there are serious problems with such an interpretation.

The composition of such a hypothetical mantle gas agrees with neither the 'planetary' ($^{20}Ne/^{36}Ar=0.4$, $^{36}Ar/^{84}Kr=74$, $^{84}Kr/^{130}Xe=8$) nor 'solar' composition ($^{20}Ne/^{36}Ar=33$, $^{36}Ar/^{84}Kr=1500$, $^{84}Kr/^{130}Xe=38$, EBERHARDT *et al.*, 1972) nor with any mixture of these two known gas patterns. Such a hypothetical mantle gas would therefore have to be a fractionated sample of a primordial pattern—probably of the 'planetary' pattern, unless one is willing to postulate a whole new primordial pattern. If this gas were a fractionated sample of 'planetary' rare gas then samples showing smaller degrees of fractionation from the planetary pattern should also exist. The mixtures of such fractionated samples with the rare gases in sea water would form an array of mixing curves whose end points would form a continuous series between the 'planetary' composition and the most fractionated sample. The region between the 'planetary' area and the right hand end of the quenched rim trend in Fig. 1 is, however, noticeably blank. I take this as an evidence that the 'quenched' rim trend is not a mixing curve between a fractionated planetary pattern which was produced in the mantle and the rare gas pattern in sea water.

Hence, I believe that the 'ridge basalts' type rare gas pattern represents a fractionated sample of the 'surface water' type rare gas. The 'ridge basalts' pattern could be produced from the 'surface water' pattern by a process which enriches Ne. Depending on the degree of fractionation (Ne enrichment) rare gases in rims of submarine basalts may fall anywhere in the 'ridge basalts' region to the right of the 'surface water' region in Fig. 1. Subsequent equilibration with sea water would merely move the points back toward the 'surface water' region. The fact that the samples which have smaller amounts of rare gases show higher $^{20}Ne/^{36}Ar$ ratios is not inconsistent with this model, because a small amount of gas could easily be subject to the most fractionation and samples which have higher amounts of gases might simply represent subsequent mixing with rare gases in sea water. The fact that $^{40}Ar/^{36}Ar$ ratios in quenched rims are usually rather low (less than 1,000) is also not inconsistent with this model since in this model most of the ^{36}Ar and ^{40}Ar is derived from the sea water. Of course, part of ^{40}Ar comes from the mantle, and this may be a reason that some of the ridge basalts show high $^{40}Ar/^{36}Ar$ ratios. High $^{3}He/^{4}He$ ratios in submarine basalts (CRAIG and LUPTON, 1976) indicate that a primordial He component still exists in quenched rims, but since $^{3}He/^{4}He$ ratios are the least sensitive to atmospheric contamination of any of the rare gases (HAMANO and OZIMA, 1978), I think high $^{3}He/^{4}He$ ratios are not a significant objection.

The explanation of the generation of the 'ridge basalts' type rare gas implies that the 'ridge basalts' may have rare gases whose elemental compositions vary continuously from the 'surface water' pattern to the highly Ne enriched 'ridge basalts' pattern. Hence, the boundary between the 'ridge basalts' region and the 'surface water' region is an artifact of still small numbers of data points.

The basic implication of this model is that the 'ridge basalts' pattern is produced by the local scale interaction of the rare gases dissolved in sea water and those originally in the magma, with the rare gases in the sea water dominant for all except He and ^{40}Ar. The 'ridge basalts' pattern is therefore trivial to the fundamental evolution of the solid earth/atmosphere system.

3.4 The 'thucholite' type rare gases

The 'thucholite' type rare gases are found in thucholites (BOGARD *et al.*, 1964), in one sample of Fig Tree shales (CANALAS *et al.*, 1969), and in carbon residues of cherts (FRICK and CHANG, 1977). A remarkable characteristic of this type of gases is a wide range of variation of $^{20}Ne/^{36}Ar$ ratio and $^{36}Ar/^{84}Kr$ ratio. In Fig. 2 all the thucholite data except for the analysis by SHERRILL (1976) fall roughly along a straight line of slope -1. The simplest explanation is that this trend is produced by a selective depletion of ^{36}Ar as compared with Ne and Kr. The trend may simply be an artifact of the small number of analysis. One noteworthy point is that all of these samples have rather high carbon concentration. Carbon may play a significant role in the generation of this type of rare gases.

3.5 The 'planetary' type rare gases in terrestrial samples

Only three samples show this type of gas. Those samples are Red Rock granite (KURODA *et al.*, 1977), a phlogopite (KANEOKA *et al.*, 1977) and a kaersutite (SAITO *et al.*, 1978). Among these samples the latter two have been derived from the mantle and may represent the rare gas in the mantle. The rare gases in Red Rock granite are anomalous and KURODA *et al.* (1977) proposed that they represent meteoritic rare gases trapped during a large impact event.

There is a good chance that shales may have this type of gas. If shales do selectively adsorb Xe, the data points will plot somewhere below the 'surface water' rare gas area. This suggests a possibility that the earth's interior could obtain excess Xe from shales, if there is a large scale circulation of Xe accompanied with plate subduction.

However, since such a large scale circulation of shales looks rather unrealistic, the 'planetary' rare gases in mantle-derived samples may represent slightly fractionated primordial rare gas.

4. Summary

1. A $^{84}Kr/^{130}Xe$ versus $^{20}Ne/^{36}Ar$ diagram can separate not only the three types of rare gases in OZIMA and ALEXANDER's classification (1976) but can also separate the 'planetary' type rare gases from the other three.

2. The data points of two mantle-derived minerals fall in the 'planetary' type rare gases area. If Xe recycling between the surface and the interior of the earth can be ruled out, this 'planetary' type rare gas is most easily accounted for as samples of the earth's primordial gases.

3. The Xe depletion in the terrestrial atmosphere may be, at least in some part, due to the preferential retention in the earth's interior.

4. Rare gases in quenched rims of submarine basalts may only be fractionated gases from rare gases in sea water.

5. In the general of the 'thucholite' type rare gases, the carbon in the samples may have played an important role.

I heartily express my thanks to Dr. E. C. Alexander, Jr. He helped me very much to improve the manuscript.

This work was supported by grant NGL 24-005-225 from the National Aeronautics and Space Administration.

REFERENCES

BENNETT, G. A. and O. K. MANUEL, Xenon in natural gases, *Geochim. Cosmochim. Acta*, **34**, 593–610, 1970.

BOGARD, D. D., M. W. ROWE, O. K. MANUEL, and P. K. KURODA, Noble gas anomalies in the mineral thucholite, *J. Geophys. Res.*, **70**, 703–708, 1965.

BOULOS, M. S. and O. K. MANUEL, The xenon record of extinct radioactivities in the earth, *Science*, **174**, 1334–1336, 1971.

BUTLER, W. A., P. M. JEFFERY, J. H. REYNOLDS, and G. J. WASSERBURG, Isotopic variations in terrestrial xenon, *J. Geophys. Res.*, **68**, 3283–3291, 1963.

CANALAS, R. A., E. C. ALEXANDER, Jr., and O. K. MANUEL, Terrestrial abundance of noble gases, *J. Geophys. Res.*, **73**, 3331–3334, 1968.

CLARKE, W. B., M. A. BEG, and H. CRAIG, Excess ^{3}He in the sea: Evidence for terrestrial primordial helium, *Earth Planet. Sci. Lett.*, **6**, 213–220, 1969.

CRAIG, H. and J. E. LUPTON, Primordial neon, helium and hydrogen in oceanic basalts, *Earth Planet. Sci. Lett.*, **31**, 369–385, 1976.

CRAIG, H., W. B. CLARKE, and M. A. BEG, Excess ^{3}He in deep water on the East Pacific Rise, *Earth Planet. Sci. Lett.*, **26**, 125–13, 21975.

DYMOND, J. and L. HOGAN, Noble gas abundance patterns in deep-sea basalts—primordial gases from the mantle, *Earth Planet. Sci. Lett.*, **20**, 131–139, 1973.

EBERHARDT, J. G., H. GRAF, N. GROGLER, M. D. MENDIA, M. MORGELL, H. SCHWALLER, and A. STETTLER, Trapped solar wind noble gases in Apollo 12 lunar fines 12001 and Apollo 11 breccia 10046, *Proc. 3rd Lunar Sci. Conf., Vol. 2, Suppl. 3, Geochim. Cosmochim. Acta*, 1821–1856, 1972.

FANALE, F. P., Review: A case for catastrophic early degassing of the earth, *Chem. Geol.*, **8**, 79–105, 1971.

FANALE, F. P. and W. A. CANON, Physical adsorption of rare gas on terrigeneous sediments, *Earth Planet. Sci. Lett.*, **11**, 362–368, 1971.

FISHER, D. E., Heavy rare gases in a Pacific seamount, *Earth Planet. Sci. Lett.*, **9**, 331–335, 1970.

FISHER, D. E., The planetary primordial component of rare gases in the deep earth, *Geophys. Res. Lett.*, 315, **1**, 161–164, 1974.

FRICK, U. and S. CHANG, Ancient carbon and noble gas fractionation, in *Lunar Science VIII*, pp. 313– Lunar Science Institute, Houston, Texas, 1977.

HAMANO, Y. and M. OZIMA, Earth-atmosphere evolution model based on Ar isotopic data, in *Terrestrial Rare Grases*, edited by E. C. Alexander, Jr. and M. Ozima, pp. 155–171, Cent. Acad. Publ. Japan, Tokyo, 1978.

HENNECKE, E. W. and O. K. MANUEL, Noble gases in CO_2 well gas, Harding County, New Mexico, *Earth Planet. Sci. Lett.*, **27**, 346–355, 1975a.

HENNECKE, E. W. and O. K. MANUEL, Noble gas in an Hawaiian xenolith, *Nature*, **257**, 778–780, 1975b.

HENNECKE, E. W. and O. K. MANUEL, Noble gases in lava rock from Mount Capuline, New Mexico, *Nature*, **256**, 284–287, 1975c.

KANEOKA, I., N. TAKAOKA, and K. AOKI, Rare gases in a phlogopite nodule and a phlogopite-bearing peridotite in South African kimberlites, *Earth Planet. Sci. Lett.*, **36**, 181–186, 1977.

KANEOKA, I., N. TAKAOKA, and K. AOKI, Rare gases in mantle-derived rocks and minerals, in *Terrestrial Rare Gases*, edited by E. C. Alexander, Jr. and M. Ozima, pp. 71–83, Cent. Acad. Publ. Japan, Tokyo, 1978.

KIRSTEN, T., Incorporation of rare gases in solidifying enstatite melts, *J. Geophys. Res.*, **73**, 2807–2810, 1968.

KONONOV, V. I., B. A. MAMYRIN, B. G. POLYAK, and L. V. KHABARIN, Helium isotopes in the thermal waters of Iceland, *Dokl. Akad. Nauk S.S.S.R.*, **217**, 41–43, 1975.

KURODA, P. K. and R. D. SHERRILL, Xenon and krypton isotope anomalies in the Besner Mine, Ontario, thucholite, *Geochem. J.*, **11**, 9–20, 1977.

KURODA, P. K., R. D. SHERRILL, and K. C. JACKSON, Abundances and isotopic compositions of rare gases in granites, *Geochem. J.*, **11**, 75–90, 1977.

LUPTON, J. E. and H. CRAIG, Excess ^{3}He in oceanic basalts, Evidence for terrestrial primordial helium, *Earth Planet. Sci. Lett.*, **26**, 133–139, 1975.

MAMYRIN, B. A., I. N. TOLSTIKHIN, G. S. ANUFRIEV, and I. L. KAMANSKIY, Anomalous isotopic composition of helium in volcanic gases, *Dokl. Akad. Nauk S.S.S.R.*, **184**, 1197–1199, 1969.

MAMYRIN, B. A., I. N. TOLSTIKHIN, G. S. ANUFRIEV, and I. L. KAMANSKIY, Helium isotopic composition in volcanic gases in Iceland, *Geochimiya,* **11**, 1369, 1972.

MAZOR, E., Paleotemperatures and other hydrological parameters deduced from noble gases dissolved in ground water: Jordan Valley, Israel, *Geochim. Cosmochim. Acta,* **36**, 1321–1336, 1972.

MAZOR, E., D. HEYMANN, and E. ANDERS, Noble gases in carbonaceous chondrites, *Geochim. Cosmochim. Acta,* **34**, 781–824, 1970.

OWEN, T., K. BIEMANN, D. R. RUSHNECK, J. E. BILLER, D. W. HOWARTH, and A. L. LAFLEUR, The composition of the atmosphere at the surface of Mars, Submitted to *J. Geophys. Res.*, 1977.

OZIMA, M., Ar isotopes and Earth-atmosphere evolution models, *Geochim. Cosmochim. Acta,* **39**, 1127–1134, 1975.

OZIMA, M. and E. C. ALEXANDER, Jr., Rare gas fractionation patterns in terrestrial samples and earth-atmosphere evolution model, *Rev. Geophys. Space Phys.*, **14**, 385–390, 1976.

OZIMA, M. and K. KUDO, Excess argon in submarine basalts and an Earth-Atmosphere evolution model, *Nature Phys. Sci.*, **239**, 23–24, 1972.

PHINNEY, D., ^{36}Ar, Kr, and Xe in terrestrial materials, *Earth Planet. Sci. Lett.*, **16**, 413–420, 1972.

RISON, W. and K. KYSER, Rare gases, oxygen and hydrogen in Hawaiian xenoliths and basalts, *EOS Trans. Am. Geophys. Union,* **58**, 537, 1977.

SAITO, K., A. R. BASU, and E. C. ALEXANDER, Jr., Planetary type rare gases in an upper mantle-derived amphibole, *Earth Planet. Sci. Lett.*, **39**, 274–280, 1978.

SHERRILL, R. D., Rare gas isotopic abundances in meteoritic and terrestrial samples, Ph.D. thesis, Univ. of Arkansas, 1976.

TOLSTIKHIN, I. N., B. A. MAMYRIN, and L. B. KHABARIN, Isotope composition of helium in ultrabasic xenoliths from volcanic rocks of Kamchatka, *Earth Planet. Sci. Lett.*, **22**, 75–84, 1974.

Earth-Atmosphere Evolution Model Based on Ar Isotopic Data

YOZO HAMANO and Minoru OZIMA

Geophysical Institute, University of Tokyo, Tokyo 113, Japan

An earth-atmosphere evolution model is discussed on the basis of Ar isotopic data. This is a second iteration of our previous attempt with more refined constraints and a more realistic dynamic earth model, where continuous K transportation from the mantle to the crust and Ar-loss from the crust are included. Validity and limitations of the earth model are also examined. New estimations of ($^{40}Ar/^{36}Ar$) and the K content in the present mantle are made. The results, and a relation between K transportation and Ar degassing rate, are used as the constraints.

The Ar inventory obtained here suggests that most of the terrestrial atmosphere was formed by an early, sudden degassing from the solid earth, very likely before about 4.0 b.y. ago.

1. Introduction

Although it is generally accepted that the terrestrial atmosphere was degassed from the interior of the earth, it is still controversial whether the degassing was continuous or catastrophic. In his classic paper, RUBEY (1951) argued that the terrestrial atmosphere has been continuously degassed via volcanic gases and hot spring emanations. Assuming continuous degassing via a first order rate process, TUREKIAN (1964) obtained a value for the degassing rate of $k=2.8\times10^{-11}\ yr^{-1}$ for a chondritic earth model. Since the degassing rate thus estimated compared favourably with the degassing rate estimated for water from the earth interior, he concluded that the model of continuous degassing to the atmosphere was valid. FANALE (1971), however, pointed out that, since few geochemical and geophysical observational data are consistent with continuous degassing, sudden, early degassing was much more likely. OZIMA and KUDO (1972) first suggested that information on the $^{40}Ar/^{36}Ar$ isotopic ratio in addition to the ^{40}Ar inventory in the earth would impose a further, severe constraint on degassing models. Assuming a value of about 2,000 for $^{40}Ar/^{36}Ar$ in the present mantle, OZIMA (1975) inferred an early catastrophic degassing. SCHWARTZMAN (1973) also favoured an early sudden degassing from the $^{40}Ar/^{36}Ar$ isotopic ratios observed in some Precambrian rocks. Contrary to the results based on the $^{40}Ar/^{36}Ar$ isotopic ratio, TOLSTIKHIN (1975), from a similar approach using the $^{3}He/^{4}He$ isotopic ratio in the earth, inferred a continuous degassing. As can be seen from the brief review above, the problem is still very much open to further investigation.

The present paper is a second iteration of our previous attempt (OZIMA, 1975) to discuss the evolution of the terrestrial atmosphere on the basis of Ar isotopes in the earth. We have improved our estimates of constraints such as the $^{40}Ar/^{36}Ar$ and K contents in the mantle and have adopted a more realistic earth evolution model.

2. Outline of a Method

There are various approaches to the problem of earth-atmosphere evolution (see

FANALE, 1971 for an excellent review of this subject). Among them, studies of the rare gas regime in the earth have provided a prime source of information. The following discussion is essentially based on the Ar inventory and its isotopic ratio regime in the earth.

We first assume a general earth-atmosphere evolution model, which should be compatible with acceptable geologic information. Such an earth-atmosphere evolution model must be quite flexible, because no definitive picture of the evolution has been worked out. The flexibility will be expressed in terms of several parameters. One such parameters will be the time of a sudden degassing, which is one of the central issues in the present discussion. Equations with these parameters are then set up to describe Ar degassing on the basis of the assumed earth-atmosphere evolution model.

To delineate more precisely the Ar degassing, we impose several constraints on the degassing equations as well as some known initial and boundary conditions. Imposition of the constraints reduces the choice of possible solutions, thus narrowing the range for the flexible parameters, which in turn yields a more specific picture of the earth-atmosphere evolution model.

3. Degassing Model and Equations

Figure 1 illustrates an earth-atmosphere evolution model in which we assume that Ar was degassed in two stages, i.e., a fraction f of the amount contained in the earth catastrophically at t_d, and subsequently by a continuous process from t_d to the present. In accord with the results inferred from Sr and Pb isotopic evolutions in the earth (e.g., RUSSELL, 1972; HART and BROOKS, 1970) we assume that K has also been continuously transported from the mantle to the crust from t_d to the present. Also, we assume that some of the ^{40}Ar produced in the crust has been degassed into the atmosphere. For all these transport processes we assume a first order rate process. The validity and limitations of the assumed earth-atmosphere evolution model will be discussed in a later section.

The equations which describe the degassing process are then: for $0<t<t_d$,

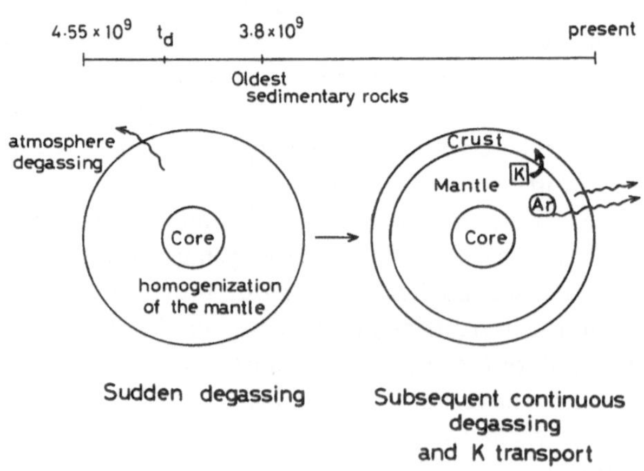

Fig. 1. Schematic representation of the earth-atmosphere evolution model employed in the present paper.

$$\frac{d}{dt}(^{40}K)_M = -\lambda(^{40}K)_M$$

$$\frac{d}{dt}(^{40}Ar)_M = \lambda_e(^{40}K)_M$$

$$\frac{d}{dt}(^{36}Ar)_M = 0,$$

in which λ and λ_e denote respectively the total ^{40}K decay constant and the electron capture decay constant of ^{40}K to ^{40}Ar.

For $t_d < t < t_p$ (present time),

$$\frac{d}{dt}(^{40}K)_M = -\lambda(^{40}K)_M - \alpha(^{40}K)_M$$

$$\frac{d}{dt}(^{40}Ar)_M = \lambda_e(^{40}K)_M - k(^{40}Ar)_M$$

$$\frac{d}{dt}(^{36}Ar)_M = -k(^{36}Ar)_M$$

$$\frac{d}{dt}(^{40}K)_C = -\lambda(^{40}K)_C + \alpha(^{40}K)_M$$

$$\frac{d}{dt}(^{40}Ar)_C = \lambda_e(^{40}K)_C - \beta(^{40}Ar)_C$$

$$\frac{d}{dt}(^{40}Ar)_A = k(^{40}Ar)_M + \beta(^{40}Ar)_C$$

$$\frac{d}{dt}(^{36}Ar)_A = k(^{36}Ar)_M,$$

where α, β, k represent respective transport coefficients and suffixes M, C, and A indicate amounts in the mantle, the crust and the atmosphere respectively. These differential equations can be solved with boundary conditions at $t = t_d$, when a fraction f of the total amounts of ^{36}Ar and ^{40}Ar then present was degassed to the atmosphere. The solutions for present amounts of ^{36}Ar, ^{40}Ar, and ^{40}K are as follows:

$$(^{40}K)_M^{t_p} = (^{40}K)_M^0 e^{-\lambda t_d} e^{-(\lambda+\alpha)(t_p - t_d)} \tag{1}$$

$$\begin{aligned}(^{40}Ar)_M^{t_p} = {} & (1-f)\left\{(^{40}Ar)_M^0 + \frac{\lambda_e}{\lambda}(^{40}K)_M^0(1 - e^{-\lambda t_d})\right\} e^{-k(t_p - t_d)} \\ & + \frac{\lambda_e}{\lambda}(^{40}K)_M^0 e^{-\lambda t_d} \frac{\lambda}{\lambda+\alpha-k} e^{-k(t_p-t_d)}(1 - e^{-(\lambda+\alpha-k)(t_p-t_d)})\end{aligned} \tag{2}$$

$$(^{36}Ar)_M^{t_p} = (1-f)(^{36}Ar)_M^0 e^{-k(t_p-t_d)} \tag{3}$$

$$(^{40}K)_C^{t_p} = (^{40}K)_M^0 e^{-\lambda t_d} e^{-\lambda(t_p-t_d)}(1 - e^{-\alpha(t_p-t_d)}) \tag{4}$$

$$\begin{aligned}(^{40}Ar)_C^{t_p} = {} & \frac{\lambda_e}{\lambda}(^{40}K)_M^0 e^{-\lambda t_d}\left\{\frac{\lambda}{\beta-\lambda} e^{-\lambda(t_p-t_d)}(1 - e^{-(\beta-\lambda)(t_p-t_d)})\right. \\ & \left. + \frac{\lambda}{\beta-\lambda-\alpha} e^{-\beta(t_p-t_d)}(1 - e^{-(\lambda+\alpha-\beta)(t_p-t_d)})\right\}\end{aligned} \tag{5}$$

$$({}^{40}\mathrm{Ar})_A^{t_p}=\left\{({}^{40}\mathrm{Ar})_M^0+\frac{\lambda_e}{\lambda}({}^{40}\mathrm{K})_M^0(1-e^{-\lambda t_d})\right\}\{f+(1-f)(1-e^{-k(t_p-t_d)})\}$$
$$+\frac{\lambda_e}{\lambda}({}^{40}\mathrm{K})_M^0 e^{-\lambda t_d}\left\{\frac{\lambda(\beta-k)}{(k-\lambda-\alpha)(\beta-\lambda-\alpha)}(1-e^{-(\lambda+\alpha)(t_p-t_d)})\right.$$
$$-\frac{\lambda}{(k-\lambda-\alpha)}(1-e^{-k(t_p-t_d)})+\frac{\beta}{\beta-\lambda}(1-e^{-\lambda(t_p-t_d)})$$
$$\left.+\frac{\lambda\alpha}{(\beta-\lambda)(\beta-\lambda-\alpha)}(1-e^{-\beta(t_p-t_d)})\right\} \tag{6}$$
$$({}^{36}\mathrm{Ar})_A^{t_p}=({}^{36}\mathrm{Ar})_M^0\{f+(1-f)(1-e^{-k(t_p-t_d)})\}. \tag{7}$$

The above system consists of seven equations with fifteen unknowns (three initial values, $({}^{40}\mathrm{K})_M^0$, $({}^{40}\mathrm{Ar})_M^0$, and $({}^{36}\mathrm{Ar})_M^0$; seven present values $({}^{40}\mathrm{K})_M^{t_p}$, $({}^{40}\mathrm{Ar})_M^{t_p}$, $({}^{40}\mathrm{Ar})_M^{t_p}$, $({}^{40}\mathrm{K})_C^{t_p}$, $({}^{40}\mathrm{Ar})_C^{t_p}$, $({}^{40}\mathrm{Ar})_A^{t_p}$, and $({}^{36}\mathrm{Ar})_A^{t_p}$; five parameters k, α, β, f, and t_d). Among these unknowns, the initial ratio $({}^{40}\mathrm{Ar}/{}^{36}\mathrm{Ar})_0$ $(=10^{-4})$ and the present amounts of ^{40}Ar and ^{36}Ar in the atmosphere (${}^{40}\mathrm{Ar}=6.6\times10^{19}$ grams, ${}^{40}\mathrm{Ar}/{}^{36}\mathrm{Ar}=295.5$) are known. Also, the present content of potassium in the sialic crust and the degassing constant β from the crust can be estimated as explained below. This reduces the number of undetermined unknowns to three. We chose f, t_d, and k as the unknown parameters. Solutions for the equations are shown in Fig. 5 by curves in a f vs. t_d diagram for various values of a parameter k. Imposition of constraints on quantities such as $({}^{40}\mathrm{Ar}/{}^{36}\mathrm{Ar})_M$ and $(\mathrm{K})_M$ (see below) narrow further the possible ranges for these solutions. Geophysical data and constraints applied in the present calculation are listed in Table 1.

Table 1. Data used for calculation.

Atmosphere abundance of ^{40}Ar	6.6×10^{19} g
Atmosphere abundance of ^{36}Ar	2.0×10^{17} g
Potassium content in the crust	1.91% (HOLLAND and LAMBERT, 1972)
Degassing rate constant, β	$3.71\times10^{-10}\mathrm{yr}^{-1}$[1)]
Potassium content in the mantle	100 ppm $\leq(\mathrm{K})_M\leq$ 400 ppm[1)]
$({}^{40}\mathrm{Ar}/{}^{36}\mathrm{Ar})$ isotopic ratio in the mantle	$({}^{40}\mathrm{Ar}/{}^{36}\mathrm{Ar})_M\geq5{,}000$[1)]
Initial $({}^{40}\mathrm{Ar}/{}^{36}\mathrm{Ar})$ ratio	10^{-4} (CAMERON, 1970)
Decay constants of ^{40}K	$\lambda=5.305\times10^{-10}\mathrm{yr}^{-1}$
	$\lambda_e=5.85\times10^{-11}\mathrm{yr}^{-1}$
Present isotopic abundance of ^{40}K	$({}^{40}\mathrm{K}/\mathrm{K})=1.19\times10^{-4}$
Mass of the crust	2.02×10^{25} g (ARMSTRONG, 1968)
Mass of the mantle	4.03×10^{27} g

1) Obtained in the present paper.

4. Constraints on a Degassing Model

The degassing history of argon can be specified by three parameters f, t_d, and k, where t_d and f represent the sudden degassing time and the fraction degassed to the atmosphere in the sudden degassing, and the degassing constant k specifies the later continuous degassing. To narrow further the range for possible values of f, t_d, and k, we impose the following constraints on the earth-atmosphere evolution model: (1) a degassing rate β of ^{40}Ar from the crust (we assume a negligible amount of ^{36}Ar in the crust), (2) the K-content in the present crust, (3, 4) the K-content and $(^{40}Ar/^{36}Ar)_M$ in the present mantle, and (5) a transportation rate α of potassium from the mantle to the crust.

4.1 Degassing rate β from the sialic crust

Degassing of ^{40}Ar from the crustal rocks can be estimated from a relation between K-Ar mineral ages and Rb-Sr whole rock ages. HURLEY and RAND (1969) compared K-Ar ages with whole rock Rb-Sr ages for cases in which both had been obtained from the same crustal rocks. They found a systematic difference between the two ages; i.e., the K-Ar ages are younger than the corresponding Rb-Sr whole rock ages and the difference between them becomes larger for older rock ages. The age difference can be most easily explained by a systematic Ar-loss from the crustal rocks.

Assuming a first order rate process with a degassing rate β for this process, the relation between the two ages is given by

$$T_{\text{K-Ar}} = \frac{1}{\lambda} \ln \left[1 + \frac{1}{1-(\beta/\lambda)} \{ e^{\lambda T_{\text{Rb-Sr}}(1-\beta/\lambda)} - 1 \} \right], \tag{8}$$

where the Rb-Sr age is assumed to represent a true age of the rocks. The derivation of this equation is given in Appendix 1. Figure 2 shows a comparison between the observed data (HURLEY and RAND, 1969) and the best fit curve of Eq. (8) to those data. The parameters of the best fit curves are $\beta = 3.7 \times 10^{-10}\ \text{yr}^{-1}$ or $\beta/\lambda = 0.7$. Although errors involved in this estimation are large, mainly due to uncertainty in the age data, the value is not very crucial in the following Ar degassing calculation and the final conclusion remains almost unchanged if a factor of two is allowed for the uncertainty in the estimated value of β.

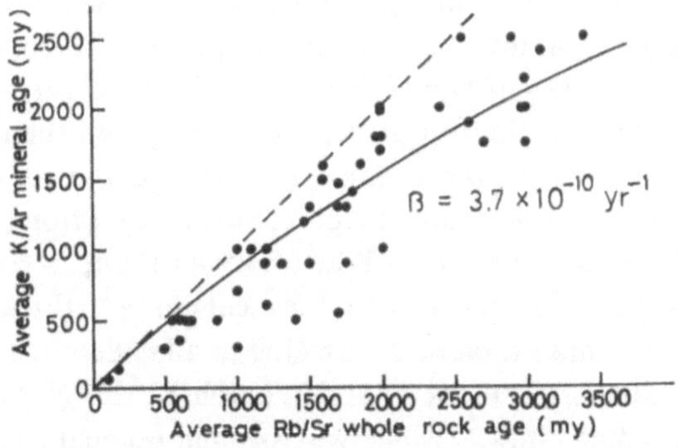

Fig. 2. Relation between K-Ar mineral ages and Rb-Sr whole rock ages (HURLEY and RAND, 1969). The solid curve, which is calculated by assuming a first order rate process for ^{40}Ar loss from crustal rocks, indicates the best fit to the observations. The dotted line gives a concordant relation between K-Ar mineral ages and Rb-Sr whole rock ages.

4.2 Potassium content in the crust

Major element composition of the continental crust has been estimated by various authors (e.g., CLARKE and WASHINGTON, 1924; GOLDSCHMIDT, 1954; TAYLOR, 1964). Estimated potassium content in the crust ranges from 2% to 3.5%. These estimates were mostly derived from analyses on surface rocks. Recently HOLLAND and LAMBERT (1972) analysed rocks from the Canadian shield and Precambrian rocks from Scotland. They found a progressive enrichment in K content toward the surface. Considering such vertical variation in K content in the crust, they estimated a value of 1.91% for the average K content in the crust. Also, heat flow data seem to favour the latter, smaller value for crustal K (CLARK and RINGWOOD, 1964). Hence, in the present paper we adopt the value of 1.91%.

4.3 *Potassium content in the present mantle*

To our knowledge, no satisfactory estimate of K content in the present mantle has been presented. Either a chondritic earth model (K=880 ppm) or an achondritic one (K=350 ppm) is commonly assumed for the total Earth. HURLEY (1969) and LARIMER (1971) estimated the minimum content of potassium in the earth to be about 85 ppm, since this much K is needed to account for ^{40}Ar in the present atmosphere if the latter were entirely derived from ^{40}K-decay in the interior of the earth. However, these estimates are too crude to serve as a useful constraint on the earth-atmosphere evolution model. Hence, we apply a new method to narrow the permissible range for $(K)_M$.

A range for the K content in the present mantle can be derived from surface heat flow data. Mean heat flow values for continental and oceanic regions are known to be about the same, 1.5 μ cal cm^{-2}s^{-1} (LEE and UYEDA, 1965). Heat generation in the continents can be estimated reasonably well from available data for U, Th, and K contents in the crustal rocks, which give a heat flow of about 1 μ cal cm^{-2}s^{-1} on the continental surface. The rest of the heat flow must come from below the crust. Heat generation in the oceanic crust is very small and the surface heat flow through the ocean floor can be attributed to heat generated in the mantle and the core. If we assume that the oceanic heat flow and the remaining continental heat flow are generated by radiogenic heat sources in the mantle, this assumption gives a maximum estimate potassium content in the mantle of about 400 ppm, assuming $K/U=10^4$ and $Th/U=3$. On the contrary, if we assume that the difference of the heat flows at the top of the mantle between the oceans and the continents results from heat sources in the oceanic mantle and the remaining oceanic and continental heat flow comes entirely from the core, this assumption gives a minimum K content in the mantle. Subtracting the heat flow of 0.5 μ cal cm^{-2}s^{-1} which is assumed to come from the core from the total heat flow of 1.5 μ cal cm^{-2}s^{-1} through the ocean floor, we have 1.0 μ cal cm^{-2}s^{-1} as a minimum heat flow due to a heat source in the oceanic mantle. Hence, if we take a value of 10^4 for K/U and 3 for Th/U, this gives a minimum K content of 250 ppm in the oceanic mantle. Since we assumed no heat source in the continental mantle, this gives a K content of about 150 ppm for the average mantle as a whole.

Lately FISHER (1975) has argued that K/U in the mantle could be as small as 3,000 from his study of $^4He/^{40}Ar$ in some ultramafic xenoliths. If a value of 5,000 is assumed for K/U in the present mantle, this reduces the minimum K content in the mantle to about 100 ppm. Considering this ambiguity for the K/U ratio in the mantle, we assume 100 ppm for the minimum K content.

4.4 *Estimation of an average* ($^{40}Ar/^{36}Ar$) *ratio in the mantle*

Rare gas isotopic compositions in the mantle may be inferred from those trapped in some mantle-derived materials, such as volcanic rocks or xenoliths in ultramafic rocks.

For this purpose glassy margins of submarine basalts seem to be the most promising, since rapidly chilled surfaces and high hydrostatic pressures would prevent rare gases from escaping from the glassy rims. Rare gases in xenoliths would be more susceptible to contamination effects because of their relatively small sample sizes. Also it is generally difficult to make a correction for in-situ decayed ^{40}Ar, as the ages of xenoliths are very difficult to estimate. Hence, the following discussions are essentially based on data obtained for young oceanic ridge basalts (ORB).

ORB are believed to be derived from an upper part of the mantle (GAST, 1968), and therefore the following discussions apply only to the upper mantle. If enough data are available for oceanic island basalts, we may possibly infer isotopic compositions in deeper regions in the mantle since ocean island basalts may be derived from deeper in the mantle via hot spots (WILSON, 1963). ($^{40}Ar/^{36}Ar$) ratios observed in submarine glasses vary from an almost atmospheric isotopic ratio (295.5) to more than a few tens of thousands. Hence, before attempting to estimate a mean value from the observed data, we must ask if the diversity in the isotopic ratio is a reflection of heterogeneity in K/Ar ratio in the mantle, or due to secondary disturbances such as contamination by crustal or atmospheric Ar, or isotopic fractionation. If the former is the case, it would be difficult to estimate an average value from the limited number of available data, whereas if the latter is the case, it may be possible, as shown below, to make a correction for disturbances and thus estimate an uncontaminated mantle isotopic ratio.

We plotted available isotopic data in a ($^{40}Ar/^{36}Ar$) vs. ($^{4}He/^{40}Ar$) diagram. Variation in ($^{40}Ar/^{36}Ar$) must result either from variation in ($K/^{36}Ar$) in the source regions for

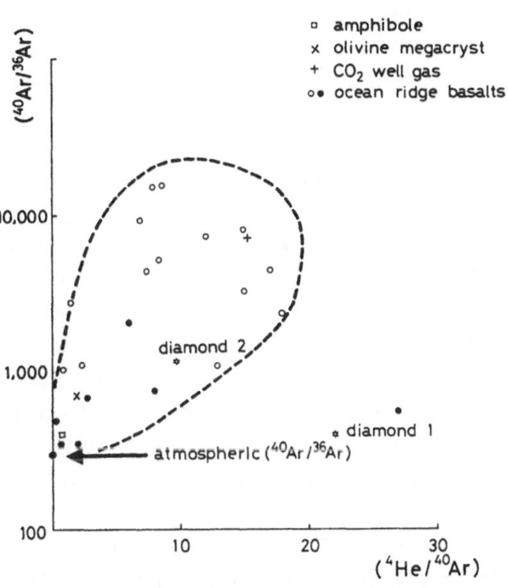

Fig. 3. $^{40}Ar/^{36}Ar$ vs. $^{4}He/^{40}Ar$ in submarine basalt glasses. ○, DYMOND and HOGAN (1973); ●, FISHER (1970). Rare gases in other mantle-derived materials are also shown: □, amphibole (SAITO *et al.*, 1977); ×, olivine megacryst (KANEOKA *et al.*, 1978); ☆, diamonds (TAKAOKA and OZIMA, 1978); +, CO_2 well gases (BOULOS and MANUEL, 1971).

oceanic ridge basalts, or from contamination by Ar of non-mantle origin, or both. The same consideration applies to the ($^{4}He/^{40}Ar$) ratio; variation in ($^{4}He/^{40}Ar$) must be a reflection of either variations in U/K, or contamination effects, or both. Variation in K content will generally change ($K/^{36}Ar$) in a direction opposite to that of U/K, giving a negative correlation in a ($^{40}Ar/^{36}Ar$) vs. ($^{4}He/^{40}Ar$) diagram. However, variation in U content and heterogeneity in rare gas contents would not result in a systematic variation between ($K/^{36}Ar$) and U/K, therefore would yield a random spread of data in the diagram. Consequently, we should expect a negative correlation or at most a random spread of data in the ($^{40}Ar/^{36}Ar$) vs. ($^{4}He/^{40}Ar$) diagram, if the spread in ($^{40}Ar/^{36}Ar$) is principally due to compositional heterogeneity in the mantle. On the other hand, if rare gas contamination—most likely atmospheric rare gases absorbed in sea water as shown below—is primarily responsible for the spread in the data, we should expect a positive correlation in the diagram, since the atmospheric rare gases, the contaminating component, have much smaller ($^{40}Ar/^{36}Ar$) and ($^{4}He/^{40}Ar$) ratios than possible upper mantle materials (see below).

As seen in Fig. 3, a positive correlation appears to exist in the ($^{40}Ar/^{36}Ar$) vs. ($^{4}He/^{40}Ar$) diagram. From the foregoing discussion, we conclude that the diversity observed in ($^{40}Ar/^{36}Ar$) ratios for submarine glasses reflects, to a first approximation, variation in the degree of atmospheric contamination, whereas heterogeneity in U, K, and rare gas contents is a secondary effect which tends to obscure the positive correlation. In Fig. 3 the region enclosed by the dashed line suggests a positive correlation between ($^{40}Ar/^{36}Ar$) and ($^{4}He/^{40}Ar$); this region includes almost all of the data for submarine glasses so far reported in the literature. The uncontaminated upper mantle value should be located at the antipode to the atmospheric value in the enclosed region, which suggests a value of about 10,000. Allowing some spread in the uncontaminated isotopic ratio in the mantle, 5,000 would be a reasonable minimum estimate for an averaged upper mantle ($^{40}Ar/^{36}Ar$) isotopic ratio. Data for other mantle-derived materials are also presented in Fig. 3.

Among possible rare gas contaminants, atmospheric rare gases introduced into submarine basalt magma via sea water are the most likely. In contrast to an almost negligible amount of He (4.1×10^{-8} ccSTP/g), the Ar content of sea water amounts to 3.3×10^{-4} ccSTP/g (VON KÖNIG, 1963). Hence, even a very small amount of sea water contamination in basalt magma would very seriously affect both the ($^{40}Ar/^{36}Ar$) and ($^{4}He/^{40}Ar$) ratios, whereas ($^{3}He/^{4}He$) would be left almost unaltered. We show the effect of sea water contamination on the isotopic composition of mantle materials in Fig. 4, in which the isotopic ratio ($^{40}Ar/^{36}Ar$) and the elemental ratio ($^{4}He/^{40}Ar$) in contaminated mantle material are plotted against the volume fraction (f) of contaminating sea water. In constructing the figure, rare gas concentrations in uncontaminated mantle materials were estimated from the data observed in xenoliths in some volcanic rocks ($^{3}He=1.4\times10^{-12}$ ccSTP/g and $^{36}Ar=2\times10^{-10}$ ccSTP/g (GRAMLICH and NAUGHTON, 1972)), the isotopic ratio of ($^{3}He/^{4}He)=1.4\times10^{-5}$ from submarine glasses (CRAIG and LUPTON, 1976), and $(^{40}Ar/^{36}Ar)_M=5{,}000$ from the discussion above. It is to be noted that sea water contamination as low as 10^{-4} volume fraction does change both ($^{40}Ar/^{36}Ar$) and ($^{4}He/^{40}Ar$) significantly, whereas even 10% contamination does not change ($^{3}He/^{4}He$) significantly from its original value. The rather uniform ($^{3}He/^{4}He$) ratio in mantle-derived materials

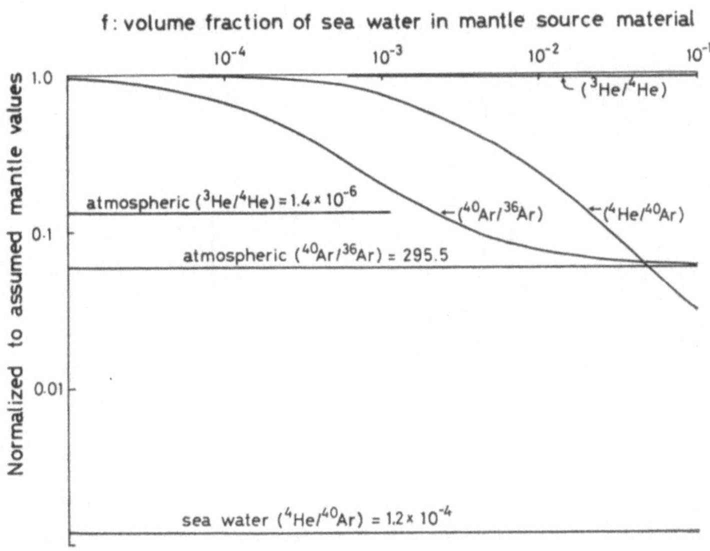

Fig. 4. Effect of sea water contamination on $^{40}Ar/^{36}Ar$, $^{4}He/^{40}Ar$, and $^{3}He/^{4}He$ in mantle material. Note that sea water contamination affects $^{40}Ar/^{36}Ar$ and $^{4}He/^{40}Ar$ very seriously, but has an almost insignificant effect on $^{3}He/^{4}He$.

(about 1.4×10^{-5}; CRAIG and LUPTON, 1976) is then easily understood as an insensitivity of this ratio to sea water contamination and suggests that the upper mantle composition is not as heterogeneous with respect to K/Ar as the observed diversity in $(^{40}Ar/^{36}Ar)$ apparently suggests.

4.5 *Transportation rate α of potassium from the mantle to the crust*

The transportation process of potassium from the mantle to the crust is specified by the rate constant α in the present model. We assume that the degassing rate constant k for Ar is larger than or equal to this rate constant α. A similar relation was used to estimate an upper limit on K in the earth (HURLEY, 1968; LARIMER, 1971). Comparing the mobility of potassium and argon, this assumption seems reasonable.

5. Results

With Eq. (1) to (7) and the geophysical data listed in Table 1, it is possible to calculate present amounts of ^{40}Ar in the crust, ^{40}Ar, ^{36}Ar, and ^{40}K in the mantle, initial amounts of ^{40}Ar, ^{36}Ar, and ^{40}K contained in the solid earth, and the transportation rate constant α for given sets of degassing parameters f, t_d, and k. If constraints are imposed on $(^{40}Ar/^{36}Ar)_M$ and $(K)_M$, the permissible ranges for f, t_d, and k are narrowed thus giving a more specific picture of the earth-atmosphere evolution process. Although the model is not unique due to ranges in the constraints, it gives a fairly restricted picture for the Ar degassing process.

The results of the calculation are presented in Figs. 5 and 6. In Fig. 5 contour lines for $(^{40}Ar/^{36}Ar)_M$ and $(K)_M$ which satisfy Eqs. (1)–(7) are plotted on a f vs. t_d diagram for three different values of the degassing constant: $k=0.53$, 1.06 and $2.65\times10^{-10}\,yr^{-1}$. For each value of k, a hatched region indicates the range in the f vs. t_d diagram which is

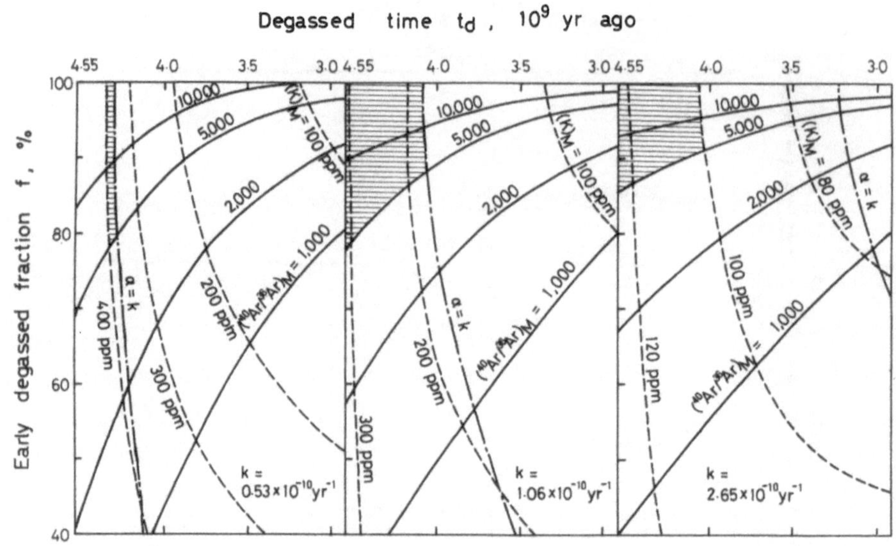

Fig. 5. Results from the earth-atmosphere evolution model. Variations of the ($^{40}Ar/^{36}Ar$) isotopic ratio and potassium content in the mantle with assumed values of f and t_d are shown by contour lines for three different values of the degassing constant k.

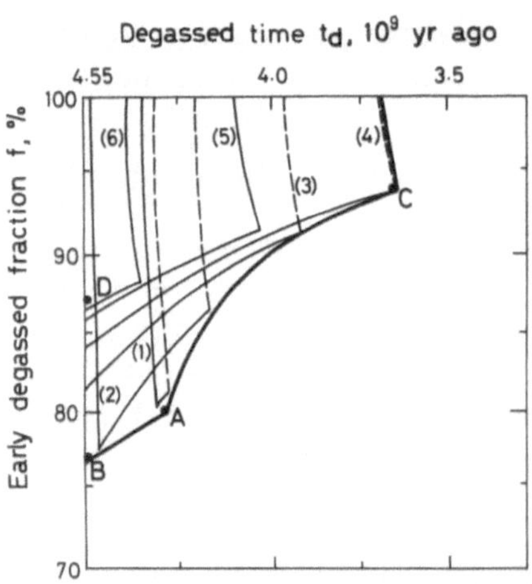

Fig. 6. Possible ranges of f and t_d are shown for degassing constant k=(1) $5.31\times10^{-11}yr^{-1}$, (2) $7.96\times10^{-11}yr^{-1}$, (3) $1.33\times10^{-10}yr^{-1}$, (4) $1.86\times10^{-10}yr^{-1}$, (5) $2.65\times10^{-10}yr^{-1}$, (6) $3.18\times10^{-10}yr^{-1}$. The solid curves indicate the range boundary by the constraints of $(^{40}Ar/^{36}Ar)_M \geqq 5{,}000$ and 100 ppm$\leq (K)_M \leq$400 ppm, and the dotted curves indicate the boundary of $k=\alpha$. The heavy curve indicates the total possible range in the present model.

Table 2. Main geophysical results from the present model.

Degassing process	
early sudden degassing	4.55 b.y. ago $< t_d <$ 3.64 b.y. ago
	$77\% < f (< 100\%)$
subsequent continuous degassing	5.04×10^{-11} yr$^{-1} < k < 3.40\times10^{-10}$yr^{-1}
Present ^{40}Ar flux from the mantle	$(5.6-14.5)\times10^{5}$ atoms cm^{-2}s^{-1}
Present ^{40}Ar flux from the crust	$(5.7-7.5)\times10^{5}$ atoms cm^{-2}s^{-1}
Transportation rate constant, α	$(4.8-18.5)\times10^{-11}$yr^{-1}
Present amounts of Ar	
(^{40}Ar) in the crust	$(6.7-8.8)\times10^{18}$ g
(^{40}Ar) in the mantle	$(3.0-18.0)\times10^{19}$ g
(^{36}Ar) in the mantle	$(0-3.4)\times10^{16}$ g

compatible with the imposed constraints: $({}^{40}Ar/{}^{36}Ar)_M \geqq 5{,}000$, $100\ \text{ppm} \leq (K)_M \leq 400$ ppm and $k \geqq \alpha$. The possible ranges for the degassing parameters f, t_d, and k are thus determined. Compatible regions in the f vs. t_d plane for six different values of the degassing constant k are presented in Fig. 6, where the total possible range in the present model is indicated by the heavy curve. These results, and other geophysical information obtained from the calculation, are summarized in Table 2 and are discussed below.

5.1 *Degassing rate constant k*

The degassing rate constant k specifying the continuous degassing process ranges in value from $5.0 \times 10^{-11}\ \text{yr}^{-1}$ to $3.4 \times 10^{-10}\ \text{yr}^{-1}$. Since the early sudden degassing has a minor effect on the present amount of atmospheric ${}^{40}Ar$, k is essentially determined by the present amount of ${}^{40}Ar$ in the atmosphere and the assumed potassium content in the earth.

The present ${}^{40}Ar$ flux from the mantle predicted by the model can be calculated from the rate constant k and the amount of ${}^{40}Ar$ in the present mantle calculated from Eq. (2). The flux averaged over the earth's surface ranges from 5.6×10^5 atoms $\text{cm}^{-2}\text{s}^{-1}$ to 14.5×10^5 atoms $\text{cm}^{-2}\text{s}^{-1}$. The estimated ${}^{40}Ar$ flux from the crust in this model is about $(5.7-7.5) \times 10^5$ atoms $\text{cm}^{-2}\text{s}^{-1}$, which is not negligible compared to the flux from the mantle.

5.2 *Early degassed fraction and t_d*

In the present model, sudden degassing must have occurred earlier than 3.64 b.y. ago, and the degassed fraction must have been greater than 77% of the argon contained in the earth at t_d. The degassed fraction increases as the degassing time becomes more recent as is evident from Fig. 6. ${}^{36}Ar$ degassed to the atmosphere in the sudden degassing constitutes more than 90% of the ${}^{36}Ar$ presently in the atmosphere, but a large fraction of the ${}^{40}Ar$ in the atmosphere was degassed through the subsequent continuous degassing process.

5.3 *${}^{36}Ar$ initially contained in the earth*

The amount of rare gases initially contained in the solid earth is necessary information for investigation of the origin of volatile elements in the earth-atmosphere system. The calculation here shows that the amount of ${}^{36}Ar$ originally contained in the earth falls in a narrow range from $(2.0\ \text{to}\ 2.4) \times 10^{17}$ g. The initial concentration of ${}^{36}Ar$ in the mantle ranged from $(3.1\ \text{to}\ 3.7) \times 10^{-8}$ ccSTP/g, which is comparable to or larger than the amounts observed in ordinary chondrites but smaller by about a factor of two than those in carbonaceous chondrites (PEPIN and SIGNER, 1965).

5.4 *Transportation rate constant α*

The transportation of potassium from the mantle to the crust is specified by the rate constant α, which ranges from $4.8 \times 10^{-11}\ \text{yr}^{-1}$ to $18.5 \times 10^{-11}\ \text{yr}^{-1}$. The variation is largely due to the range assumed for the potassium content in the mantle, i.e., 100 to 400 ppm.

5.5 *Time evolution of (${}^{40}Ar/{}^{36}Ar$)* in the atmosphere and the mantle

After obtaining the parameters controlling the degassing process, it is possible to calculate the time evolution of the ${}^{40}Ar/{}^{36}Ar$ isotopic ratio in the atmosphere and the mantle. The evolution for some special cases is shown in Fig. 7, where the governing parameters for these cases are summarized in Table 3 and indicated in Fig. 5. The variation is rather insensitive to the potassium content in the mantle and the degassed fraction f, but sensitive to the time t_d of sudden degassing.

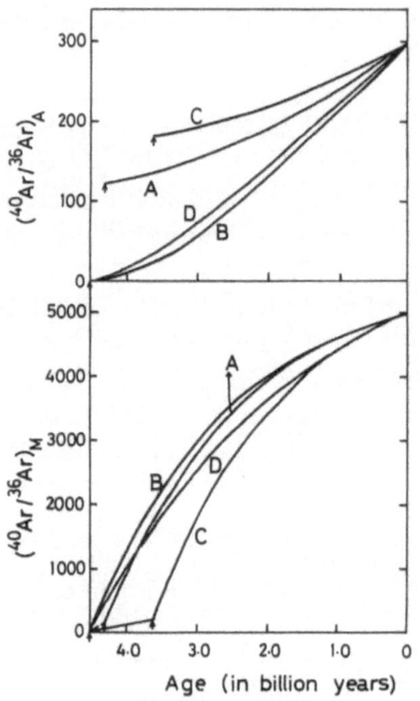

Fig. 7. Time variation of the $(^{40}Ar/^{36}Ar)$ isotopic ratio in the atmosphere (upper) and in the mantle (lower). Assigned parameters for these four models are listed in Table 3. Arrows indicate the early degassed time t_d.

Table 3. Parameters in special models used for the calculation of the time evolution of Ar isotopic ratios.

Model	A	B	C	D
$k(10^{-11}yr^{-1})$	5.0	8.5	18.6	34.0
$f(\%)$	81	77	94	86
t_d (in b.y. before present)	4.23	4.55	3.64	4.55
$(^{40}Ar/^{36}Ar)_M^{tp}$	5,000	5,000	5,000	5,000
$(K)_M$ (ppm)	400	400	100	100

6. Discussion

6.1 *Crustal evolution*

In the foregoing Ar degassing discussions, Eqs. (1)–(7) were set up for an earth-atmosphere evolution model which assumes a two stage degassing of the atmosphere; that is, catastrophic degassing at t_d and subsequent continuous degassing. Here, it is also implicitly assumed that the subsequent continuous degassing accompanied a continuous evolution of the crust. We now examine this latter assumption for compatibility with observed geophysical and geochemical evidence. Numerous models for crustal evolution have been proposed by various authors with a variety of approaches. Among them, radiogenic isotopic ratios provide the prime source of information. On the basis of Sr and Pb isotopic ratios, ARMSTRONG (1968) concluded that the volumes and the bulk compositions of ocean, continent and mantle have remained essentially constant for at least the last 2.5 b.y. This result corresponds to a recycling crust model, where a continental crust of nearly the present size had differentiated from the mantle in an early stage of earth history and henceforth has been continuously recycling within a mantle-crust system. This is one of the extreme cases of a crustal evolution model.

At the other extreme, many authors believe that the continental crust has been growing continuously (e.g., HURLEY *et al.*, 1963). In the latter interpretation, the commencement of crustal evolution is generally assumed to predate the age of the oldest crustal rocks. From the fact that initial $^{87}Sr/^{86}Sr$ ratios of granitic rocks in the continents are generally low and do not correlate with their ages, HURLEY *et al.* (1962, 1963) concluded that these granitic rocks were derived directly from the mantle, thus favouring a continuous evolution of the crust. In addition, both ore lead isotopic evolution (CUMMING and RICHARDS, 1975; STACEY and KRAMERS, 1975) and strontium isotopic evolution observed in oceanic basalts (FAURE and POWELL, 1972) suggest a continuous evolution of the crust-mantle system.

To specify further the mode of crustal development, we extend the approach originally proposed by HURLEY and RAND (1969). Hurley and Rand compiled almost all the available Rb-Sr whole rock age data of continental basement rocks. This compilation shows an apparent increase in the volume of younger rock. From this result they postulated an accelerating growth of continents. However, this apparently accelerating continental growth may be an artifact, since in their calculation they did not consider the effects of weathering and erosion of crustal rocks. Hence, we re-examine the problem, taking into account the erosion and weathering effects.

We assume a production rate $P(t)$ of continental crust, of the form

$$P(t)\mathrm{d}t = P_0 \mathrm{e}^{-pt}\mathrm{d}t,$$

where time t is measured from the commencement of crustal formation (or an initial time) and p is a parameter specifying the growth rate of continents. This assumption for continental evolution is very general and should cover almost all conceivable cases occurring in nature simply by adjusting the value of p. $p=0$ corresponds to a steady-state continuous growth. A large positive value of p corresponds to sudden formation of continents. Accelerating growth is represented by a negative value of p. We represent the decrease of crustal material due to weathering and erosion by a first order rate process, that is, the amount of eroded material is proportional to the total volume of the contemporaneous crust. Then an age distribution function $D(t)$ in the crust observed at present can be expressed as

$$D(t) = P_0 \mathrm{e}^{-pt}\mathrm{e}^{-q(t_p - t)},$$

where q is a rate constant representing the decrease of continents due to erosion and weathering and t_p represents the present time measured from the initial time defined above. The initial time and the rate constant q must be found in order to specify the mode of continental growth. We assume that the initial time predates the age of the oldest crustal rocks. We take the age of the earth (4.55×10^9 yr) for the initial time. The choice of initial time is, however, rather insensitive in the following calculation and the conclusion are unchanged if we assume 3.8×10^9 yr ago for the initial time.

We can estimate the rate constant q from the amount of sediments under the sea. The total mass of sediments, including suboceanic and pelagic sediments, was estimated to be about 1.2×10^{24} g, of which about 80% (1×10^{24} g) can be attributed to a continental origin (POLDERVAART, 1955). No oceanic sediments older than 150 m.y. are observed and sediments older than 80 m.y. are quite rare. These facts suggest that most of the sediments observed at present were produced during the last 100 m.y. Hence, q may be estimated as $1 \times 10^{24}/10^8 \times M_c \text{ yr}^{-1}$, where M_c is the mass of the present continental crust, giving $q = 0.5 \times 10^{-9} \text{ yr}^{-1}$. This is a minimum estimate, since we neglect recent sediments on land. However, if we take recent continental sediments on land into account, this would not change the estimated value of q by more than a factor of two. Using the above value for q, age distributions $D(t)$ for various values of the parameter p are calculated. These curves are compared with the age distribution compiled by HURLEY and RAND (1969) in Fig. 8. As evident from the figure, the curve with $p=0$ shows the best fit to the observed data; curves for larger positive values of p ("sudden" formation models) deviate markedly from the observed distribution. From this we conclude that a model of continuous growth of the continental crust is more likely than a recycling model.

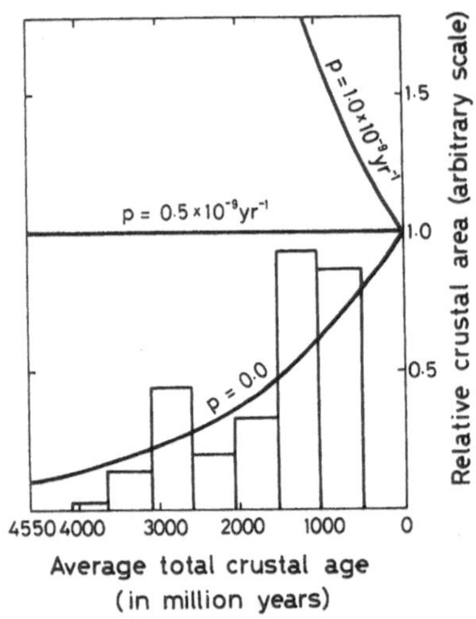

Fig. 8. Age distribution of the continental crust compiled by HURLEY and RAND (1969). Curves indicate calculated distributions based on the model of crustal evolution derived in the text.

6.2 Sudden degassing event

Since the $\mu(\equiv {}^{238}U/{}^{204}Pb)$ value in the present mantle ($\mu \cong 10.0$) estimated from Pb isotopic data is entirely different from that observed in chondrites ($\mu \sim 0.4$), TUREKIAN and CLARK (1975) suggested that high temperature early condensates ($\mu \simeq \infty$) must have been mixed with low temperature later condensates to give the observed μ value of ~ 10.0. They therefore proposed a two stage earth evolution model, in which the earth was first formed heterogeneously and subsequently underwent a secondary remelting which resulted in redistribution of U, Th, Pb, and other elements in the earth. They further suggested that the redistribution event predates the age of the oldest crustal rock. A similar conclusion was reached by HUTCHINSON (1974) on the basis of the observed deficiency of

Ti and Cr in the mantle. Since such global redistribution of elements is very likely to cause liberation of rare gases and other volatiles from the solid earth, we identify the sudden degassing time t_d in the earth-atmosphere evolution model as the element redistribution event. It is very remarkable that both the geochemical and Ar isotopic data indicate quite independently an early global catastrophic event.

6.3 Degassing of other volatiles

We have assumed implicitly that rare gas degassing was accompanied by degassing of other volatiles to the atmosphere, and that the evolution of the atmosphere can be described reasonably well in terms of rare gas degassing. Proof of this assumption of contemporaneous degassing of both rare gases and other volatiles can not be developed until we know the exact mechanisms of degassing from the solid earth. However, ^{36}Ar degassing enables us to make reasonable estimates of the degassing of other volatiles. (We note that a model of ^{40}Ar degassing cannot be used to estimate degassing of other volatiles, since ^{40}Ar has been accumulating in the earth due to radioactive decay of ^{40}K, unlike ^{36}Ar and other volatiles (FANALE, 1976)).

As shown in the previous section, more than $\sim 80\%$ of the ^{36}Ar originally contained in the earth was degassed at t_d, while $\sim 80\%$ of the remaining ^{36}Ar has been degassing continuously from t_d to the present. This result shows that the activity of the sudden degassing, defined as the ratio f of the degassed ^{36}Ar to the initial ^{36}Ar content of the earth, is about the same or even larger than the degassing activity for the period from t_d to the present, where we define degassing activity for this subsequent continuous process as the ratio of the continuously degassed fraction to the amount remaining after the sudden degassing. Unless we assume that the degassing of other volatiles was entirely independent from degassing of rare gases (hardly an acceptable assumption), an inescapable conclusion of the ^{36}Ar degassing history is that the degassing activity of other volatiles for the sudden degassing period was at least as large as the degassing activity for the rest of the period. Therefore, more than half of the present atmospheric constituents were degassed during the early sudden degassing period.

An early sudden degassing of the atmosphere seems to be compatible with other geological evidence. In this respect the recent discovery of 3.7 b.y. old gneiss in west Greenland, metamorphosed from both volcanic and sedimentary source rocks, is very important since the existence of the well-developed sediments indicates the existence of non-trivial ocean prior to 3.7 b.y. In addition, we would like to emphasize again that geochemical considerations (i.e., redistribution of elements in the mantle (TUREKIAN and CLARK, 1975; HUTCHINSON, 1974)) independently lead to a conclusion of an early global catastrophic event, which would very likely have resulted in degassing of the earth.

APPENDIX I

A relation between K-Ar ages and Rb-Sr whole rock ages compiled by HURLEY and RAND (1969) is shown in Fig. 2. Younger ages obtained by the K-Ar method are most likely due to Ar-loss from the crustal rocks. The systematic difference between the two ages suggests that we can approximate the Ar-loss by a first order rate process. Assuming this relation, ^{40}Ar contained in a rock with a formation age T is expressed as

$$({}^{40}\mathrm{Ar})=\frac{\lambda_e}{\beta-\lambda}({}^{40}\mathrm{K})_0(e^{-\lambda T}-e^{-\beta T}), \tag{A. 1}$$

where β is the degassing constant defined in the text. Then the present ratio of the amount of ^{40}Ar to that of ^{40}K is given by

$$({}^{40}\mathrm{Ar}/{}^{40}\mathrm{K})=\frac{\lambda_e}{\beta-\lambda}(1-e^{(\lambda-\beta)T}), \tag{A. 2}$$

and the K-Ar age $T_{\mathrm{K\text{-}Ar}}$ for the rock is

$$T_{\mathrm{K\text{-}Ar}}=\frac{1}{\lambda}\ln\left\{\frac{\lambda}{\lambda_e}\left(\frac{{}^{40}\mathrm{Ar}}{{}^{40}\mathrm{K}}\right)+1\right\}. \tag{A. 3}$$

If we assume that a Rb-Sr age obtained for the same rock gives the true formation age T, Eq. (8) in the text is obtained by combining Eqs. (A. 2) and (A. 3).

REFERENCES

ARMSTRONG, R. L., A model for the evolution of strontium and lead isotopes in a dynamic earth, *Rev. Geophys.*, **6**, 175–199, 1968.

BOULOS, M. S. and O. K. MANUEL, The Xenon record of extinct radioactivities in the earth, *Science*, **174**, 1334–1336, 1971.

CAMERON, A. G. W., Abundances of the elements in the solar system, *Space Sci. Rev.*, **15**, 121–146, 1970.

CLARK, S. P. and A. E. RINGWOOD, Density distribution and constitution of the mantle, *Rev. Geophys.*, **2**, 35–88, 1964.

CLARKE, F. W. and H. S. WASHINGTON, The composition of the Earth's crust, U.S.G.S. Prof Paper 127, 1924.

CRAIG, H. and J. E. LUPTON, Primordial neon, helium and hydrogen in oceanic basalts, *Earth Planet. Sci. Lett.*, **31**, 369–385, 1976.

CUMMING, G. L. and J. R. RICHARDS, Ore lead isotope ratios in a continuously changing earth, *Earth Planet. Sci. Lett.*, **26**, 207–221, 1975.

DYMOND, J. and L. HOGAN, Noble gas abundance patterns in deep sea basalts—Primordial gases from the mantle, *Earth Planet. Sci. Lett.*, **20**, 131–139, 1973.

FANALE, F. P., A case for catastrophic early degassing of the earth, *Chem. Geol.*, **8**, 79–105, 1971.

FANALE, F. P., Martian volatiles: Their degassing history and geochemical fate, *Icarus*, **28**, 179–202, 1976.

FAURE, G. and J. L. POWELL, *Strontium Isotope Geology*, Springer-Verlag, 1972.

FISHER, D. E., Heavy rare gases in a Pacific seamount, *Earth Planet. Sci. Lett.*, **9**, 331–335, 1970.

FISHER, D. E., Trapped helium and argon and the formation of the atmosphere by degassing, *Nature*, **256**, 113–114, 1975.

GAST, P. W., Trace element fractionation and the origin of tholeiitic and alkaline magma types, *Geochim. Cosmochim. Acta*, **32**, 1057–1086, 1968.

GOLDSCHMIDT, V. M., *Geochemistry*, edited by A. Muir, Oxford University Press, London, 1954.

GRAMLICH, J. W. and J. J. NAUGHTON, Nature of source material for ultramafic minerals from Salt Lake Crater, Hawaii, from measurement of helium and argon diffusion, *J. Geophys. Res.*, **77**, 3032–3042, 1972.

HART, S. R. and C. BROOKS, Rb-Sr mantle evolution models, Ann. Report of the Director, Dept. Terr. Magnetism, Carnegie Institution 1968–1969, 426–429, 1970.

HOLLAND, J. G. and R. St. J. LAMBERT, Major elemental chemical composition of shields and the continental crust, *Geochim. Cosmochim. Acta*, **36**, 673–683, 1972.

HURLEY, P. M., Absolute abundances and distribution of Rb, K and Sr in the earth, *Geochim. Cosmochim.*

Acta, **32**, 273–284, 1968.

HURLEY, P. M. and J. R. RAND, Pre-drift continental nuclei, *Science*, **164**, 1229–1242, 1969.

HURLEY, P. M., H. HUGHES, G. FAURE, H. W. FAIRBAIRN, and W. H. PINSON, Radiogenic strontium-87 model of continent formation, *J. Geophys. Res.*, **67**, 5313–5334, 1962.

HURLEY, P. M., G. FAURE, H. HUGHES, H. W. FAIRBAIRN, and W. H. PINSON, Evidence of continuing separation of sial from the mantle from the isotopic composition of common strontium, in Nuclear geophysics, Nuclear Sci. Ser., Rept. No. 38, Nat. Acad. Sci.-Nat. Res. Council, Publ. 1075, 83–92, 1963.

HUTCHINSON, R., The formation of the earth, *Nature*, **250**, 556–558, 1974.

KANEOKA, I., N. TAKAOKA, and K. AOKI, Rare gases in the mantle-derived rocks and minerals, in *Terrestrial Rare Gases*, edited by E. C. Alexander, Jr. and M. Ozima, pp. 71–83, Cent. Acad. Publ. Japan, Tokyo, 1978.

LARIMER, J. W., Composition of the earth: chondritic or achondritic? *Geochim. Cosmochim. Acta*, **35**, 769–786, 1971.

LEE, W. H. K. and S. UYEDA, Review of heat flow data, in *Terrestrial Heat Flow, American Geophysical Union Monograph*, *8*, pp. 87–190, 1965.

OZIMA, M., Ar isotopes and earth-atmosphere evolution models, *Geochim. Cosmochim. Acta*, **39**, 1127–1134, 1975.

OZIMA, M. and K. KUDO, Excess argon in submarine basalts and an earth-atmosphere evolution model, *Nature Phys. Sci.*, **239**, 23–24, 1972.

PEPIN, R. O. and P. SIGNER, Primordial rare gases in meteorites, *Science*, **149**, 253–264, 1965.

POLDERVAART, A., The chemistry of the earth's crust, *Geol. Soc. Am. Spec. Pap.*, **62**, 119–144, 1955.

RUBEY, W. W., Geologic history of sea water, An attempt to state the problem, *Bull. Geol. Soc. Am.*, **62**, 1111–1147, 1951.

RUSSELL, R. D., Evolutionary model for lead isotopes in conformable ores and in ocean volcanics, *Rev. Geophys. Space Phys.*, **10**, 529–549, 1972.

SAITO, K., A. R. BASU, and E. C. ALEXANDER, Jr., Planetary rare gas in a mantle derived amphibole, Talk given at: U.S.-Japan Seminar on "Rare Gas Abundance and Isotopic Constraints on the Origin and Evolution of the Earth's Atmosphere", held at Hakone, Japan, June 28–July 1, 1977.

SCHWARTZMAN, D. W., Ar degassing and the origin of the sialic crust, *Geochim. Cosmochim. Acta*, **37**, 2479–2495, 1973.

STACEY, J. S. and J. D. KRAMERS, Approximation of terrestrial lead isotope evolution by a two-stage model, *Earth Planet. Sci. Lett.*, **26**, 207–221, 1975.

TAKAOKA, N. and M. OZIMA, Rare gases in diamond, in *Terrestrial Rare Gases*, edited by E. C. Alexander, Jr. and M. Ozima, pp. 65–70, Cent. Acad. Publ. Japan, Tokyo, 1978.

TAYLOR, S. R., Abundance of chemical elements in the continental crust: A new table, *Geochim. Cosmochim. Acta*, **28**, 1273–1285, 1964.

TOLSTIKHIN, I. N., Helium isotopes in the earth's interior and in the atmosphere: A degassing model of the earth, *Earth Planet. Sci. Lett.*, **26**, 88–96, 1975.

TUREKIAN, K. K., Degassing of argon and helium from the earth, in *The Origin and Evolution of Atmosphere and Oceans*, edited by P. J. Brancazio and A. G. W. Cameron, pp. 74–82, Wiley, 1964.

TUREKIAN, K. K. and S. P. CLARK, Jr., The non-homogeneous accumulation model for terrestrial planet formation and the consequences for the atmosphere of Venus, *J. Atmos. Sci.*, **32**, 1257–1261, 1975.

VON KÖNIG, H., Über die Löslichkeit der Edelgase in Meerwasser, *Z. Naturforschg*, **18a**, 363–367, 1963.

WILSON, J. T., A possible origin of the Hawaiian Islands, *Can. J. Phys.*, **41**, 863–870, 1963.

Terrestrial Potassium and Argon Abundances as Limits to Models of Atmospheric Evolution

D. E. FISHER

Rosenstiel School of Marine and Atmospheric Science, University of Miami, Miami, Florida 33149, U. S. A.

Internal-earth abundances of Potassium and Argon are discussed, and calculations are presented which set limits to models of atmospheric evolution. The primary uncertainty in such calculations seems to be in the transport of K from mantle to crust.

1. Introduction

Terrestrial abundances of the rare gases have been used to establish boundary conditions on the origin and evolution of the earth's atmosphere ever since BROWN (1952) recognized that the extremely low abundances found in our atmosphere were an indication that the accreting Earth was incapable of retaining a primary atmosphere and that, therefore, the atmosphere we see today must have resulted from a later degassing of the solid Earth.

In particular the isotopic composition of Ar has been useful, beginning with DAMON and KULP's (1958) analysis of atmospheric $^{40}Ar/^{36}Ar$ which asserted that a major degassing must have occurred about 3.5 billion years ago. TUREKIAN (1959) showed that the observed ratio could have resulted instead from steady state degassing with a first order degassing constant. The argument vis-à-vis the continuous (RUBEY, 1951; TUREKIAN, 1963; OZIMA and KUDO, 1972) and catastrophic (SCHWARTZMAN, 1973a; FANALE, 1972; CHASE and PERRY, 1972) degassing models has raged unabated since then, although some combination of the two may be most reasonable. OZIMA has concentrated the discussion on the isotopic composition of Ar *within* the earth, first showing (1973) that the $^{40}Ar/^{36}Ar$ value then obtained for deep-sea basalts ($\sim 1{,}000$) ruled out the catastrophic model if the time of the catastrophe was constrained to be older than that of the oldest surface rocks (~ 3.9 b.y.) and if the K content of the earth is $<1{,}500$ ppm. He noted, however, following SCHWARTZMAN (1973b) that if ultramafic values of $^{40}Ar/^{36}Ar \sim 10^4$ were indicative of the deep-earth ratio his conclusion was invalid.

FISHER (1975) subsequently reported deep-sea basaltic values for this ratio of up to 1.5×10^4. A later analysis by OZIMA (1975) concluded that when the transportation of K from the mantle to the crust is taken into account, models of catastrophic degassing are compatible with *any* value of $^{40}Ar/^{36}Ar$ within the earth ≥ 295.5 while the continuous degassing model is disallowed if the K content of the mantle is ≥ 50 ppm and the rate constant for K transport is $\gg 10^{-10}$ yr^{-1}.

In this paper we reinvestigate this question by ascertaining whether values calculated by the various models are indeed compatible with what we know of present-day K and Ar abundances within the earth (i.e., mantle abundances) and with what we may estimate as a rate constant for K transport. There are large uncertainties associated with our knowledge of such parameters: the very concept of mantle-wide abundances may be chimerical, with measured abundances reflecting only local conditions, while a constant rate of K transport from mantle to crust over the whole 4.5 aeons of earth history is certainly a fiction. However these concepts seem likely to provide boundary conditions for atmospheric evolution models that are at least interesting, and may even be useful.

If there is any real meaning to the concept of a mantle-wide $^{40}Ar/^{36}Ar$ ratio, the highest values obtained in the glass rims of oceanic basalts ($\sim 1.5\times 10^4$) are probably our best approximation to it (OZIMA and KUDO, 1972; OZIMA, 1973, 1975; FISHER, 1970a, 1973, 1974).

The K content of the earth is not known. Limits can be specified by assuming that the earth formed from material of roughly chondritic composition (K$\sim$850 ppm), but may have lost K either through volatilization or through partitioning into the core where it is effectively lost insofar as atmospheric or mantle ^{40}Ar production is concerned (RINGWOOD, 1966; GOETTEL, 1976). A lower limit can be established by noting that if all the ^{40}Ar ever produced in the earth is currently in the atmosphere, 85 ppm K were necessary to produce it. Therefore, $85<K_e<850$ ppm are safe limits.

2. The Simple Continuous Model

Following OZIMA (1975) we describe the continuous evolution of radiogenic ^{40}Ar within the earth ($^{40}Ar^e$) as:

$$\frac{d}{dt}{}^{40}Ar^e=\lambda R'{}^{40}K_0e^{-\lambda t}-k{}^{40}Ar^e,$$
$$ {}^{40}Ar^e=\frac{R'\lambda{}^{40}K_0}{k-\lambda}(e^{-\lambda t}-e^{-kt}), \tag{1}$$

where λ is the ^{40}K decay constant, $^{40}K_0$ is the ^{40}K content of an internally homogeneous earth 4.55 b.y. ago, $R'=R/1+R$ where R is the branching ratio for K-decay, and k is the rate constant for the degassing of Ar from the earth to the atmosphere. Similarly,

$$\frac{d}{dt}{}^{36}Ar^e=-k{}^{36}Ar^e,$$
$${}^{36}Ar^e={}^{36}Ar^0e^{-kt}. \tag{2}$$

The atmospheric components ($^iAr^a$) are given by:

$$\frac{d}{dt}{}^{40}Ar^a=k{}^{40}Ar^e,$$
$${}^{40}Ar^a=\frac{R'{}^{40}K_0}{k-\lambda}[k(1-e^{-\lambda t})-\lambda(1-e^{-kt})], \tag{3}$$

and

$$\frac{d}{dt}{}^{36}Ar^a=k{}^{36}Ar^e,$$
$${}^{36}Ar^a={}^{36}Ar^o(1-e^{-kt}). \tag{4}$$

Combining these we arrive at equations describing the evolution of Ar within the earth in terms of atmospheric Ar, the K content of the earth, and the decay and degassing constants:

$$\frac{^{40}\mathrm{Ar^e}}{^{36}\mathrm{Ar^e}}=\frac{R'\lambda^{40}\mathrm{K}_0\mathrm{e}^{kt}}{(k-\lambda)^{36}\mathrm{Ar^a}}[(\mathrm{e}^{-\lambda t}-\mathrm{e}^{-kt})(1-\mathrm{e}^{-kt})],$$
$$^{40}\mathrm{K}_0=\frac{^{40}\mathrm{Ar^a}(k-\lambda)}{R'}[k(1-\mathrm{e}^{-\lambda t})-(1-\mathrm{e}^{-kt})]^{-1}. \qquad (5)$$

Equations (5) are plotted vs. k in Fig. 1, for the known values of R', λ, t, $^{36}\mathrm{Ar^a}$, and $^{40}\mathrm{Ar^a}$. The oceanic basalt value for $^{40}\mathrm{Ar}/^{36}\mathrm{Ar}$ ($\sim 1.5\times 10^4$) is reached at $k \sim 1.5\times 10^{-9}\ \mathrm{yr}^{-1}$, and this indicates $\mathrm{K_e} \sim 90$ ppm. The model is therefore compatible with the known values of the $\mathrm{Ar^e}$ and $\mathrm{K_e}$ parameters.

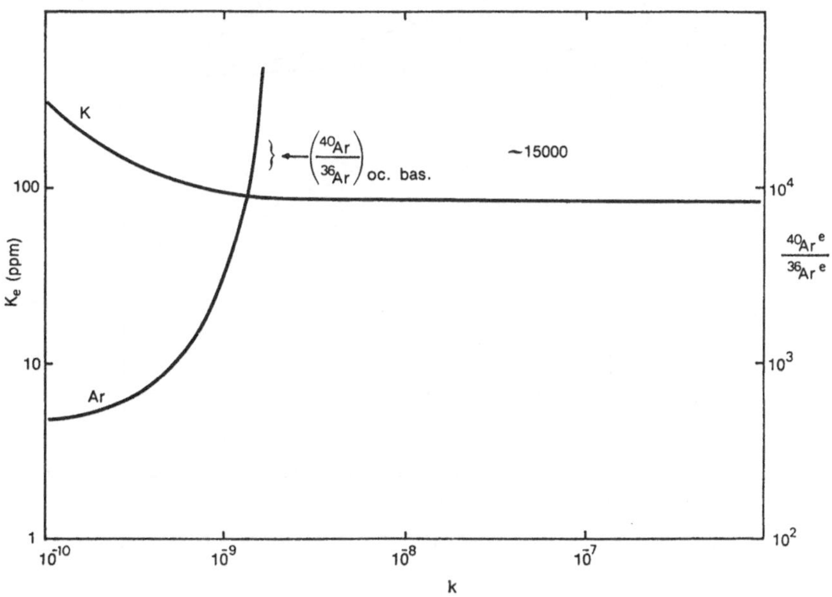

Fig. 1. $^{40}\mathrm{Ar^e}/^{36}\mathrm{Ar^e}$ and $\mathrm{K_e}$ plotted vs. k, the Ar degassing constant, according to Eqs. (5).

3. Time-Variable Models

The next approximation to reality is to allow the rate of degassing to vary with time, i.e., $k=k(t)$. We first assume that the rate of degassing decreases with time in proportion to the half lives of U, Th, and K. The time-dependence of heat production due to these elements is shown in Fig. 2, for assumed values of K=93 ppm, U=50 ppb, Th/U=3.6. The final curve is not significantly affected by any reasonable change in these values.

We fit k to this curve in a stepwise fashion and denote the evolution of nuclide X in the time interval $t_i \rightarrow t_j$ as X_i^j, so that now

$$(^{40}\mathrm{Ar^e})_i^j=\int_i^j(\lambda R'^{40}\mathrm{K}_0)\mathrm{e}^{-\lambda t}-k^{40}\mathrm{Ar^e}\mathrm{d}t,$$

$$(^{40}\mathrm{Ar^e})_0^j=\frac{(^{40}\mathrm{Ar^e})_0^i+\dfrac{R'\lambda^{40}\mathrm{K}_0}{k-\lambda}[\mathrm{e}^{(k-\lambda)t_j}-\mathrm{e}^{(k-\lambda)t_i}]}{\mathrm{e}^{kt_j}}, \qquad (6)$$

$$(^{36}\mathrm{Ar^e})_0^j=(^{36}\mathrm{Ar^e})_0^i\mathrm{e}^{-k(t_j-t_i)}, \qquad (7)$$

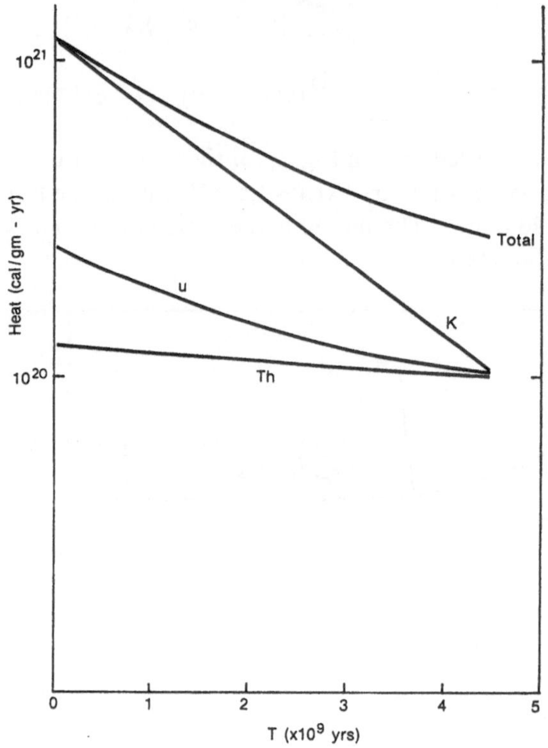

Fig. 2. Heat production via K, U, and Th decay. The final curve is the sum calculated for assumed abundances given in the text.

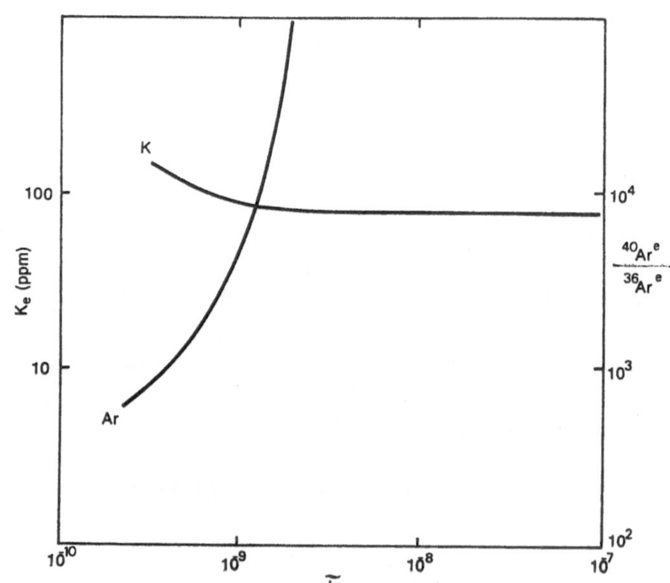

Fig. 3. $^{40}Ar^e/^{36}Ar^e$ and K_e vs. k (average k fitted stepwise to Fig. 2) according to Eqs. (6)–(8).

$$^{40}K_0 = \frac{(^{40}Ar^a)_i^j(k-\lambda)}{R'}[\lambda(e^{-kt_j}-e^{-kt_i})-k(e^{-\lambda t_j}-e^{-\lambda t_i})]^{-1}, \tag{8}$$

with k constant in each ij time interval (taken as 0.5 b.y.). The results of Eqs. (6)–(8) are plotted as before in Fig. 3. The results are identical with the case of a constant k except that the most likely K content has decreased to ~ 80 ppm.

The effect of the time-dependence has also been exaggerated by taking k equal to the square of the heat production curve; the only effect (not shown) is to lower the most probable k from $\sim 1.5\times 10^{-9}$ to $\sim 1.3\times 10^{-9}$ and the K_e content from ~ 80 to ~ 76 ppm.

4. Catastrophic Degassing

We now introduce a contribution from a catastrophic degassing event. We assume first that by $t=0.5$ b.y., 95% of the earth's internal Ar atmosphere has been degassed, i.e., that:

$$\begin{aligned}(^{40}Ar^a)^{0.5} &= 0.95(^{40}Ar^{total})^{0.5}\\ &= 0.95(R'^{40}K_0e^{-0.5})\\ &= 0.0238^{40}K_0,\\ (^{36}Ar^a)^{0.5} &= 0.95(^{36}Ar^e)^{\circ},\end{aligned}$$

and that since then degassing has continued with k proportional to the declining heat production as previously. The results are shown in Fig. 4: the most probable $\bar{k}$ is now $\sim 0.9\times 10^{-9}$ with a corresponding $K_e \sim 150$.

A similar calculation with 95% degassing taking place within the first 10^8 years, corresponding more closely to estimates of core formation (OVERSBY and RINGWOOD,

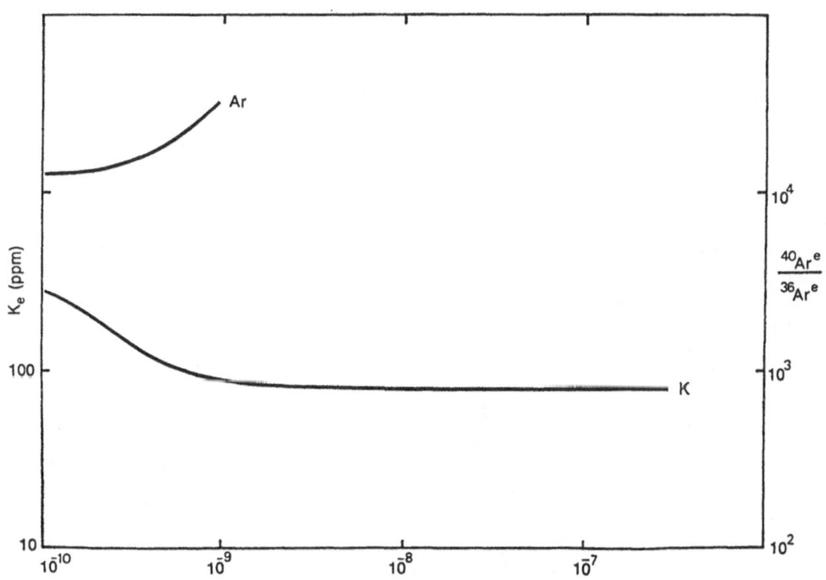

Fig. 4. $^{40}Ar^e/^{36}Ar^e$ and K_e vs. k, in the event of 95% of the earth's internal Ar atmosphere having degassed within the first 0.5 billion years.

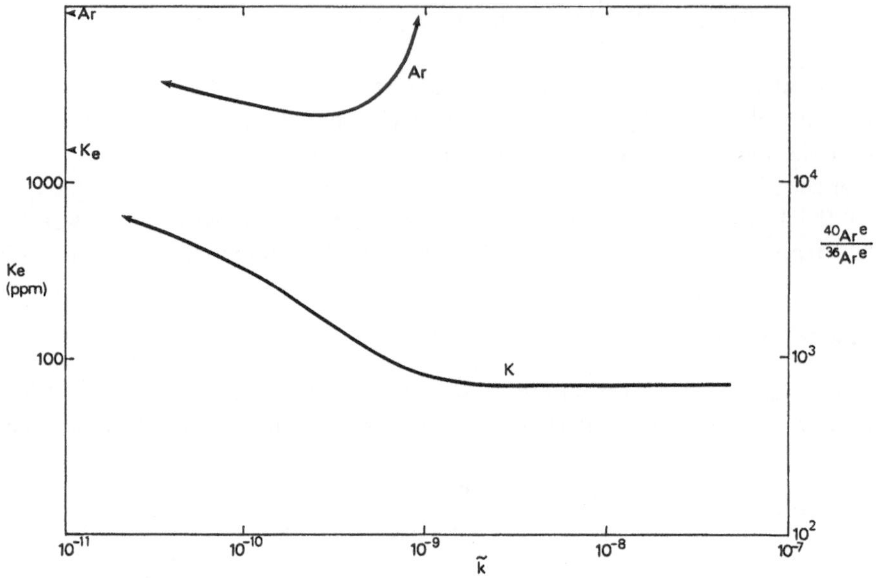

Fig. 5. $^{40}Ar^{e}/^{36}A_{re}$ and K_e vs. K, in the event of 95% of the earth's internal Ar atmosphere having degassed with the first 10^8 years.

1971), leads to the results of Fig. 5: the $^{40}Ar^{e}/^{36}Ar^{e}$ values nowhere reach low enough to match the oceanic basalt data. It appears that such an early degassing event is not compatible with the data; either a later time of degassing or a lesser degassed fraction is mandatory.

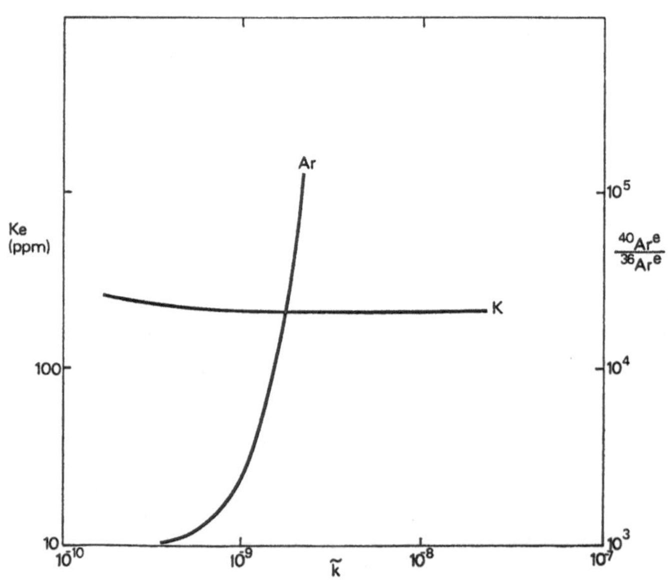

Fig. 6. $^{40}Ar^{e}/^{36}Ar^{e}$ and K_e vs. k, according to Eqs. (6)–(8), with k fitted stepwise to a linearly-increasing curve.

The arrows shown on the left-hand border of Fig. 5 indicate values for $k=0$, i.e., no contribution from the continuous degassing process. Both the Ar and K values thus calculated are unrealistically high, ruling out a totally catastrophic origin of the atmosphere.

5. Degassing Increasing with Time

GREGOR (1977) has pointed out that HURLEY and RAND's (1969) analysis of continental radiometric age data may indicate an *accelerating* generation of crustal material. This might be interpreted as evidence for a degassing process that is accelerating—increasing with time. To investigate this effect on model calculations we fit HURLEY and RAND's (1969) linearly-increasing curve of crustal area vs. time with a stepwise k-function and apply Eqs. (6)–(8). The results are shown in Fig. 6. The results are again compatible with the data, with $\bar{k} \sim 1.5\times10^{-9}$ and $K_e \sim 190$ ppm.

6. Transport of K from Mantle to Crust

We now allow the transport of K from the mantle to the crust, following OZIMA's (1975) suggestion, with a first-order transport constant:

$$\frac{d}{dt}{}^{40}Ar^e = R'\lambda^{40}K_0 e^{-(\lambda+\alpha)t} - k^{40}Ar^e, \tag{9}$$

$$\begin{aligned} {}^{40}Ar^e &= \frac{R'\lambda^{40}K_0}{(k-\lambda-\alpha)}[e^{-(\lambda+\alpha)t} - e^{-kt}], \\ {}^{40}K_0 &= \frac{{}^{40}Ar^a(\lambda+\alpha)(k-\lambda-\alpha)}{R'\lambda[k(1-e^{-(\lambda+\alpha)t})-(\lambda+\alpha)\{1-e^{-kt}\}]}. \end{aligned} \tag{10}$$

The evolution of $^{36}Ar^e$ is unaffected, but the values for $^{40}Ar^e$ and K are strongly dependent on the value of α. A reasonable choice for this parameter is not immediately obvious. It might be estimated from today's values of K in the crust and mantle, assuming initially homogeneous conditions, but although the K content of the earth's crust is reasonably well known, that of the mantle (or whole earth) is not (as discussed previously). UREY (1956), HURLEY (1957), and BIRCH (1958) have pointed out that the observed heat flow from the surface of the earth matches that calculated from chondritic abundances of K, U, and Th, but measured abundances of these elements in terrestrial rocks have since shown that the straightforward interpretation of chondritic earth abundances is an oversimplification. The K/U ratio measured on a variety of crustal rocks is $\sim 1\times10^4$ (WASSERBURG *et al.*, 1964), the chondritic ratio is $\sim 5\times10^4$ (FISHER, 1972), and the values measured on a variety of deep-earth materials are shown in Table 1: they average $\sim 0.3\times10^4$. It

Table 1. K/U ratios in deep-earth materials.

Sample	K/U (10^3)	Reference
Basalt peridotite inclusions	3±2	FISHER, 1970b
Kimberlite peridotite inclusions	2	FISHER, 1970b
Peridotite nodules	3	WAKITA *et al.*, 1967
Lherzolite nodules	1–5	GREEN *et al.*, 1968

therefore seems impossible for the mantle-crust system to have chondritic values for these elements.

Two solutions have been proposed: (1) A substantial fraction of the earth's K must be in the very deep interior (i.e., the core), as suggested by LEWIS (1971), HALL and MURTHY (1971), and GOETTEL (1976). (2) The earth as a whole is depleted in K relative to chondrites (RINGWOOD, 1966; GAST, 1960; WASSERBURG *et al.*, 1964; HURLEY, 1968; FISHER, 1970b), most likely by volatilization during earth formation.

K that has been incorporated in Fe-FeS melts and thus in the core early in the earth's history has probably remained trapped there, and thus is as lost to the mantle-crust transportation process as if it has never existed. It is likely that radiogenic ^{40}Ar associated with it may also remain trapped, and therefore whether K has been lost from the mantle-crust system by entrapment in the core or from the entire earth by a volatility process during planetary formation is immaterial: either way we do not know the present or original K content of the mantle and therefore cannot calculate α from abundance arguments.

Instead we turn our attention to U because (1) the K/U ratio is known and reasonably constant for a variety of geologic materials within different environments (meteorites, crust, deep-earth), (2) U is not a volatile element, nor is it likely to have been substantially fractionated into the core, and (3) there is an abundance of U data and estimates on pertinent materials.

The evolution of U in the crust and mantle is given by (after TOLSTIKHIN *et al.*, 1975):

$$\frac{d}{dt}U_cM_c = -\lambda U_cM_c + \alpha U_mM_m,$$

$$U_c = U_c^0\left[\frac{M_m}{M_c}(e^{-\lambda t} - e^{-(\lambda+\alpha)t}) + e^{-\lambda t}\right], \tag{12}$$

$$\frac{d}{dt}U_mM_m = -\lambda U_mM_m - \alpha U_mM_m, \quad U_m = U_m^0 e^{-(\lambda+\alpha)t}, \tag{13}$$

where U_i is the concentration of U in phase i, M_i is the mass of that phase, and λ and α are the decay and transport constants respectively. We are assuming that initially the earth was homogeneous with respect to its U content, that the masses of mantle and crust are time-independent, and that the present-day variation in U between mantle and crust is due to transport and decay as described in these equations. Then $U_m^0 = U_c^0$ and

$$\frac{U_c}{U_m} = e^{\alpha t}\left(\frac{M_m}{M_c} + 1\right) - \frac{M_m}{M_c}. \tag{14}$$

Table 2. U concentration in the earth's mantle.

U(ppb)	Basis	K/U (assumed)	Reference
56	Nucleosynthesis	—	HOYLE and FOWLER, 1963
40–50	Heat flow	1×10^4	MACDONALD, 1964
50–150	Basalt chemistry	1×10^4	CLARK and RINGWOOD, 1964
100	Basalt chemistry	1×10^4	MASUDA, 1965
50–200	Ultramafic inclusions	—	FISHER, 1970b
50–130	Ultramafic inclusions	—	KLEEMAN *et al.*, 1969
<17	Ar degassing and K/U ratio	1×10^4	LARIMER, 1971

The abundance of U in the crust today is $\sim$1–2 ppm (LAMBERT and HEIER, 1968); $M_m=4.1\times10^{27}$ g, $M_c=2.4\times10^{25}$ g. Various estimates of the mantle concentration are given in Table 2. Aside from Larimer's estimate all the values lie in the range 50–200 ppb. Larimer's estimate is based on an oversimplified model of Ar degassing and an erroneous K/U ratio, and can be ignored. Values of α calculated for mantle concentrations of 50 and 200 ppb are 3.44×10^{-11} and 0.82×10^{-11}; the average is 2.1×10^{-11} yr^{-1}.

The K/U ratio of most deep-earth materials is $\sim3\times10^3$ (Table 1); the K/U ratio in most crustal materials is $\sim1\times10^4$ (WASSERBURG *et al.*, 1964). Therefore the transport constant for K is probably $\sim$3 times that for U; from the above discussion we estimate $\alpha_K\sim6\times10^{-11}$ yr^{-1}. Since OZIMA's (1975) dismissal of the continuous Ar degassing model was based on a value of $\alpha\gg10^{-10}$ yr^{-1} his conclusion must be reexamined. In Fig. 7 we plot the values of $^{40}Ar^e/^{36}Ar^e$ and K^e vs. k, as determined from Eqs. (9) and (10) with $\alpha=6\times10^{-11}$. The results are seen to be virtually unchanged from those with $\alpha=0$, and therefore the continuous degassing model with such a rate of K transport *can* satisfy the parameters.

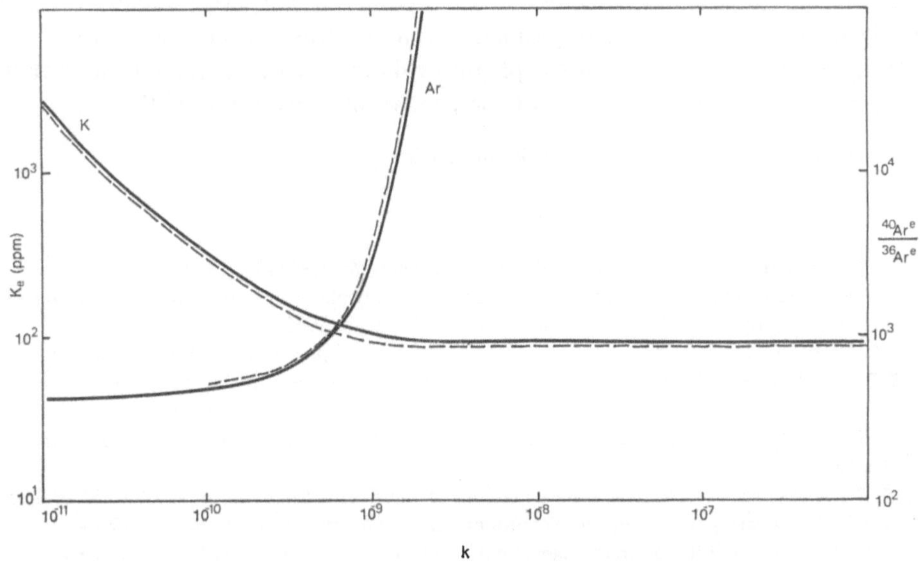

Fig. 7. $^{40}Ar^e/^{36}Ar^e$ and K_e vs. k, according to Eqs. (9) and (10), with$=6\times10^{-11}$ yr^{-1}.

The reality of the K transport process and the estimate of α given here are open to question, as indicated above. If we accept the estimate of K_e given here as a first approximation and combine this with crustal K ($\sim$1.9%) values, we calculate (assuming a first order constant transport process) $\alpha\sim1$–2×10^{-10}. Equations (9) and (10) have been reevaluated using this value of α; the Ar vs. k and K vs. k graphs (not shown) are not significantly different from those calculated on the basis of $\alpha\sim6\times10^{-11}$.

RUSSELL and BIRNIE (1974) have published a model for lead isotope evolution in the earth which includes U transport from mantle to crust within only a limited time interval. If K transport has been similarly spasmodic it could affect to a great extent the calculations described above.

7. Conclusions

(a) Assuming that there is meaning to the concept of world-wide mantle abundances, these are best approximated by $^{40}Ar/^{36}Ar \sim 1.5\times 10^4$ and $85<K<850$ ppm.

(b) All models of continuous degassing (with k constant, increasing, or decreasing) give results compatible with such values so long as K transport from mantle to crust is assumed to be a constant first-order process with $\alpha \lesssim 10^{-10}$. Values resulting from all such models are $k \sim 1.5\times 10^{-9}$ and $K^e \sim 100$ ppm.

(c) The catastrophic degassing model is not compatible with such values if $\gtrsim 95\%$ degassing occurred very early in the earth's history ($\lesssim 10^8$ yrs.). However, a later catastrophe or a combination of catastrophic with continuous degassing do give compatible values. OZIMA (1975) has pointed out that catastrophic degasssing alone will give compatible values if $\alpha \gg 10^{-10}$.

(d) The fulcrum on which all these calculations are balanced is the manner and rate of K transport from mantle to crust, and this is very imperfectly known. The evaluation of α is certainly model-dependent, and even the assumption of a first-order transport equation for K is questionable. OZIMA (1975) has shown that larger values of α lead to completely different results, as could spasmodic or multi-stage transport processes. Future refinements in the calculation of atmospheric evolution based on radiogenic ^{40}Ar must await improvement in our knowledge of the geochemical evolution of K.

I am grateful to John Southam for his help and advice.

REFERENCES

BIRCH, F., Differentiation of the mantle, *Bull. Geol. Soc. Am.*, **69**, 483–485, 1958.

BROWN, H., Rare gases and the formation of the Earth's atmosphere, in *Atmospheres of the Earth and Planets*, 2nd. ed., pp. 258–266, Univ. Chicago Press, Chicago, 1952.

CHASE, C. G. and E. C. PERRY, The oceans growth and oxygen isotope evolution, *Science*, **161**, 1132 1972.

CLARK, S. P. and A. E. RINGWOOD, Density distribution and constitution of the mantle, *Rev. Geophys*, **2**, 35, 1964.

DAMON, P. E. and J. L. KULP, Inert gases and the evolution of the atmosphere, *Geochim. Cosmochim. Acta*, **13**, 280–292, 1958.

FANALE, F. P., A case for catastrophic early degassing of the Earth, *Chem. Geol.*, **8**, 79–105, 1972.

FISHER, D. E., Heavy rare gases in Pacific Seamount, *Earth Planet. Sci. Lett.*, **9**, 331, 1970a.

FISHER, D. E., Homogenized fission track determination of uranium in whole rock geologic samples, *Anal. Chem.*, **42**, 414–416, 1970b.

FISHER, D. E., Uranium content and radiogenic ages of hypersthene, bronzite, amphoterite and carbonaceous chondrites, *Geochim. Cosmochim. Acta*, **36**, 15–33, 1972.

FISHER, D. E., Primordial rare gases IN the deep Earth, *Nature*, **244**, 344, 1973.

FISHER, D. E., The planetary primordial component of rare gases in the deep Earth, *Geophys. Res. Lett.*, **1**, 161, 1974.

FISHER, D. E., Trapped He and Ar and the formation of the atmosphere by degassing, *Nature*, **256**, 113, 1975.

GAST, P. W., Limitations on the composition of the upper mantle, *J. Geophys. Res.*, **65**, 1287–1297, 1960.

GOETTEL, K. A., Potassium in the earth's core: Evidence and implications, in *The Physics and Chemistry of Minerals and Rocks*, pp. 479–488, John Wiley, New York, 1976.

GREEN, D. H., J. W. MORGAN, and K. S. HEIER, Thorium, uranium, and potassium abundances in peridotite inclusions and their hosts, *Earth Planet. Sci. Lett.*, **4**, 155, 1968.

GREGOR, B., Mass-age distribution of crystalline shield rocks, *EOS* (Abstract), **58**, 537, 1977.

HALL, H. T. and V. R. MURTHY, The early chemical history of the earth: Some critical elemental fractionations, *Earth Planet. Sci. Lett.*, **11**, 39–244, 1971.

HOYLE, F. and W. A. FOWLER, On the abundances of uranium and thorium in solar system material, in *Isotopic and Cosmic Chemistry*, pp. 516–529, 1963.

HURLEY, P. M., Test on the possible chondritic composition of the earth's mantle and its abundances of uranium, thorium, and potassium, *Bull. Geol. Soc. Am.*, **68**, 379, 1957.

HURLEY, P. M., Correction to: Absolute abundance and distribution of Rb, K and Sr in the earth, *Geochim. Cosmochim. Acta*, **32**, 1025–1030, 1968.

HURLEY, P. M. and J. R. RAND, Pre-drift continental nuclei, *Science*, **164**, 1229–1242, 1969.

HUTCHISON, R., Strontium and lead isotopic ratios, heterogeneous accretion of the earth, and mantle plumes, *Geochim. Cosmochim. Acta*, **40**, 482–485, 1976.

KLEEMAN, J. D., D. H. GREEN, and J. F. LOVERING, Uranium distribution in ultramafic inclusions from Victorian basalts, *Earth Planet. Sci. Lett.*, **5**, 449–458, 1969.

LAMBERT, I. B. and K. S. HEIER, Estimates of the crustal abundances of Th, U, and K, *Chem. Geol.*, **3**, 233–238, 1968.

LARIMER, J. W., Composition of the earth: Chondritic or achondritic? *Geochim. Cosmochim. Acta*, **35**, 769–786, 1971.

LEWIS, J. S., Consequences of the presence of sulfur in the core of the earth. *Earth Planet. Sci. Lett.*, **11**, 130–134, 1971.

MACDONALD, G. J. F., Dependence of the surface heat flow on the radioactivity of the earth, *J. Geophys. Res.*, **69**, 2933–2946, 1964.

MASUDA, A., Geothermal and petrogenetic implications of the analysis of the distributional relationship between thorium and uranium, *Tectonophysics*, **2**, 69–82, 1965.

OVERSBY, V. M. and A. E. RINGWOOD, Time of formation of the Earth's core, *Nature*, **234**, 463–464, 1971.

OZIMA, M., Was the evolution of the atmosphere continuous or catastrophic? *Nature Phys. Sci.*, **246**, 41, 1973.

OZIMA, M., Ar isotopes and Earth-atmosphere evolution models, *Geochim. Cosmochim. Acta*, **39**, 1127–1134, 1975.

OZIMA, M. and K. KUDO, Excess argon in submarine basalts and an earth-atmosphere evolution model, *Nature Phys. Sci.*, **239**, 23–24, 1972.

RINGWOOD, A. E., Chemical evolution of the terrestrial planets, *Geochim. Cosmochim. Acta*, **30**, 41–104, 1966.

RUBEY, W. W., Geologic history of sea water, *Bull. Geol. Soc. Am.*, **62**, 1111, 1951.

RUSSELL, R. D. and D. J. BIRNIE, A bi-directional mixing model for lead isotope evolution, *Phys. Earth Planet. Inter.*, **8**, 158–166, 1974.

SCHWARTZMAN, D. W., Ar degassing and the origin of the sialic crust, *Geochim. Cosmochim. Acta*, **37**, 2479–2496, 1973a.

SCHWARTZMAN, D. W., On argon degassing models of the Earth, *Nature Phys. Sci.*, **245**, 20, 1973b.

TOLSTIKHIN, I. N., I. Y. ASBEL, and L. V. KHABARIN, The isotopes of the light inert gases in the earth's mantle, crust and atmosphere, *Geokhimiya*, No. 5, 653–666, 1975.

TUREKIAN, K. K., The terrestrial economy of helium and argon, *Geochim. Cosmochim. Acta*, **17**, 37–43, 1959.

TUREKIAN, K. K., Degassing of argon and helium from the Earth, in *The Origin and Evolution of Atmospheres and Oceans*, edited by P. J. Brancazio and A. G. W. Cameron, pp. 74–82, Wiley, New York, 1963.

WAKITA, H., H. NAGASAWA, S. UYEDA, and H. KUNO, Uranium, thorium, and potassium contents of possible mantle materials, *Geochem. J.*, **1**, 183, 1967.

WASSERBURG, G. J., G. J. F. MACDONALD, F. HOYLE, and W. A. FOULER, Relative contributions of uranium, thorium, and potassium to heat production in the Earth, *Science*, **143**, 465–467, 1964.

On the Ambient Mantle $^4He/^{40}Ar$ Ratio and the Coherent Model of Degassing of the Earth

D. W. Schwartzman

Department of Geology and Geography, Howard University, Washington, D.C. 20001, U. S. A.

Deep-sea basalt glasses are highly depleted in ^{40}Ar rad. and ^4He relative to concentrations predicted by a coherent model (Schwartzman, 1973a). A plausible mechanism for depletion is the partitioning of the rare gases between the magma and a vesiculating gas phase. This mechanism would also result in enrichment of He relative to Ar in the magma because of the greater He solubility. Thus measured $^4He/^{40}Ar$ rad. ratios are not necessarily minimum values for the mantle source.

A sizeable group of published data from the glasses appear to define a partial equilibration line on a ^{40}Ar total versus ^{36}Ar total plot, indicating addition of an atmospheric component with simultaneous loss of ^{40}Ar rad. and ^4He. These data show a significant inverse correlation between ^4He and ^{36}Ar concentrations supporting the reality of the partial equilibration line. When the $^4He/^{40}Ar$ ratios are corrected for this effect they are consistent with the proposed depletion mechanism, assuming coherent concentrations of ^{40}Ar rad. and ^4He and a "crustal" K/U ratio (1×10^4). A lower K/U ratio (Fisher, 1975) would remove the requirement for relative He enrichment in the magma. The above interpretation indicates a minimum $^{40}Ar/^{36}Ar$ ratio of 16,000 for the mantle source of deep-sea basalt. If the much lower ratios recently reported are real, they imply the existence of primitive mantle sources which have retained their primordial rare gases.

The ambient mantle $^4He/^{40}Ar$ ratio is an important boundary condition to theories of degassing of the Earth and the geochemical evolution of the mantle. Observed $^4He/^{40}Ar$ ratios in glassy margins of deep-sea basalts are higher than can be generated by in situ decay of K, U, and Th in mantle parent material for crustal or chondritic K/U abundances. Fisher (1975, 1976) assumed that these high ratios are lower limits to the mantle-ambient values, because upon gas loss on emplacement or cooling He would be expected to be lost preferentially to Ar. He concludes that the data rule out all models of mantle degassing which are chondritic or crustal with respect to the K/U ratio. Schwartzman (1973b) showed that a coherent model of degassing predicts an ambient mantle $^4He/^{40}Ar$ ratio of about 1.8* compared to observed ratios in glasses of 0.06–27. He explained the high values by preferential leaking of He into the magma from wall rocks. Ozima and Alexander (1976) and Alexander (1976) have a similar explanation for this pattern except that the preferential diffusion of He takes place in the mantle during partial fusion episodes. Their argument is based on extrapolation of rare gas diffusion data in silica glass. However the diffusivities of the rare gases in minerals at mantle

* Using the new decay constants for U (Jaffey *et al.*, 1971) and ^{40}K (Beckinsale and Gale, 1969), the ratio ranges from 1.82 (Th/U=2) to 2.18 (Th/U=3.7), assuming $K/U=1\times10^4$. The new constants will be used in all model calculations.

P, T conditions could be much lower.

One aspect of magma emplacement has thus far been ignored in relation to possible alteration of relative rare gas abundances; the partitioning of rare gases between a separate volatile phase and the magma. KIRSTEN (1968) measured He, Ne and Ar solubilities in an enstatite melt. The ratio of solubilities for He to Ne and to Ar is 1.7 and 6 respectively. He and to a lesser extent Ne would be expected to be enriched in the melt relative to Ar in the presence of a separate volatile phase. Before this effect can be evaluated for quenched rims of submarine basalts, the possible contribution of atmospheric Ar must be taken into account. Data from FISHER (1975), and DYMOND and HOGAN (1973) (samples 1–5) are plotted on a ^{40}Ar total versus ^{36}Ar total diagram (Fig. 1). Three

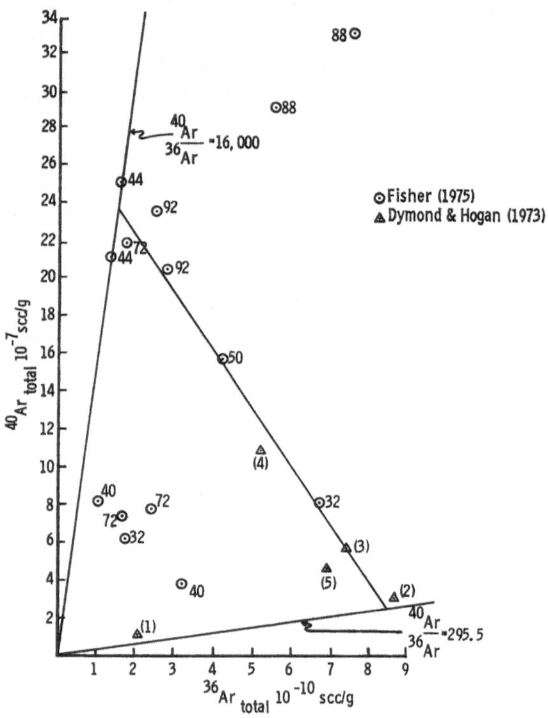

Fig. 1. ^{40}Ar total versus ^{36}Ar total for quenched rims of deep-sea basalts.

of Dymond and Hogan's samples with high ^{36}Ar contents, nearly atmospheric $^{40}Ar/^{36}Ar$ ratios and low He/^{40}Ar ratios were not plotted (they would fall near the atmospheric line far removed to the right from the other samples). 11 analyses (representing 9 samples) fall near an apparent "mixing" line. Strictly speaking it is not a true mixing line, but can be interpreted to express a linear relation between ^{40}Ar rad. loss and atmospheric Ar gain in different samples with the same initial Ar. It is thus a partial equilibration line rather than a true mixing line. Only 3 of a total of 12 samples (88, 40, 1) have no analyses falling near the "mixing" line. There is no obvious systematic correlation of these samples with H_2O, K content or other parameters including $^{4}He/^{40}Ar$ ratios. The apparent "mixing" line has a radiogenic end member with a minimum $^{40}Ar/^{36}Ar$ ratio of 16,000 which can be taken as a mantle minimum. The atmospheric end member apparently represents a dissolved sea water Ar component in an equilibrated volume of the basalt. This value (2.5×10^{-7} scc $^{40}Ar/g$) corresponds to a volume partition coefficient (basalt/sea water) of 3×10^{-3}. This is lower than the coefficient for sea water Ar/atmosphere (2.7×10^{-2}) but larger than the coefficient for Ar mafic melt/gas (3.7×10^{-4}). Perhaps this bracketing reflects the equilibration of the molten outer rim with a steam phase enriched in Ar relative to sea water. The apparent radiogenic end member (2.4×10^{-6} scc/g for $^{40}Ar/^{36}Ar = 16,000$) is lower by a factor of about 0.03 from the value predicted from the coherent model of degassing (assumes a primary magma with 0.10% K**). This model (SCHWARTZMAN, 1973a) assumes ^{40}Ar rad. and ^{4}He enter the crust-atmosphere upon mantle differentiation along with parent K and U, Th respectively. As a limiting model it assumes no ^{40}Ar or ^{4}He leaks into the magma, apart from the coherent supply. If the primary basalt magma is assumed to have a K content of 0.10% (with a K/U of 1×10^{4}, Th/U=2) then it would have accumulated 7.9×10^{-5} scc $^{40}Ar/g$ and 1.4×10^{-4} scc $^{4}He/g$ in 4.55 b.y. (for Th/U=3.7, the He concentration$=1.7 \times 10^{-4}$ scc/g). Note that this model calculation does not assume a single-stage mantle history, only that coherency is maintained for any mantle melting episode. The coherent model thus predicts a $^{4}He/^{40}Ar$ rad. ratio of 1.8 (2.2 for Th/U=3.7). FISHER's (1978) recent estimate of K/U in deep-sea basalts and other "deep Earth materials" is $2 \pm 1 \times 10^{-3}$ giving a coherent model $^{4}He/^{40}Ar$ rad. ratio of 7^{+11}_{-1}.

The observed $^{4}He/^{40}Ar$ total ratios are 2.2 to 18 for samples on the "mixing" line. ^{4}He would be enriched by a factor of 6 relative to Ar if a mechanism involving depletion of rare gases into a separate volatile phase occurred, assuming a one-stage depletion episode. Using an assumed K/U ratio of 1×10^{4}, all but 3 "mixing" line analyses (falling near the atmospheric end member) are less depleted in He than Ar relative to the coherent model initial He and Ar contents. The ratio (He observed/He coherent mode)/(^{40}Ar rad. end member/^{40}Ar coherent model) varies from 0.1 (sample (2)) to 6 (sample (50)), with the predicted value being 6 (see Table 1). For Fisher's estimate of K/U the case for the postulated depletion mechanism leading to relative enrichment of He is much weaker; for the lower limit of 1×10^{3} it collapses entirely.

Table 1

Sample	Ratio depletion factors $^{4}He/^{40}Ar$[1)]
(2)	0.1
(3)	1
(4)	2
(5)	0.3
(92)	4.5
(50)	6
(32)	0.4
(44)	4.5

1) Assumes coherent model with K/U$=1 \times 10^{4}$, Th/U=2. No ^{4}He analysis was given for the "mixing" line analysis of sample 72 (FISHER, 1975).

** The available data on K contents of 9 samples plotted in Fig. 1 range from 0.06 to 0.117%, with one sample (32) at 0.26% (data on Fisher's samples in FUNKHOUSER *et al.*, 1968).

A plot of ^4He versus ^{36}Ar (Fig. 2) shows an inverse correlation for "mixing" line analyses (circled) while the other analyses scatter. A least squares linear regression gives $m=-0.319$, $b=2.702$ (correlation coefficient $=-0.794$) for the "mixing" line analyses. This correlation supports the reality of the "mixing" line and suggests diffusive loss of He along with ^{40}Ar rad. during the process which added atmospheric Ar. Further, the computed depletion factor ratio of 5.1 for the ^4He end member (corresponding to the ^{36}Ar value of the radiogenic ^{40}Ar end member in Fig. 1) is in good agreement with the predicted value.

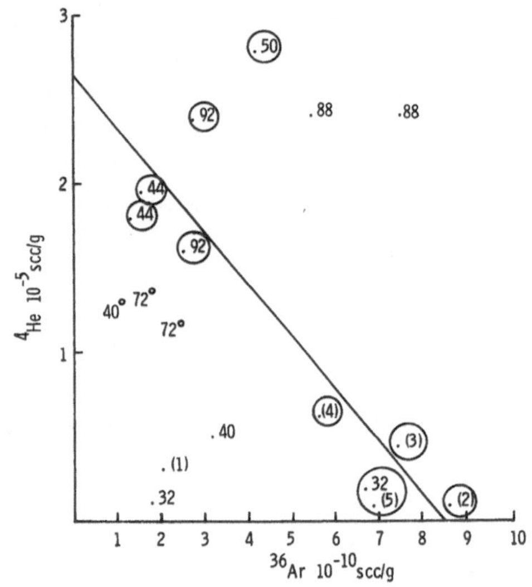

Fig. 2. ^4He versus ^{36}Ar. Circled analyses group on "mixing" line in Fig. 1.

The samples not grouping on the "mixing" line can be interpreted as being disturbed, having lost Ar and He (except sample 88) by an additional mechanism. Analyses of a few samples are split; e.g., one analysis of 32 lines on the "mixing" line and another is depleted in ^{40}Ar and ^4He, suggesting inhomogeneous gas loss on a small scale. Incipient weathering might result in this observed inhomogeneity. The proposed mechanism for depletion of ^{40}Ar and ^4He from the magma, and the enrichment in He relative to ^{40}Ar rad. is vesiculation of CO_2 during upwelling and emplacement, since the magmas are undersaturated in H_2O (MOORE, 1970). For exsolution of CO_2 bubbles at 100 atm. 0.1 of the CO_2 in an original melt with 0.1% is required for an Ar depletion of 0.003, assuming a volume partition coefficient of 3.7×10^{-4} for Ar. Some of the scatter may be an expression of different initial Ar, K and CO_2 contents. The interpretation of the "mixing" line samples depends on their having similar K, Ar and CO_2 contents in their mantle differentiates and at least a similar depth of emplacement. The relative constancy of K has already been mentioned. The depths of origin of 7 samples range from 2,800 to 3,215 m (all happen to be "mixing" line samples) with 1 sample from, 3,713 (50) and 2,260 m (1). The 7 samples from 2,800 to 3,215 m depth are also all mid-ocean ridge basalts, indicating similar pressures at the time of emplacement. Sample 50 is from a sea mount some 700 km from the crest of the East Pacific Rise. The narrow limits in the isotopic composition of Sr and Pb in mid-ocean ridge basalts supports their common mantle parentage (e.g., SUN and HANSON, 1975).

Large Ne enrichments relative to planetary abundances were reported in some samples ((3), (4)) by DYMOND and HOGAN (1973). This was not confirmed by CRAIG and LUPTON (1976). Some Ne enrichment relative to primordial ^{36}Ar (by a factor of 3.5) is to be expected based on the relative solubilities of Ne and Ar, and the above depletion mechanism.

Clearly, much more data is needed on rare gas abundances in submarine basalts to evaluate the proposed model. In my view, the coherent model should be tested to establish the boundary conditions to the process of mantle degassing. Coherency, as a limiting condition, can be applied to computing quantitative models of degassing corresponding to definite rates of crustal development (SCHWARTZMAN, 1973b). An upper limit to the extent of non-coherent degassing can be established if we assume K in the sialic crust is conserved; for a 4.6 b.y. Earth, with a 1.9% K sialic crust the maximum non-coherent contribution to the atmosphere is 0.2 of the total (SCHWARTZMAN, 1973b). However, the K-Ar age of the sialic crust determines the ratio of coherent to non-coherent Ar ("R") since by definition neither include Ar degassed from the crust by reworking.*** R would set limits on the openness of the mantle to Ar and He degassing and its history, particularly the importance of the proposed catastrophic degassing event.

GRAMLICH and NAUGHTON's (1972) data on ultramafic xenoliths from Hawaii give an inferred $^{4}He/^{40}Ar$ ratio for the mantle source of 1.2 ± 0.2 (corrected for diffusive loss in the magma). This ratio is consistent with a coherent degassing for both ^{4}He and Ar (SCHWARTZMAN, 1973b). The rare gases are apparently present in liquid CO_2 inclusions (FUNKHOUSER and NAUGHTON, 1968). Radiogenic ^{129}Xe was reported in similar xenoliths from Hawaii (HENNECKE and MANUEL, 1975) indicating a mantle source. The abundance pattern of Ne, ^{36}Ar and Kr in this sample is very close to a Kirsten pattern. The later authors conclude that the Ne, Ar and Kr in the atmosphere is a fair representation of noble gases which equilibrated with the mantle. Thus primordial noble gases in the mantle, represented by the Hawaiian xenolith, may be a residue of an early catastrophic degassing producing the atmospheric inventory, enriched in Ne and depleted in Kr relative to the atmosphere. The coherent model is consistent with an early catastrophic event (close to 4.55 b.y. ago) releasing the bulk of primordial rare gases into the atmosphere, followed by more or less continuous degassing associated with mantle to crust differentiation and crustal reworking. While coherent degassing does not require a single-stage mantle history, it may require the absence of a volatile phase in the mantle during melting episodes to prevent non-coherent gas from being strongly partitioned into the magma. However, Gramlich and Naughton's data on xenoliths from nephelinite-melilite lava suggest that coherency may be expected event in the case of generation of volatile-rich alkaline magmas, providing the xenoliths are cognate.

The inferred minimum $^{40}Ar/^{36}Ar$ ratio for the mantle source of the deep-sea basalt glasses is 16,000. Others have recently reported much lower ratios (less than 500) in

*** This contribution was ignored by OZIMA (1975). This neglect however strengthens the case for a high $^{40}Ar/^{36}Ar$ mantle early in Earth history, resulting from catastrophic degassing, assuming 1st order degassing. The 1st order models are equivalent to a coherent model when the transport coefficient for Ar is the same as K (mantle to crust).

mantle-derived materials (MANUEL, 1978; SAITO *et al.*, 1978). If these observations are confirmed, it would require a mantle with either of the following properties:

1) inhomogeneous, with very low K systems, isolated since the age of the Earth.

2) non-equilibrium partial melting ($^{40}Ar/^{36}Ar$ basalt greater than $^{40}Ar/^{36}Ar$ partial melting residue).

The second alternative is quite unlikely considering diffusion rates at mantle conditions (HOFMANN and HART, 1975). However, if primitive mantle still exists, it could have on the order of 100 ppm K *if* it still has a "primitive" ^{36}Ar concentration (i.e., mass ^{36}Ar atmosphere/mass Earth) and have an $^{40}Ar/^{36}Ar$ ratio of about 300. If the mantle as a whole has not been substantially degassed of primordial rare gases, then the K content could be still higher and give $^{40}Ar/^{36}Ar$ ratios of about 300.

The participants in the Hakone conference at which a preliminary version of this paper was presented are thanked for their discussion. In particular, I wish to thank Dr. Alexander, Dr. Fisher and Dr. Ozima. Dr. A. Hofmann reviewed a first draft of the Hakone paper and made valuable suggestions. Two reviewers made strengthening criticisms of a first draft of this paper.

REFERENCES

ALEXANDER, E. E., Jr., Comments on Trapped helium and argon and the formation of the atmosphere, *Nature*, **261**, 77, 1976.

BECKINSALE, R. D. and N. H. GALE, A reappraisal of the decay constants and branching ratio of ^{40}K, *Earth Planet. Sci. Lett.*, **6**, 289–294, 1969.

CRAIG, H. and J. E. LUPTON, Primordial neon, helium, and hydrogen in oceanic basalts, *Earth Planet. Sci. Lett.*, **31**, 369–385, 1976.

DYMOND, J. and L. HOGAN, Noble gas abundance patterns in deep-sea basalts—primordial gases from the mantle, *Earth Planet. Sci. Lett.*, **20**, 131–139, 1973.

FISHER, D. E., Trapped helium and argon and the formation of the atmosphere by degassing, *Nature*, **256**, 113–114, 1975.

FISHER, D. E., Reply to comments on Trapped helium and argon and the formation of the atmosphere, *Nature*, **261**, 77, 1976.

FISHER, D. E., K, U and the rare gases in oceanic basalts: Their relation to atmospheric evolution, This volume, 1978.

FUNKHOUSER, J. G., D. E. FISHER, and E. BONATTI, Excess argon in deep-sea rocks, *Earth Planet. Sci. Lett.*, **5**, 95–100, 1968.

FUNKHOUSER, J. G. and J. J. NAUGHTON, Radiogenic helium and argon in ultramafic inclusions from Hawaii, *J. Geophys. Res.*, **73**, 4601–4607, 1968.

GRAMLICH, J. W. and J. J. NAUGHTON, Nature of source material for ultramafic minerals from Salt Lake Crater, Hawaii, from measurement of helium and argon diffusion, *J. Geophys. Res.*, **77**, 3032–3042, 1972.

HENNECKE, E. W. and O. K. MANUEL, Noble gases in an Hawaiian xenolith, *Nature*, **257**, 778–780, 1975.

HOFMANN, A. W. and S. R. HART, An assessment of local and regional isotopic equilibrium in a partially molten mantle, *Carnegie Inst. Washington, Year Book*, **74**, 195–210, 1975.

JAFFEY, A. H., K. F. FLYNN, L. E. GENEDENIN, W. C. BENTLEY, and A. M. ESSLING, Precision measurement of half-lives and specific activities of ^{235}U and ^{238}U, *Phys. Rev.*, **C4**, 1889–1906, 1971.

KIRSTEN, T., Incorporation of rare gases in solidifying enstatite melts, *J. Geophys. Res.*, **73**, 2807–2810, 1968.

MANUEL, O. K., A comparison of terrestrial and meteoritic noble gases, This volume, 1978.

MOORE, J. G., Water content of basalt erupted on the ocean floor, *Contrib. Mineral. Petrol.*, **28**, 272–279, 1970.

OZIMA, M., Ar isotopes and Earth-atmosphere evolution models, *Geochim. Cosmochim. Acta*, **39**, 1127–1134, 1975.

OZIMA, M. and E. C. ALEXANDER, Jr., Rare gas fractionation patterns in terrestrial samples and the Earth-atmosphere evolution model, *Rev. Geophys. Space Phys.*, **14**, 385–390, 1976.

SAITO, K., A. R. BASU, and E. C. ALEXANDER Jr., Planetary rare gas in a mantle derived amphibole, This volume, 1978.

SCHWARTZMAN, D. W., Argon degassing models of the Earth, *Nature Phys. Sci.*, **245**, 20–21, 1973a.

SCHWARTZMAN, D. W., Ar degassing and the origin of the sialic crust, *Geochim. Cosmochim. Acta*, **37**, 2479–2495, 1973b.

SUN, S. S. and G. N. HANSON, Evolution of the mantle: Geochemical evidence from alkali basalt, *Geology*, **3**, 297–302, 1975.

Note added in proof

MOORE *et al.* (1977) recently confirmed that vesicles in mid-ocean basalts quenched on the sea floor are filled with CO_2 and that the melts were saturated with the same gas.

MOORE, J. G., J. N. BATCHELDER, and C. G. CUNNINGHAM, CO_2-filled vesicled in mid-ocean basalts, *J. Volcanol. Geotherm. Res.*, **2**, 309–327, 1977.

Earth Degassing Models, and the Heterogeneous vs. Homogeneous Mantle

R. HART and L. HOGAN

School of Oceanography, Oregon State University, Corvallis, Oregon 97331, U. S. A.

The likelihood of early catastrophic degassing of ^{36}Ar as well as the likelihood of compositional heterogeneities in the mantle increases with the average K content of the mantle. A continuously degassed homogeneous mantle is allowed only if its original K conent is in the range 75–150 ppm. A mantle with K content in the range 150–400 ppm requires either early catastrophic (or time skewed) degassing of ^{36}Ar or heterogeneities in the Ar content of the mantle. A mantle with average K content greater than 400 ppm might have undergone early catastrophic degassing of ^{36}Ar, must be heterogeneous in ^{40}Ar, and is probably heterogeneous in K. While a present day mantle heterogeneous in K could be the result of heterogeneous accretion, heterogeneities in Ar are more likely the result of preferentially faster degassing of the upper mantle relative to the deeper mantle.

1. Introduction

The discovery of radiogenic ^{40}Ar and ^{4}He in the earth's atmosphere first led to the suggestion that these gases were produced in the interior of the earth by radioactive decay of K and U and subsequently degassed to the atmosphere (TATEL, 1950; BIRCH, 1951; KULP, 1951; RANKAMA, 1954; SHILLIBEER and RUSSELL, 1955; NICOLET, 1956; DAMON and KULP, 1958; and TUREKIAN, 1959). The early models of earth degassing assumed a first order rate process and a degassing constant was calculated by stipulating a chondritic potassium content for the earth (TUREKIAN, 1959, 1964). FANALE (1971) pointed out that continuous degassing of radiogenic gases does not preclude an early catastrophic degassing of non-radiogenic gases. He suggested catastrophic degassing accompanied large scale melting of the earth, segregation of the core from the mantle, and the concentration of U, Th, and K in the outermost portion of its mass. In addition, FANALE (1971) also noted the total mass of ^{36}Ar in the atmosphere if originally distributed throughout the entire mass of the earth, would result in a concentration of ^{36}Ar similar to that found in chondrites and suggested that ^{36}Ar has not been lost from the earth to outer space. OZIMA and KUDO (1972) assuming the mantle is homogeneous and the ^{40}Ar/^{36}Ar ratio measured in submarine basalts and ultramafic rocks represent the entire mantle were able to simultaneously solve for a K concentration of the mantle as well as a degassing constant in a continuous first order degassing process. SCHWARTZMAN (1973) applied the principle of coherency which states only the ^{40}Ar associated with parent K is degassed as K is transferred from mantle to Ar retentive crust to the problem of earth degassing. He showed if coherency holds, the maximum ^{36}Ar/Si ratio of the Stillwater complex indicates at least 70% of ^{36}Ar in the Stillwater magma source was degassed 2.7 b.y. ago. He also pointed out a continuous degassing earth requires a continuous increase in the $^{40}Ar/^{36}Ar$ of the mantle.

OZIMA (1975), and HAMANO and OZIMA (1978) introduced first order equations for the transfer of potassium from the mantle to an argon retentive sialic crust. They also suggested, if the first order degassing constant of ^{40}Ar is uniform for the entire earth and less than 0.37×10^{-9} year^{-1}, early catastrophic degassing of ^{36}Ar is required to produce $^{40}Ar/^{36}Ar$ ratios greater than 2,000 in the mantle.

All of the previous discussions of earth degassing models are strongly based on the assumption that the earth's mantle is homogeneous, well-mixed, and uniformly degassed. Furthermore, most models assume the ^{36}Ar and ^{40}Ar of the atmosphere degassed uniformly from the entire mantle rather than preferentially from a small portion of it. In this paper we consider two separate K-Ar systems and attempt to define the conditions in which the K and Ar concentrations and the rate of degassing are similar and the conditions where they are different. The two K-Ar systems we use are:

1) atmospheric ^{36}Ar and ^{40}Ar and their source which is taken to be some portion of the mantle-crust system.

2) the ^{36}Ar and ^{40}Ar contained in the glassy margins of submarine pillow basalts and their source which is taken to be the upper mantle under oceanic ridges.

We discuss the following two possible types of heterogeneities that could exist in the earth's mantle today:

1) heterogeneity of argon which could have resulted if atmospheric argon degassed from only a small portion of the mantle or preferentially faster from one portion than another.

2) Heterogeneity of potassium resulting from either inhomogeneous accretion or transfer of potassium from the upper mantle to sialic crust faster than from the deep mantle.

2. Data Evaluation

The available data on ^{40}Ar and ^{36}Ar in submarine basalts are displayed in Fig. 1, a plot of total ^{36}Ar vs. apparent K-Ar age derived from the total ^{40}Ar. DYMOND and HOGAN (1978) have suggested the apparent K-Ar age is a sensitive indicator of degassing with samples of high apparent age having lost less of their ^{40}Ar than samples of younger apparent age. The data fall into four discrete groups defined by their $^{40}Ar/^{36}Ar$ ratios, and each individual group appears to trend towards lower ^{36}Ar with decreasing apparent K-Ar age. Most likely these trends have been produced by the degassing of Ar either in the magma chamber or during the process of eruption onto the sea floor. The samples with $^{40}Ar/^{36}Ar$ ratio most similar to the atmospheric ratio of 295 also have the highest ^{36}Ar contents suggesting atmospheric argon has contaminated them. This interpretation was favored by DYMOND and HOGAN (1978) to explain the diversity of Ar concentration in submarine basalts, and is fortified by comparison of the Ar concentration in the glassy margin of a basalt pillow with those in the central holocrystalline portions which presumably had more time to interact with sea water during cooling of the lava. This comparative study is represented by an arrow and dotted line in Fig. 1.

It is possible the data shown in Fig. 1 have resulted from the effects of degassing and atmospheric contamination of samples from a homogeneous mantle of low ^{36}Ar and high

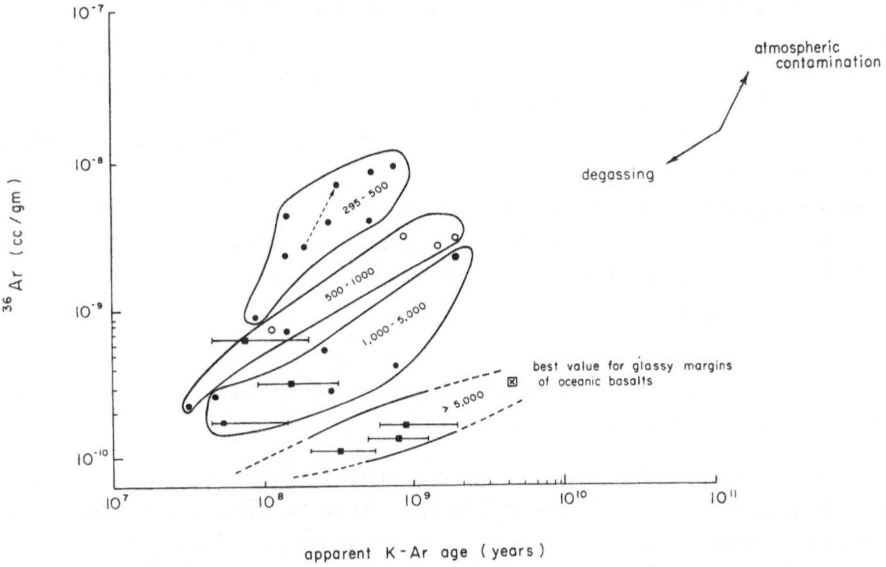

Fig. 1. Plot of ^{36}Ar vs. apparent K-Ar age in submarine basalts. The K-Ar age is calculated from the total ^{40}Ar. Data is from DYMOND and HOGAN (1978), and FISHER (1975). Open circles represent samples dredged from the Juan de Fuca Ridge. Closed squares with error bars indicate samples of Fisher in which the potassium was estimated from an analysis of the same sample by FUNKHOUSER *et al.* (1968). The envelopes enclose points of similar $^{40}Ar/^{36}Ar$ ratios which are indicated by the numbers in each envelope. The dotted line and small arrow connect an analysis made on the glassy margin of a basalt pillow with an analysis made on a holocrystalline interior and demonstrates the effect of atmospheric contamination. High apparent K-Ar age represents samples that have undergone little or no loss of ^{40}Ar by degassing. A negative correlation exists between apparent K-Ar age and ^{36}Ar for samples of the same $^{40}Ar/^{36}Ar$ ratio. A maximum apparent K-Ar age of 4.5 b.y. would be given if a sample had never undergone degassing of ^{40}Ar in earth history. The square and cross represents the maximum values of ^{36}Ar and apparent K-Ar age of samples with $^{40}Ar/^{36}Ar$ in the range 5,000–15,000. This value of ^{36}Ar is taken as a maximum for the upper mantle today if it is homogeneous with respect to ^{36}Ar.

$^{40}Ar/^{36}Ar$. By projecting the "degassing trend" to an apparent maximum K-Ar age of 4.5 b.y. it is possible to estimate the maximum ^{36}Ar content of the mantle to be 3×10^{-10} cc/gm. This is a maximum for the mantle because partially melting and crystal fractionation would have concentrated ^{36}Ar in the melt. Furthermore, a $^{40}Ar/^{36}Ar$ of 15,000 most likely represents the ratio least contaminated by atmospheric argon and a value of 4.5×10^{-6} cc/gm can be calculated as a likely concentration for mantle derived ^{40}Ar in submarine basalts. DYMOND and HOGAN (1977) suggested a somewhat lower value of 2.3×10^{-6} cc/gm for the concentration of mantle derived ^{40}Ar in submarine basalts and point out this value is surprisingly uniform among many of the samples they studied.

A second interpretation of the data follows from the suggestion that many rocks of supposed mantle origin have $^{40}Ar/^{36}Ar$ ratios similar to that of the present day atmosphere, and rather than being a result of atmospheric contamination these ratios actually represent the mantle (BROWN *et al.*, 1974; SAITO *et al.*, 1978). Most of the submarine basalts in

Fig. 1 were dredged from spreading ridges where MORB or LIL depleted basalts are common. Four samples are from the Juan de Fuca ridge where FeTi or "plume type" basalts are often encountered (MELSON *et al.*, 1976). None of the Juan de Fuca basalts (open circles in Fig. 1) have as a low ^{36}Ar or as high $^{40}Ar/^{36}Ar$ as some basalts dredged from regions of MORB. The available data are inadequate but are consistent with the possibility of two different concentrations of Ar in the mantle. The mantle beneath normal ridges has been depleted in ^{36}Ar relative to the mantle source for plume or hot spot material. If the depletion event occurred early in earth history the $^{40}Ar/^{36}Ar$ of depleted mantle could have increased more rapidly than non-depleted mantle by the production of radiogenic ^{40}Ar through time.

3. ^{36}Ar Inventory of the Earth and a Comparison of Degassing Constants for the Atmosphere and Upper Mantle

An estimate of the ^{36}Ar inventory of the earth is given in Table 1. The quantity of ^{36}Ar in the atmosphere is well established, the ^{36}Ar of the sialic crust is determined from the range of measured ^{36}Ar concentrations in continental rocks (DALRYMPLE and LANPHERE, 1969), and multiplied by the volume of sialic crust (ARMSTRONG, 1968). The two values of ^{36}Ar in the mantle given in Table 1 correspond to the two interpretations of ^{36}Ar concentration in submarine basalts as discussed in section 2. One assumes a homogeneous mantle in which the lowest ^{36}Ar concentration in submarine basalts represents the total mantle and the other assumes a heterogeneous mantle in which the maximum ^{36}Ar reported in all basalts and ultramafic rocks (3×10^{-8} cc/gm, DALRYMPLE and LANPHERE, 1968) represents the maximum for non-depleted mantle. It is assumed the ^{36}Ar in the core is similar to and definitely not greater than the mantle. The volume of the mantle and the core are from ARMSTRONG (1968). The case of a homogeneous mantle represents a minimum ^{36}Ar for the earth with more than 95% of the total ^{36}Ar now in the atmosphere. The heterogeneous mantle case requires a higher total ^{36}Ar and up to half of the total ^{36}Ar may still be present in the mantle.

Table 1. Argon-36 inventory of the earth.

	$^{36}Ar\times10^{19}$ cc
Atmosphere	13.68
Sialic crust	0.01–0.04[1)]
Mantle	0.04[2)]–12.23[3)]
Core	0.01[2)]–5.77[3)]
	13.92–31.90

1) Based on ^{36}Ar content of terrestrial rocks (DALRYMPLE and LANPHERE, 1969).
3) Based on ^{36}Ar content of submarine basalts assuming a homogeneous mantle (see Fig. 1 and text).
2) Based on maximum ^{36}Ar content of basalts and ultramafic rocks (DALRYMPLE and LANPHERE, 1969).

If all the ^{36}Ar now in the atmosphere was originally occluded and uniformly distributed in the interior of the earth, the initial ^{36}Ar concentration of the mantle can be calculated by dividing the total ^{36}Ar by the mass of the earth. The calculated concentrations are: $^{36}Ar_{m0}=2.32\times10^{-8}$ cc/gm for a homogeneous mantle and $^{36}Ar_{m0}=5.29\times10^{-8}$ cc/gm for a heterogeneous mantle. If a portion of the present day atmospheric ^{36}Ar was not occluded in the interior of the earth or degassed in an early catastrophic event $^{36}Ar_{m0}$ would be proportionally lower. If the present day atmospheric ^{36}Ar degassed from only a portion of the earth, then the mantle is heterogeneous and $^{36}Ar_{m0}$ would be higher than in the case of uniform degassing.

The initial concentration of ^{36}Ar in the mantle ($^{36}Ar_{m0}$) and the present day concentration ($^{36}Ar_{m}$) can now be used to calculate degassing constants in first order degassing equations. The derivations of all equations used in this paper are given in the appendix. The equation for first order degassing of the mantle is given by:

$$^{36}Ar_{m}={}^{36}Ar_{m0}e^{-\mu_m t}.$$

The value for t used throughout this paper is 4.5 b.y. Assuming $^{36}Ar_{m}$ is similar to that of the submarine basalts least contaminated by atmospheric argon (4×10^{-10} cc/gm) and taking the range of possible values of $^{36}Ar_{m0}$ (2.32–5.29$\times10^{-8}$ cc/gm) solutions of μ_m fall in the range, 0.8–2$\times10^{-9}$ year^{-1}.

First order degassing of the atmosphere is given by:

$$^{36}Ar_{at}={}^{36}Ar_{at_0}+{}^{36}A_{m0}\,(1-e^{-\mu_a t}).$$

The term $^{36}Ar_{at_0}$ represents the amount of ^{36}Ar initially in the atmosphere and for continuous degassing is zero. The amount of ^{36}Ar in the atmosphere is well established and by assuming it was initially evenly distributed throughout the earth (average concentration$=2.3\times10^{-8}$ cc/gm) solutions of μ_a fall in the range 0.1–1$\times10^{-9}$ year^{-1}. The higher values corresponding to a present day mantle homogeneous in ^{36}Ar. Both a heterogeneous mantle (with $^{36}Ar_{m}$ higher than a homogeneous mantle), and a catastrophically degassed earth (with $^{36}Ar_{m0}$ lower than a continuously degassed earth) give degassing constants lower than a continuously degassed homogeneous mantle.

4. Limitations on the K Transfer Constant (α) of the Mantle

If K has been systematically transferred from the mantle to sialic crust the original K content of the mantle would have been greater than it is today. Ozima (1975) considered the possibility of a first order transferal process:

$$K_m=K_{m0}e^{-\alpha t},$$

where K_m and K_{m0} are the present day and original K content of the mantle and α is the K transfer constant. The total loss of ^{40}K in the mantle must combine the transfer constant and decay constants:

$$^{40}K_m={}^{40}K_{m0}e^{-(\lambda'+\alpha)t},$$

where λ' is the total decay constant for ^{40}K, and t is taken to be 4.5 b.y.

The case for heterogeneous accretion of the sialic crust and constant K content for the mantle is represented by $\alpha=0$. A maximum value for α can be estimated by assuming, on the basis of thermal requirement, that the K content of the present day mantle is not lower than 100 ppm. If the original K content of the mantle was similar to chondrites (800 ppm), α could not be greater than 0.4×10^{-9} year^{-1} and still allow a present K content greater than 100 ppm. For initial K contents lower than the chondritic concentration, values of below 0.4×10^{-9} year^{-1} are required.

The introduction of K transfer into the degassing equations does not change their basic nature but broadens the range of possible solutions of initial K in the mantle with the larger values of α producing the higher value of K.

5. Calculation of the K Content of the Source of Atmospheric Argon from the Atmospheric $^{40}Ar/^{36}Ar$ Ratio

As shown in Eq. (A. 12), if the original ^{40}Ar is negligible, then the $^{40}Ar/^{36}Ar$ of the atmosphere is proportional to the potassium and inversely proportional to the original ^{36}Ar of the atmosphere and its mantle source. The $^{40}Ar/^{36}Ar$ of the atmosphere is well

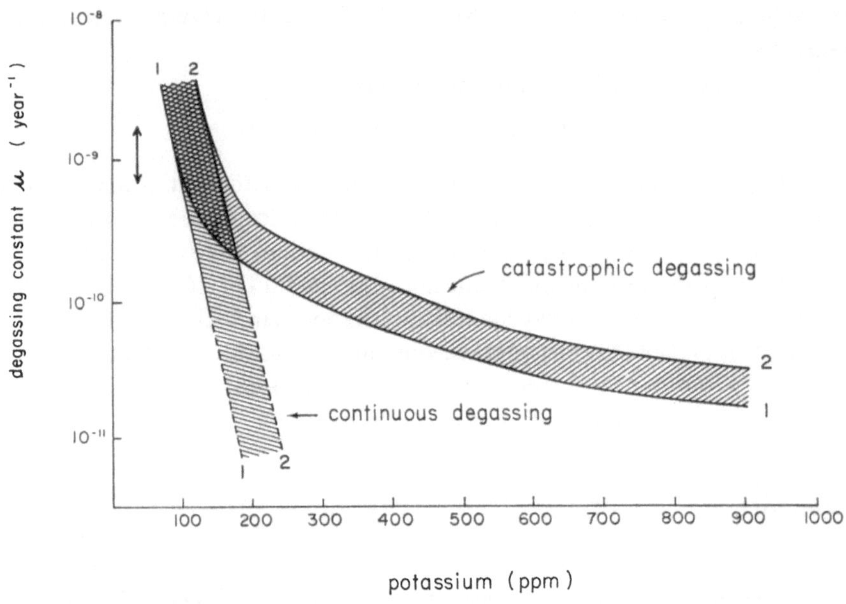

Fig. 2. Graphic representation of the possible solutions of the K content of the mantle source of ^{40}Ar as a function of the degassing constant μ_a. The solutions were derived from Eq. (A. 12) in the appendix using an atmospheric $^{40}Ar/^{36}Ar$ of 295 and an original ^{36}Ar content of mantle ($^{36}Ar_{m0}$) of 2.32×10^{-8} gm/cc. In the case of continuous degassing the original ^{36}Ar content of the atmosphere was taken as zero ($^{36}Ar_{at_0}=0$). In the case of catastrophic degassing the original ^{36}Ar content of the atmosphere was taken as 90% of the total earth inventory for a homogeneous earth ($^{36}Ar_{at_0}=2.09\times10^{-8}$ cc/gm). If the original ^{36}Ar of the mantle ($^{36}Ar_m$) was higher as in the case of a heterogeneous mantle the possible value of K is proportionally higher but probably not more than twice the values plotted here, unless more than half the earth's total ^{36}Ar is still in the mantle. The curves labeled 1 correspond to K transfer constant (α) equal to zero. The curves labeled 2 correspond to K transfer constants of 0.4×10^{-9} year^{-1}. The double headed arrow indicates the range of μ allowed for a continuously degassed mantle.

established to be 295, therefore this equation relates the potassium content of the mantle directly to the original amount of ^{36}Ar in the mantle plus atmosphere with low original ^{36}Ar requiring low potassium. The K-transfer constant α, and the atmosphere degassing constant, μ_a, also enter into the equation. As mentioned in section 3, the original potassium content of the mantle increases with the value of the K-transfer constant. The relationship between K content and degassing constent is displayed graphically in Fig. 2 which shows an inverse relation with the lowest degassing constants requiring the high potassium contents.

The K content for a homogeneous continuously degassed mantle can be calculated assuming the original ^{36}Ar of the atmosphere was zero ($^{36}Ar_{at_0}=0$). As argued in section 2 the $^{36}Ar_{m0}$ of a homogeneous mantle must be close to 2.38×10^{-8} cc/gm and the degassing constants must fall in the range $0.8\text{–}2\times10^{-9}$ year^{-1} for continuous first order degassing throughout 4.5 b.y. As shown in Fig. 2 the possible potassium contents for this range of degassing contents fall in the range 75–150 ppm.

For the case of early catastrophic degassing large amounts of ^{36}Ar would be injected into the early atmosphere and require proportionally high K contents in the mantle to produce the $^{40}Ar/^{36}Ar$ ratio of the atmosphere. The possible solutions for early catastrophic degassing of 90% of the earth's total ^{36}Ar, assuming a present day mantle homogeneous in ^{36}Ar, are shown in Fig. 2. Potassium contents as high as 900 ppm are well within reason for an early catastrophic or time skewed degassing. The possible degassing constants allowed for ^{40}Ar depend upon the extent of the early catastrophic event but are always smaller than in the case of continuous degassing.

A mantle heterogeneous in ^{36}Ar leads to solution of higher K contents than a mantle of homogeneous ^{36}Ar content. This result follows from both the higher $^{36}Ar_{m0}$ allowed as well as the lower degassing constants permitted (see section 3). If atmospheric ^{40}Ar was degassed from only a portion of the mantle proportionally higher values of K are permitted. For example if the atmospheric ^{40}Ar degassed from only half of the mantle, the K content would be twice as high as if it had degassed from the entire mantle. It should be pointed out that if the atmosphere did degass from only a portion of the mantle, then it must be heterogeneous with respect to its present day argon content.

6. Calculation of K Content of Upper Mantle from ^{40}Ar Content of Submarine Basalts

Equation (A. 3) in the appendix can be used to calculate the K content of the upper mantle source of submarine basalt from the ^{40}Ar content of the mantle using the values of degassing constant and K transfer constant presented in previous sections. In Eq. (A. 3) the ^{40}Ar content of the mantle is proportional to the K content of the mantle and inversely related to the degassing content. For a known ^{40}Ar of the mantle then the ^{40}K is directly related to the degassing constant with the lowest degassing constant requiring the lowest potassium contents. This is opposite the relation between K and μ_a in the equation for the $^{40}Ar/^{36}Ar$ of the atmosphere used in previous section. Using Dymond and Hogan's (1978) value for $^{40}Ar_m$ (2.3×10^{-6} cc/gm) and the degassing constants for a homogeneous mantle ($0.8\text{–}2\times10^{-9}$ year^{-1}) the calculated values of K fall in the same range (75–150 ppm) as those calculated from the atmospheric $^{40}Ar/^{36}Ar$ using the model of a homogeneous, continuously degassed mantle. This result should be emphasized, because it constitutes and independent check on the model of a continuously degassed homogeneous earth. Progressively lower degassing constants yield progressively divergent solutions of K and values of degassing constant below 0.2×10^{-9} give distinctly discordant solutions for K and therefore are consistent with the concept of a mantle heterogeneous in K.

Degassing constants below 0.2×10^{-9} are allowed in two cases:

A) The present day mantle is heterogeneous in ^{36}Ar with 20–50% of the total earth inventory still in the interior (see Table 1 and Fig. 1).

B) The earth catastrophically degassed ^{36}Ar early in its history (see Eq. (A. 11) in the appendix).

However, these are not the only conditions that lead to a solution of a mantle heterogeneous in K. As pointed out in section 3, the values of K calculated from the $^{40}Ar/^{36}Ar$ of the atmosphere assumes uniform degassing from the entire mantle. If the atmosphere ^{40}Ar degassed from only a portion of the mantle discordant solutions for K would result over a wide range of degassing constants.

7. Discussion

The fact that two independent solutions to the K content of the mantle for degassing constants in the range $0.8–2\times10^{-9}$ year^{-1} give results that agree must be given further consideration. The result may be fortuitous but certainly deserves careful analysis, particularly when combined with the observation of DYMOND and HOGAN (1978) that the mantle component of ^{40}Ar is uniform in submarine basalts.

One possible restriction on a continuously degassed homogeneous mantle is its low K content (75–150 ppm). Generally, theories of heat flow in the oceanic basalt call for K contents in the range of 400–800 ppm, although values as low as 150 ppm have been cited (OZIMA and HAMANO, 1977). It is possible a substantial amount of the heat flow results from chemical reactions between sea water and the oceanic crust (HART, 1973). If the U/K ratio of the deep mantle were higher than that for crustal rocks, heat flow from radioactive decay would be compatible with K content below 150 ppm.

Another serious objection to a homogeneous mantle is the observation that Sr and Pb isotope isochron ages are concordant in MORB type basalts (TATSUMOTO, 1966; BROOKS *et al.*, 1977). This observation strongly suggests heterogeneities have existed between Sr-Rb and U-Pb in the mantle for 2 b.y. The Sr and Pb mantle isochrons in oceanic regions although convincing, do not completely rule out the possibility of homogeneous Ar concentration. For example, the compositional heterogeneities may exist only between mineral phases and Ar because its high mobility may reach a homogeneous equilibrium concentration between all phases while heterogeneities continue to exist for less mobile elements. A more extensive comparison of Ar contents of plume type basalts with MORB types may offer a useful insight into the question of the scale of heterogeneities in the mantle.

At this point the implications of a mantle heterogeneous in K on Ar degassing are discussed. As mentioned in section 5, the likelihood of a mantle heterogeneous in K increases with an increase in the average K content. Therefore, the problem of a mantle high in K is the same as the problem of a mantle heterogeneous in K. While a continuously degassed earth is restricted to K concentrations of 75–150 and could be homogeneous, small heterogeneities are also allowed because of uncertainties in the ^{40}Ar of the upper mantle, the degassing constant, and the K-transfer constant. Furthermore, catastrophic degassing of a mantle with an average K content in the range 75–150 ppm cannot be ruled out because of the uncertainty in the original ^{36}Ar of the mantle.

In the case of a mantle with K greater than 400 ppm the uncertainties in the data become less prominent and several points can be made. As shown in Fig. 2, K contents greater than 400 ppm are allowed for a uniformly degassed mantle only for a degassing constant, μ_a, lower than 0.1×10^{-9} year^{-1}. On the other hand, the ^{40}Ar of submarine basalts restricts the original K content of the upper mantle under oceanic ridges to less than 100 ppm for degassing constants lower than 0.1×10^{-9} year. Thus, original K content of the upper mantle is likely to be different from the original K content of the total mantle if the average K content of the mantle is greater than 400 ppm. This condition breaks down if the upper mantle degassed faster than the total mantle. Continuous degassing of ^{36}Ar with degassing constants below 0.1×10^{-9} year^{-1} is not compatible with the ^{36}Ar contents of submarine basalts (see Fig. 2), or most terrestrial basalts and ultramafic rocks (DALRYMPLE and LANPHERE, 1968). However, the possibility cannot be ruled out that the mantle is heterogeneous in ^{36}Ar and high concentrations exist in the deep mantle that have not yet been represented in samples collected at the surface. A more likely condition that would allow degassing constants below 0.1×10^{-9} year^{-1} is early catastrophic degassing of ^{36}Ar. For example a process in which 90% of the ^{36}Ar of the atmosphere was degassed in an early catastrophic event, and followed by first order degassing with a degassing constant μ_a, lower than 0.1×10^{-9} would result in the ^{36}Ar concentrations for both the mantle and atmosphere consistent with the available data. Early catastrophic degassing of ^{36}Ar is not the only condition leading to a heterogeneous mantle with average K greater than 400 ppm but it is certainly a good possibility.

Table 2. Estimate of the earth's ^{40}Ar inventory assuming a mantle homogeneous in ^{40}Ar.

	$^{40}Ar \times 10^{21}$ cc
Atmosphere	36.78
Sialic crust	4.67–14[1)]
Mantle	0.09–93[2)]
Core	0.00–43[3)]
	41.54–186.78
Range of K=89–400[4)] ppm	

[1)] Estimated by assuming age of sialic crust is 2 b.y. and K composition is in the range 1–3%. The average age of the sialic crust is probably less than 2 b.y.

[2)] Estimated from the range of ^{40}Ar concentration measured in glassy margins of submarine basalts (see Fig. 1 and text).

[3)] Minimum assumes K content of core is negligible, maximum assumes ^{40}Ar concentration of core is not greater than the mantle.

[4)] Assumes an age of 4.5 b.y. for the earth and the original ^{40}Ar content was negligible compared to present day value.

We now end this paper with a discussion of non-uniform degassing of ^{40}Ar and the implications of heterogeneities in the ^{40}Ar content of the mantle. If the earth's mantle is heterogeneous in K or if the upper mantle degassed at a rate faster than the deep mantle, then heterogeneities in ^{40}Ar may exist in the mantle. The existence of such heterogeneities require that convective overturn of the mantle as well as movement of ^{40}Ar by diffusion and bubble transfer is negligible. In Table 2 we present an attempt at estimating the total ^{40}Ar of the earth based on the assumption that the mantle is homogeneous in ^{40}Ar and that its concentration of ^{40}Ar can be approximated by the ^{40}Ar content of submarine basalts. Allowing for possible effects of partial melting, crystal fractional, degassing, and air contamination on the concentration of ^{40}Ar in submarine basalts, we took what seems like a safe range of $0.23–23\times10^{-6}$ cc/gm to represent the mantle concentration of ^{40}Ar.

According to our inventory the mass of ^{40}Ar associated with the earth is between 41.54 and 187×10^{21} cc. Assuming the age of the earth is 4.5 b.y. and the original ^{40}Ar of the earth was negligible (the solar $^{40}Ar/^{36}Ar$ ratio is generally believed to be lower than 10^{-4}) the corresponding range of K for the earth is 89–400 ppm. The minimum value could actually be lower because we assumed a value of 2 b.y. in calculating the ^{40}Ar of the sialic crust.

It is possible to conclude that a mantle of K much greater than 400 ppm requires a reservoir of ^{40}Ar we have not accounted for in our inventory. If the mantle is not homogeneous in ^{40}Ar and the deep mantle contains concentrations of ^{40}Ar higher than observed in submarine basalts, then average K contents greater than 400 ppm are permitted.

The probability of ^{40}Ar of the atmosphere degassing from a portion of rather than the total mantle depends on the K and ^{40}Ar content of the present day mantle. The lower the K and higher the ^{40}Ar of the mantle the more likely the atmospheric ^{40}Ar degassed from the entire mantle. For example, if the ^{40}Ar of the mantle is 2.3×10^{-6} cc/

Table 3. Summary of models of degassing and types of mantle compositional heterogeneities allowed for different ranges of K composition.

	K composition of mantle (ppm)		
	75–150	150–400	greater than 400
Continuous degassing of ^{36}Ar	Permitted	Permitted only if mantle heterogeneous in Ar	Not permitted unless ^{40}Ar degassed from upper mantle more rapidly than deep mantle
Early catastrophic degassing of ^{36}Ar	Permitted	Permitted	Permitted
Mantle homogeneous in:			
^{36}Ar	Permitted	Permitted only with early catastrophic degassing of ^{36}Ar	
^{40}Ar	Permitted		
K	Permitted	Permitted only if ^{40}Ar degassed preferentially from the upper mantle	Possible but unlikely
Mantle heterogenous in:			
^{36}Ar	Compositions must be within a factor of 2	Permitted	Permitted
^{40}Ar		Permitted	Permitted
K		Permitted	Permitted

gm and the K content 100 ppm the total mantle must be called upon to produce all the ^{40}Ar in the mantle plus atmosphere. If instead the concentration of K in the mantle is 200 ppm then only half of the mantle is required to produce the total ^{40}Ar observed in the mantle plus atmosphere.

8. Conclusions

Some principle conclusions of this paper are presented in Table 3 and others are outlined below.

1) Two interpretations are possible for the observed range of ^{36}Ar and ^{40}Ar in submarine basalts. One interpretation stipulates the array of argon data is due to the effects of degassing and atmospheric contamination on magma derived from a homogeneous source low in ^{36}Ar and with a high $^{40}Ar/^{36}Ar$ ratio. The second interpretation stipulates a heterogeneous mantle with the upper mantle source of MORB basalts depleted in ^{36}Ar relative to the deep mantle early in earth history resulting in a subsequently high $^{40}Ar/^{36}Ar$ ratio in the upper mantle.

2) If the mantle is homogeneous with respect to ^{36}Ar, more than 95% of the total inventory of ^{36}Ar is now in the atmosphere. Assuming the ^{36}Ar of the atmosphere was originally evenly distributed in the mantle and subsequently degassed in a first order process, the initial ^{36}Ar concentration of the mantle can be estimated and degassing constants for both the upper mantle source of ridge basalt and the mantle source of atmospheric argon can be calculated. These calculations show the degassing constants are similar and in the range of 0.8–2×10^{-9} for both portions of the mantle. If large amounts of ^{36}Ar were degassed in an early catastrophic event subsequent degassing of the remaining ^{36}Ar (and ^{40}Ar) must have taken place at a relatively slower rate than in the case of continuous degassing.

3) The present day atmospheric $^{40}Ar/^{30}Ar$ ratio of the atmosphere limits the K content of a continuously degassed homogeneous mantle to 75–150 ppm.

4) The ^{40}Ar contents of the glassy margins of submarine basalts permits K content of the upper mantle in the same range (75–150 ppm) as those calculated for the total mantle from the atmospheric $^{40}Ar/^{36}Ar$ and therefore are consistent with a mantle originally homogeneous in K.

5) K contents for the mantle greater than 150 ppm are only permitted if the ^{40}Ar of the atmosphere degassed from a small portion of the total mantle or also if the degassing constant for ^{40}Ar is lower than 0.1×10^{-9} year^{-1}. Low degassing constants are only permitted in two cases: 1) the ^{36}Ar was catastrophically degassed early in earth history, 2) the ^{36}Ar concentration of the present day mantle is greater than observed in most terrestrial rocks. In either case, mantle of greater than 150 ppm K should be heterogeneous in Ar or K or both.

6) Preferential degassing of the upper mantle relative to the deep mantle can be invoked to produce a mantle heterogeneous in ^{40}Ar or ^{36}Ar. Such preferential degassing must be time skewed toward early earth history in order to provide the high $^{40}Ar/^{36}Ar$ ratios observed in MORB type submarine basalts.

The work in this paper was supported by NSF Grant OCE 76-82015. We thank Jack Dymond for a critical reading of the manuscript and many helpful discussions.

APPENDIX I

In the upper mantle ^{40}Ar is produced by the decay of ^{40}K and lost to the atmosphere by degassing. If potassium is transferred from the mantle to a retentive crust, then we may write two differential equations which describe this process

$$\frac{d^{40}Ar}{dt} = -\mu^{40}Ar + \lambda_\varepsilon{}^{40}K, \tag{A. 1}$$

and

$$\frac{d^{40}K}{dt} = -\alpha^{40}K - \lambda'^{40}K, \tag{A. 2}$$

where, μ, is the degassing constant, λ_ε is the decay constant for production of ^{40}K from ^{40}Ar, and α is the K transference constant.

The simultaneous solution of these equations yield for ^{40}Ar in the upper mantle

$$^{40}Ar_m = {}^{40}Ar_{m0}e^{-\mu t} + \frac{\lambda_\varepsilon K_0}{\lambda' + \alpha - \mu}(e^{-\mu t} - e^{-(\alpha+\lambda')t}), \tag{A. 3}$$

where $^{40}Ar_{m0}$ and $^{40}K_0$ respectively represent the initial ^{40}Ar and ^{40}K in the upper mantle. Since there is no production of ^{36}Ar in the upper mantle, only a loss by degassing, we may write for ^{36}Ar

$$\frac{d^{36}Ar}{dt} = -\mu_m{}^{36}Ar. \tag{A. 4}$$

The solution of this is

$$^{36}Ar_m = {}^{36}Ar_{m0}e^{-\mu_m t}. \tag{A. 5}$$

The $^{40}Ar_m/^{36}Ar_m$ ratio is then

$$\frac{^{40}Ar_m}{^{36}Ar_m} = \frac{^{40}Ar_{m0}}{^{36}Ar_{m0}} + \frac{\lambda_\varepsilon{}^{40}K_0}{(\lambda' + \alpha - \mu)^{36}Ar_{m0}}(1 - e^{-(\alpha+\lambda'-\mu_m)t}). \tag{A. 6}$$

Based upon the previous assumption for the degassing of ^{40}Ar, we may write for the ^{40}Ar release to the atmosphere

$$\frac{d^{40}Ar_{at}}{dt} = \mu_a{}^{40}Ar_m, \tag{A. 7}$$

where $^{40}Ar_m$ is given above. The solution for this equation leads to

$$^{40}Ar_{at} = {}^{40}Ar_{at_0} + {}^{40}Ar_{m0}(1 - e^{-\mu_a t}) + \frac{\mu^{40}K_0\lambda_\varepsilon}{\lambda' + \alpha - \mu_a}\frac{1}{\mu_a}(1 - e^{-\mu_a t}) - \frac{1}{\alpha + \lambda'}(1 - e^{-(\alpha+\lambda')t}). \tag{A. 8}$$

Similarly, we may write for ^{36}Ar,

$$\frac{d^{36}Ar_t}{dt} = \mu_a{}^{36}Ar_m, \tag{A. 9}$$

which becomes

$$\frac{d^{36}Ar_{at}}{dt} = \mu^{36}Ar_{m0}e^{-\mu t}, \tag{A. 10}$$

and leads to the solution

$$^{36}Ar_{at} = {}^{36}Ar_{at_0} + {}^{36}Ar_{m0}(1 - e^{-\mu_a t}). \qquad (A. 11)$$

We may now write the $^{40}Ar/^{36}Ar$ ratio for the atmosphere

$$\frac{^{40}Ar_{at}}{^{36}Ar_{at}} = \frac{^{40}Ar_{at_0} + {}^{40}Ar_{m0}(1 - e^{-\mu_a t}) + \dfrac{\mu^{40}K_0\lambda_s}{\lambda' + \alpha - \mu_a}\left[\dfrac{(1 - e^{-\mu_a t})}{\mu_a} - \dfrac{(1 - e^{-(\alpha + \lambda')t})}{\alpha + \lambda'}\right]}{^{36}Ar_{at_0} + {}^{36}Ar_{m0}(1 - e^{-\mu_a t})}. \qquad (A. 12)$$

In this paper we assume the original ^{40}Ar of the earth is negligible. This assumption is based on the commonly accepted values of 10^{-4}–10^{-6} for the solar $^{40}Ar/^{36}Ar$ ratio.

REFERENCES

Armstrong, R. L., A model for evolution of strontium and lead isotopes in a dynamic earth, *Rev. Geophys.*, **6**, 175–199, 1968.

Birch, F., Recent work on the radioactivity of potassium and some related geophysical problems, *J. Geophys. Res.*, **56**, 107–126, 1951.

Brooks, C., S. R. Hart, A. Hofman, and D. E. James, Pb-Sr mantle isochrons from oceanic regions, *Earth Planet. Sci. Lett.*, **32**, 51–61, 1977.

Brown, J. F., C. T. Harper, and A. L. Odom, Petrogenetic implications of argon isotopic evolution in the upper mantle, *Nature*, **250**, 130–132, 1974.

Dalrymple, G. B. and M. A. Lanphere, *Potassium Argon Dating*, W. H. Freeman and Co., San Francisco, 1969.

Damon, P. E. and J. L. Kulp, Inert gases and the evolution of the atmosphere, *Geochim. Cosmochim. Acta*, **13**, 280–292, 1958.

Dymond, J. and L. Hogan, Factors controlling the noble gas abundance patterns of deep sea basalts, in *Earth Planet. Sci. Lett.*, **38**, 117–128, 1978.

Fanale, F. P., A case for catastrophic degassing of the earth, *Chem. Geol.*, **8**, 79–105, 1971.

Fisher, D. E., Trapped helium and argon and the formation of the atmosphere by degassing, *Nature*, **256**, 113–114, 1975.

Funkhouser, J. G., D. E. Fisher, and E. Bonatti, Excess argon in deep sea rocks, *Earth Planet. Sci. Lett.*, **5**, 95–100, 1968.

Hamano, Y. and M. Ozima, Earth-atmosphere evolution model based on Ar isotopic data, in *Terrestrial Rare Gases*, edited by E. C. Alexander, Jr. and M. Ozima, pp. 155–171, Cent. Acad. Publ. Japan, Tokyo, 1978.

Hart, R. A., A model for chemical exchange in the basalt-seawater system of oceanic layer II, *Can. J. Earth Sci.*, **10**, 799–816, 1973.

Kulp, J. L., Origin of the hydrosphere, *Bull. Geol. Soc. Am.*, **22**, 326–329, 1951.

Melson, W. G., T. L. Vallier, T. L. Wright, G. Byerly, and J. Nelen, Chemical diversity of abyssal volcanic glass erupted along Pacific, Atlantic, and Indian Ocean sea-floor spreading centers, in *The Geophysics of the Pacific Ocean Basin and its Margin* (*Wollard Volume*), p. 51, Am. Geophys. Union, Washington, D. C., 1976.

Nicolet, M., Sur l'origine de l'argun atmospherique, *Bull. Acad. R. Belg.*, **42**, 482–489, 1956.

Ozima, M. and K. Kudo, Excess argon in submarine basalts and an earth-atmosphere evolution model, *Nature Phys. Sci.*, **234**, 23–24, 1972.

Ozima, M., Ar isotopes and earth-atmosphere evolution models, *Geochim. Cosmochim. Acta*, **39**, 1127–1134, 1975.

Rankama, K., A calculation of the amount of weathered rock, *Geochim. Cosmochim. Acta*, **5**, 81–84, 1954.

Saito, K., A. R. Basu, and E. C. Alexander, Planetary type rare gases in an upper mantle derived amphibole, *Earth Planet. Sci. Lett.*, **39**, 274–280, 1978.

SCHWARTZMAN, D. W., Argon degassing models of the earth, *Nature*, **245**, 20–21, 1973a.

SCHILLIBEER, H. A. and R. D. RUSSEL, The Argon-40 content of the atmosphere and the age of the earth, *Geochim. Cosmochim. Acta*, **8**, 16–21, 1955.

TATEL, H. E., Argon 40 and the age of the earth, *Geophys. Res.*, **55**, 329–336, 1950.

TATSUMOTO, M., Genetic relations of oceanic basalts as indicated by lead isotopes, *Science*, **153**, 1094–1101, 1966.

TUREKIAN, K. K., The terrestrial economy of helium and argon, *Geochim. Cosmochim. Acta*, **17**, 37–43, 1959.

TUREKIAN, K. K., Degassing of argon and helium from the earth, in *The Origin and Evolution of Atmospheres and Ocean Water*, edited by P. J. Brancazio and A. G. W. Cameron, pp. 74–82, New York, 1964.

Lead Isotope Constraints on the Early History of the Earth

R. D. RUSSELL

Department of Geophysics and Astronomy, The University of British Columbia, Vancouver, B.C. V6T 1W5, Canada

Isotopic analyses of common lead samples have revealed several distinctive patterns. These include, for example, the primary growth curve for terrestrial leads, the secondary isochrons of ore leads, the primary or meteoritic isochron, and the characteristic distribution of ocean-basalt leads. From each of these patterns can be determined useful information about the history of the earth.

The present study focuses on the relationship between the primary growth curve and the ocean-basalt lead array. Galenas from conformable lead ore deposits represent a good approximation to the locus of the primary growth curve. The curve can be labelled with a timescale, such that it indicates the approximate ages of deposition of the various deposits. RUSSELL and BIRNIE (1973) have proposed that lead from ocean island basalts can be represented by a secondary isochron and have estimated the parameters for that isochron. The fact that the oceanic isochron lies distinctly below the young end of the primary growth curve is considered to be a characteristic of extreme importance, because it shows that no sample from one of the populations can be made by simply averaging any number of samples from the other.

Mathematical models are formulated to give quantitive explanations of the two patterns described, and these are used to infer early earth history. The models represent the evolution of two quasi-isolated systems from an early primitive earth. Both unidirectional and bidirectional models are considered. Exact agreement between the theory and the isochron parameters of Russell and Birnie can be obtained from one or the other of these models for a beginning of differentiation anywhere in the range 2,400 Ma to 4,000 Ma ago. The differentiation process for lead extended well beyond the first quarter of the earth's history. Geochemical parameters for uranium, thorium and lead are inferred for the two evolving systems, as well as rate constants for differentiation.

The principal conclusion is that it is possible to reproduce exactly the oceanic isochron by a simple two reservoir model and that, in particular, such a model can explain quantitatively the lead-207 deficiency.

1. Introduction

Writing in 1975, John Sutton remarked that

"One of the most intriguing problems in Geology at the present is the investigation of the first 2,000 million years of the earth's history. We have now a detailed knowledge of small areas of the earth's crust dating from the latter part of the first 1,000 million years of geological time, but the remnants of crust formed more than 3,500 million years ago are no more than a few tens of kilometers across at the most and are widely separated from one another.

"What we need is information on large scale activities within the primitive Earth, for this might link our fragmentary scraps of knowledge obtained from the ancient rocks exposed in Minnesota, West Greenland, South Africa and Antarctica."

(Sutton, 1975)

In the above paragraphs Sutton has proposed a problem of major proportions, which will ultimately be solved only by bringing together many lines of available evidence. In this present paper, I wish to show that a proper interpretation of lead isotope abundances makes useful contributions to its solution. The discussion that follows will reflect a personal approach to lead isotope interpretations, in that it focuses on some of the larger isotopic features and on the broad scale, and attempts to relate these to very generalized models. That the models are severely oversimplified is freely admitted.

To start with, let us review the "hard data" with respect to the early history of the earth. These data are, in the first place, the ages of the oldest rocks and the age of the earth itself.

Of the very old areas to be recognized, Africa was one of the first. The old rocks there are near the town of Barberton in subgroups of the Swaziland sequences. Model lead studies by RUSSELL and FARQUHAR (1960) were interpreted to indicate ages of about 3,300 Ma. ULRYCH *et al.* (1967) extended those studies and estimated as a result an age of 3,460 Ma. ROBERTSON (1973) obtained data delineating a secondary lead isochron which indicated a primary event at 3,300$\pm$100 Ma when a secondary event at 2,700 Ma was assumed. ALLSOPP *et al.* (1968) found that rubidium-strontium age data were essentially uninterpretable, but JAHN and SHIH (1974) found an age of 3,500$\pm$200 Ma, with an initial strontium isotopic ratio of 0.70048$\pm$0.00005. These data are of variable reliability and the accepted values for the decay constants have changed since many of them were published, but an overall estimate of about 3,370$\pm$80 Ma seems to be a reasonable conclusion from this evidence when modern decay constants are substituted.

For North America, SLAWSON *et al.* (1963) used secondary lead isochrons as evidence in support of ages in excess of 3,100 Ma for the Canadian Shield. More specifically, a uranium lead age of 3,560 Ma was found by GOLDICH *et al.* (1970) by interpreting discordant data for the Morton/Montevideo Gneiss of Minnesota. GOLDICH and HEDGE (1974) obtained an age of 3,800 Ma by rudibium/strontium methods, with an initial ratio of about 0.700. With modern constants, these data suggest an age of approximately 3,640$\pm$112 Ma.

The most dramatic example of such ages is, of course, the Amitsoq Gneiss and Isua iron deposits of the Gotthaab District of south-western Greenland. MOORBATH *et al.* (1972) published whole rock rubidium/strontium ages which, with $\lambda(^{87}Rb)=1.42\times 10^{-11}\ yr^{-1}$, give an age of about 3,620 to 3,670 Ma. Interpretation of common lead secondary isochrons by BLACK *et al.* (1971) and MOORBATH *et al.* (1973) indicated 3,740 $\pm$70 Ma. BAADSGAARD (1973) obtained figures of 3,650$\pm$50 Ma and 3,646$\pm$85 Ma from discordant uranium-lead measurements. A very detailed study of the mineral isotopic age relationships from the Amitsoq Gneiss by BAADSGAARD *et al.* (1976) led him to the conclusion that the microcline, biotite, anorthosite, quartz, oxides of iron and titanium, apatite and zircon all record, directly or indirectly, the consequences of an event of about 3,600 Ma (using the newer constants). He also concluded that evidence for an older event is not yet convincing.

The above summary does not include all of the very old areas of the earth and is not inclusive of all the data available for even these areas. But, it does give a good impression of the confidence with which we can state the earth's surface retains the record of 3,600 Ma events and of the extreme difficulty of finding reliable records of much older events.

Now let's turn to the other benchmark, the age of the earth. There have been many calculations of this quantity, the more precise of which have been based on common lead data from meteorites and from the earth. These have yielded ages of about 4,500 Ma since about 1954. Because of the importance of this value to many model calculations, it is worth briefly reviewing its origin.

It seems to be possible to group all of the various methods for determining this quantity under one of two basic ideas. The first uses the slope of the meteoritic isochron (after PATTERSON, 1956) and the second uses the shape of the primary growth curve for the earth (after ALLAN *et al.*, 1953).

Figure 1 shows some measured isotope ratios of lead from various meteorites, as plotted by CUMMING and RICHARDS (1975). To the data have been fitted straight lines. If it is assumed that all the meteorites experienced only one common geochemical dispersion of U/Pb ratios, and that each has remained an isolated closed system since that dispersion, one can calculate from the average slope a unique time for that event which, in the case of the figure, turns out to be about 4,400 Ma. To equate this number to the age of the earth requires, in addition, the conclusion that the earth participated in the same event that altered the chemical ratios of the meteorites. One way of doing this is to show that modern terrestrial lead lies on the meteoritic isochron. In practice this is difficult to do. In the first place, it is difficult to establish precisely the average isotopic composition of modern lead in the outer part of the earth (compare, for example, with the difficulty of determining the present average mantle ratio of the argon isotopes). Moreover, as was pointed out by RUSSELL (1957), the scatter about the meteoritic isochron is very significant. Many meteorite leads are in fact "anomalous" to a degree that would certainly attract attention in the case of terrestrial samples. Also, often it is not possible to establish that the radiogenic lead observed in meteorites is supported by the necessary amounts of thorium and uranium. In spite of these difficulties, the result looks rather precise.

In the second case, one establishes a more or less unique and independent time scale for the primary lead isotope growth curve, from terrestrial lead isotope measurements. In practice, essentially all methods for establishing the shape of such a curve rely on the measurements reported by STANTON and RUSSELL (1959), corrected and extended, for example by CUMMING and RICHARDS (1975). If the isotopic ratios of lead are assumed to be known for a time at which the earth was formed, these isotopic ratios point to unique time on the primary growth curve, and hence a unique age for the earth. It has been found that things work well if for the primordial ratios, the least radiogenic natural leads are chosen, namely those from the troilite phase of iron meteorites. Figure 2 shows isotopic ratios of some conformable ore leads superimposed on hypothetical closed system growth curves passing through the troilite abundances, but for which the troilite leads are assumed widely differing ages.

The point is that there are two procedures which involve fundamentally different concepts, and that the critical assumptions in each are quite different. In their simplest

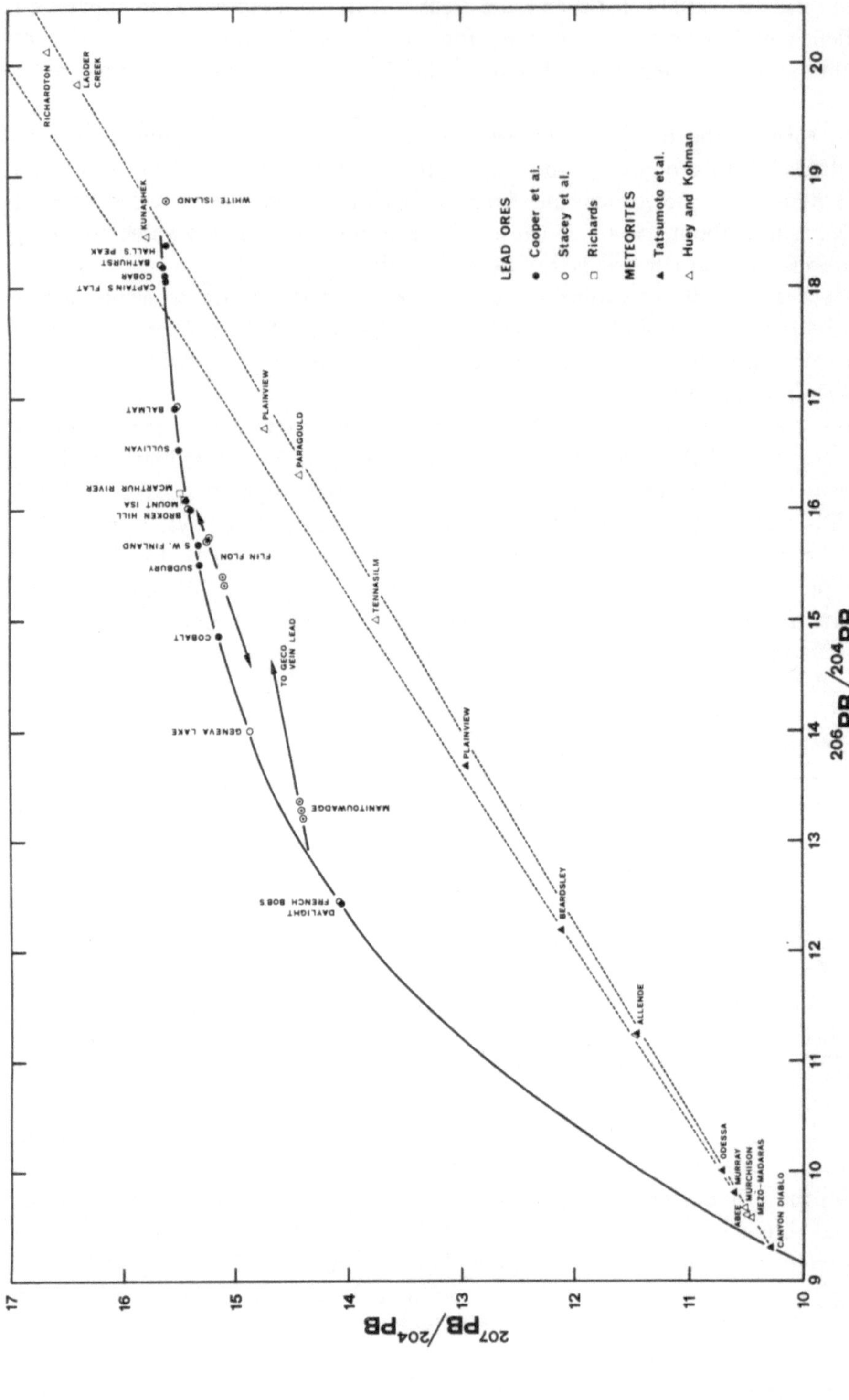

Fig. 1. Isotopic patterns for certain natural leads, reproduced with permission from CUMMING and RICHARDS (1975). The solid curved line indicates the approximate position of the primary growth curve for terrestrial leads. The dotted lines delineate the region of meteoritic lead isotopic ratios. The arrows indicate representative secondary isochrons for multistage leads.

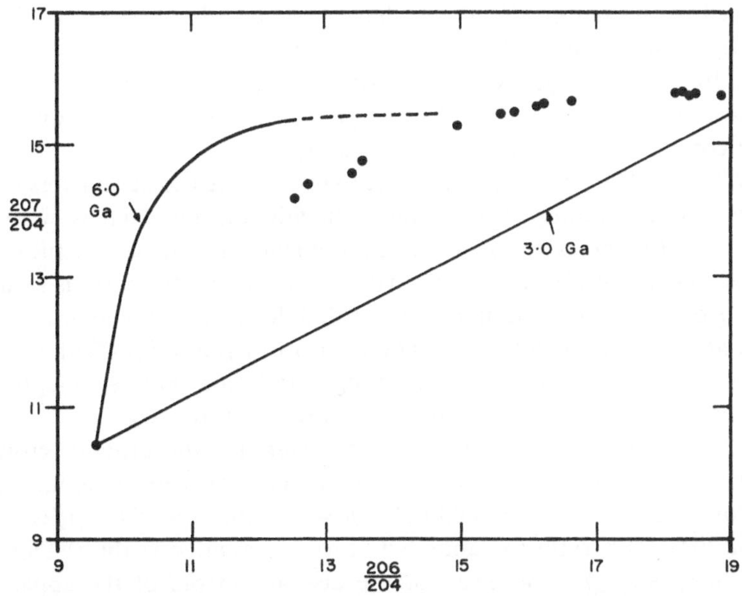

Fig. 2. Single-stage lead isotopic growth curves drawn through the meteoritic lead isotopic ratios and passing as nearly as possible to the isotopic ratios of the conformable lead ores. For the left-hand curve the age of the meteoritic leads is assumed to be 6,000 Ma. For the right-hand curve (nearly a straight line) the age is assumed to be 3,000 Ma. Clearly an intermediate value is required.

forms, both use equations for evolution in a closed chemical system. Fortunately, the two approaches lead to results that are in reasonably good accord, and therefore the numerical values obtained have justifiably acquired considerable credibility.

2. The Primary Growth Curve for Terrestrial Leads

The calculations that follow are based on the assumption that the total of the two evolving systems together constitute a closed system and that the evolution of that closed system corresponds to the primary lead isotope growth curve for the earth. For internal consistency, parameters for such a system are calculated here.

Until recently, the best estimates for the primary terrestrial lead isotope growth curve probably were those by STACEY *et al.* (1969) and by COOPER *et al.* (1969). Those authors used quite different numerical techniques, rather different assumptions, and somewhat different sets of data for the conformable ores. Nevertheless, their conclusions were in impressive agreement. KANASEWICH (1968) has given a general review of the problem of estimating the primary growth curve.

New, and apparently more precise, estimates have been reported for the constants essential to such calculations. TATSUMOTO *et al.* (1973) have made new analyses of the isotope ratios of lead extracted from the meteorite Canyon Diablo, from which they estimate the isotopic composition of primordial lead to be: $^{206}Pb/^{204}Pb=9.307\pm0.006$, $^{207}Pb/^{204}Pb=10.294\pm0.006$, and $^{208}Pb/^{204}Pb=29.476\pm0.018$. These values differ significantly from those by PATERSON (1956) from which many lead isotope studies including those mentioned in the paragraph above, have been based. JAFFREY *et al.* (1971) have redetermined the decay constants for ^{238}U and ^{235}U, obtaining values of $0.155125 \times 10^{-9} yr^{-1}$, and $0.98485 \times 10^{-9} yr^{-1}$, respectively. LE ROUX and GLENDENIN (1963) reported the decay constant of ^{232}Th to be $0.049475 \times 10^{-9} yr^{-1}$. These values also differ significantly from those in common use before their publication.

V. M. OVERSBY (1974) was among the first to re-evaluate the parameters for the growth curve with the new constants. The assumptions adopted by Oversby were, firstly, that the parent material for the galena met the requirements of a single, chemical closed U-Pb system and, secondly, that the age of the earth (at which time the earth is assumed to have had the lead isotopic composition of primordial lead) was the oldest age found by TATSUMOTO *et al.* (1973) for meteorites. These a priori assumptions, although reasonable, lead to the contradictions that there is a strong correlation between μ-apparent and the age of the samples, and that the apparent ages are much too young.

DOE and STACEY (1974) evaluated the parameters for the primary growth curve by minimizing the sum of the squares of the differences between apparent and "known" ages. They found in this way t_0=4,430 Ma, μ=9.56 and κ=4.04. (t_0 is the age of the earth, μ is the $^{238}U/^{204}Pb$ ratio extrapolated to the present, κ is the $^{232}Th/^{238}U$ ratio extrapolated to the present.) The mean of the absolute errors of the apparent ages was $\pm$80 Ma. The above t_0-value is significantly younger than the value of 4,570 Ma found by TATSUMOTO *et al.* (1973) for the meteorite system. Moreover, the assumed primordial $^{208}Pb/^{204}Pb$ ratio appears too small to be consistent with the other data points. To circumvent these criticisms, STACEY and KRAMERS (1975) fit a two-stage model for which t_0 was constrained to be 4,570 Ma. They also minimized age residuals to show that the first stage terminated at 3,700 Ma. The mean of the absolute values of the apparent age discrepancies was reduced to an impressive $\pm$45 Ma.

The present paper takes the point of view that it is dangerous to fit apparent ages because it would be much too optimistic to believe that any suite of natural leads reflect the idealized single stage model and even slight departures from the ideal give significant age discrepancies. This view is also taken by CUMMING and RICHARDS (1975) who discuss this point in detail and elaborate the difficulties of assigning reliable ages to the galenas. The latter authors adopted the procedure of RUSSELL and REYNOLDS (1969) that evaluates the parameters of primary growth curve from the *shape* of the trend of the data points and makes no assumptions about the ages of the samples. They found (model II) that t_0=4,470 Ma, μ=9.15, κ=3.94. The same authors extended their analysis to include time varying parameters (effectively abandoning a single-stage model) to obtain benefits similar to those obtained by STACEY and KRAMERS (1975).

In this paper a primary growth curve is constructed rather differently. As in Oversby's paper, it is assumed that the primary $^{207}Pb/^{204}Pb$-$^{206}Pb/^{204}Pb$ growth curve originates at the primary lead abundances as determined by TATSUMOTO *et al.* (1973), but the age of the earth and the ages of the conformable lead ores are assumed to be entirely unknown. The second assumption is that the *shape* of the primary growth curve is determined by the isotopic ratios of the conformable ore leads. Estimates of the parameters for primary growth curve based on curvature have been used previously by ALLAN *et al.* (1953), RUSSELL and REYNOLDS (1965), and by COOPER *et al.* (1969), as well as by CUMMING and RICHARDS (1975).

Applications of the curvature criterion can be applied by asserting that the apparent μ-values for the various conformable ores should be as nearly as possible the same. With the decay constants and primordial abundances assumed known, the apparent age and μ-value for each conformable ore can be determined uniquely if a t_0 value is postulated. (By definition t_0 is the time at which the troilite abundances fit the primary growth curve.) A series of t_0 values were selected in turn, and the apparent μ and t_0 values calculated for each of the conformable ores listed in Table 1. For each t_0 value, the sum of the squares of the deviations from the mean of the individual μ-values was evaluated. A sharp minimum was obtained at t_0=4,473 Ma (Fig. 4). Using the method of Cumming and Richards, the uncertainty in the value of this parameter is found to be less than 10 Ma.

The data for the individual samples are illustrated in Table 1. This includes all the data used in the present calculation, and the parameter values for the various samples. It is clear that the μ-values derived in this paper are very constant and hence consistent with the hypothesis of a closed system, but this is hardly surprising in view of the fact that the whole calculation is based on making these values constant. One can, though, consider the model ages, which were *not* introduced as data into the calculation. The ages based on the constants preferred in this paper seem in reasonable accord with the geological age of the surrounding rocks.

To obtain the parameter for lead-208, the same data were fitted except that the troilite lead was excluded. This seemed desirable because it has been amply demonstrated that the lead-208 abundance in troilites is not consistent with proposed closed system models for the primary growth curve. This fact is generally attributed to changing

Table 1

Analytical data[1] (values used in the present calculation)				This paper t_0=4,473			Apparent geological age
				t_{app}	μ	κ	
Halls' Peak	18.360	15.600	38.347	13	9.05	3.97	
Bathurst, N.B. (T800)	18.204	15.655	38.122	202	9.17	3.97	
Bathurst, N.B.	18.163	15.627	38.090	197	9.12	3.97	460
Cobar, N.S.W.	18.077	15.617	38.107	249	9.11	4.03	
Captain's Flat, N.S.W.	18.051	15.612	38.128	263	9.10	4.05	
Balmat, N.Y. (F-19)	16.935	15.505	36.423	960	9.06	3.85	1,100
Balmat, N.Y.	16.900	15.524	36.440	1,005	9.11	3.89	
Sullivan, B.C.	16.522	15.503	36.186	1,252	9.16	3.99	
Mt. Isa, Queensland	16.096	15.441	35.809	1,494	9.16	4.05	
Broken Hill, N.S.W.	16.007	15.397	35.675	1,515	9.09	4.02	1,500–1,700
Broken Hill, N.S.W.	16.003	15.390	35.660	1,512	9.07	4.01	
South West Finland	15.678	15.329	35.238	1,688	9.06	3.96	
Sudbury, Ontario	15.491	15.305	35.342	1,800	9.09	4.17	
Cobalt, Ontario	14.876	15.162	34.420	2,123	9.10	3.97	
Geneva Lake, Ontario	14.002	14.870	33.716	2,545	9.06	4.12	2,500–2,700
Barberton, S.A. (Fr.)	12.461	14.077	32.285	3,247	9.09	4.21	3,100
Barberton, S.A. (Ca.)	12.456	14.078	32.308	3,253	9.11	4.25	
				Means:	9.10	4.03	
				Sigmas:	0.03	0.11	

[1] From RUSSELL (1972).

chemistry in the very early part of the earth's history. Provided that the effects are largely confined to the period before 4,000 Ma, they do not affect our calculations.

Table 2 summarized the numerical values used in the model calculations in this paper.

Table 2. Comparison of recent proposals for single-stage closed system primary growth curve for the earth.

	DOE and STACEY (1974)	CUMMINGS and RICHARDS (1975)	This paper
t_0 (Ma)	4,430	4,469	4,473
μ	9.63	9.157 ± 0.12	9.10
$\kappa\mu$	39.0	36.058 ± 0.13	36.7
Modern lead	18.823	18.465 ± 0.074	18.430
ratios	15.706	15.642 ± 0.095	15.632
	39.033	38.507 ± 0.095	38.554

3. Assumptions

The following discussion is intended to provide a physical basis for the models that will be developed later in the paper. For the purposes of the numerical calculations, some features are more important than others, but nevertheless it is useful to have in mind a general physical environment for a mathematical model.

First, it is assumed that the earth was entirely molten at some time early in its history. This is in accord with calculations by WETHERILL (1972) and with those of MIZUTANI *et al.* (1972). The latter authors estimated that 83 per cent of the earth was initially molten, compared with only ten per cent of the moon. Work of SHARPE *et al.* (1976) and of HSUI *et al.* (1976) support the same conclusions.

There is assumed to be a significant period of time after the formation of the earth until the surface became sufficiently solid to preserve thereafter large-scale heterogeneities. After solidification, it became possible to preserve large-scale systems between which material mixing was rather limited. There is beginning to be a rather large body of evidence in support of this idea. For example, MOORBATH (1975) said "... the process of repeated reworking of older sialic crust must be rejected." He was arguing on the basis of initial strontium isotope ratios. SUN and HANSON (1975) said "... the variations in the $^{238}U/^{204}Pb$ and Rb/Sr ratios seem to be in the mantle source and seem to have existed for 1,000–3,000 m.yr." DZIEWONSKI (1976) made a critical assessment of seismological earth models and concluded that "the data require that differences between the two models (continental and oceanic) persist to at least four hundred kilometres." SIPKIN and JORDAN (1976) reached a similar conclusion.

Finally, the models to follow will suppose that there is a mechanism for transport between residual mantle and proto-crust, and that this mechanism has operated since solidification. For example, the transport process of present plate tectonics would seem to operate like a simple "conveyor belt" and if a simple partition law governs loading and unloading of the conveyor, the overall process can be described by simple first-order rate constants. (This last point has also been made by M. Ozima.) Current opinion seems to be moving in the direction that the processes of plate tectonics in early pre-Cambrian times must have been rather different from what we now observe. Nevertheless, this paper will use the assumption that whatever forms of evolutionary process were involved, first order rate constants may be appropriate. It is rather more difficult to defend assumptions about early processes, but the view adopted here seems to be concordant with that of DICKINSON and LUTH (1971) who reached similar conclusions from geothermal-petrological evidence.

The primary objective of the model calculations is to explain the observed fact that lead from oceanic basalts lies, in general, along a trend that is distinctly below the primary growth curve for the earth. This is illustrated by Fig. 3, taken from RUSSELL and BIRNIE (1973). It will be shown that this lead-207 discrepancy can be explained without doing violence to most of the other accepted relationships for lead isotope abundances.

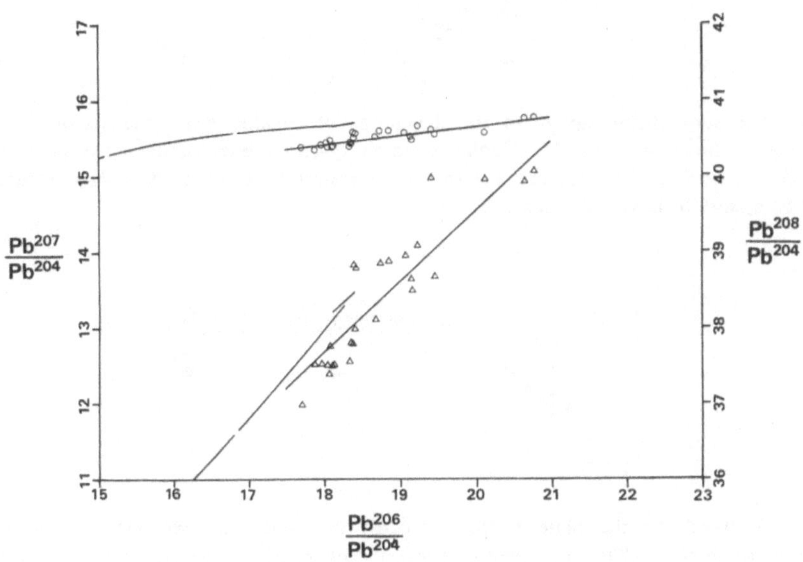

Fig. 3. Oceanic basalt leads, showing their isotopic ratios in relation to the primary growth curve. The primary growth curve is interrupted at intervals of 1,000 Ma. (After RUSSELL and BIRNIE, 1973).

For the purpose of this paper we are assuming that the oceanic basalts can be represented by the regression of Russell and Birnie, namely

$$y=0.1158\ (x-18.322)+15.543,$$

where

$$x={}^{206}\text{Pb}/{}^{204}\text{Pb} \text{ and } y={}^{207}\text{Pb}/{}^{204}\text{Pb}.$$

There are now more data and an improved estimate could undoubtedly be obtained. But, in view of the much greater uncertainties of the assumptions in the model and in view of the fact that many of the newer data seem concordant with the above regression, it is reasonable to stay with it. Moreover, to adjust the parameters would seem to add unnecessary confusion when comparing the several papers on this topic.

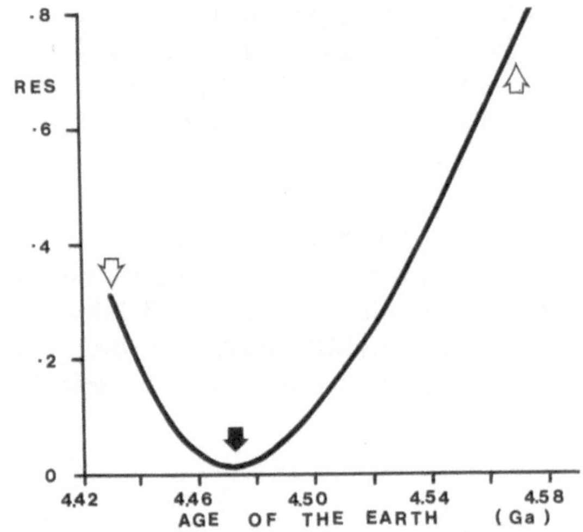

Fig. 4. The sum of the square of the deviations of μ-value from the mean for the samples shown in Table 1. Ideally a closed system model should correspond to a constant μ-value and therefore the minimum residual. Open arrows show values of oversby and of Doe and Stacey.

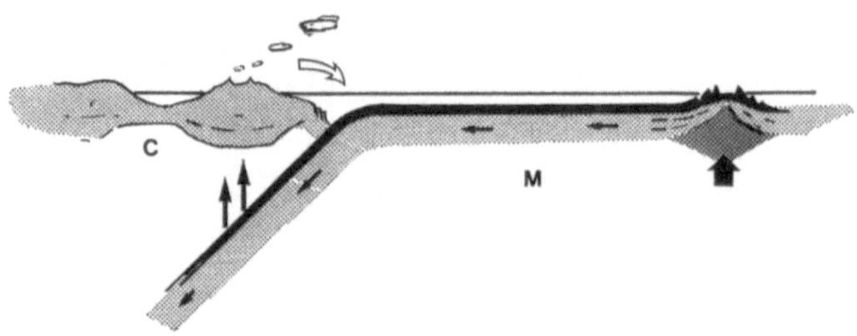

Fig. 5. A sketch of the plate tectonic process on which is indicated the presumed exchange modes. The solid arrows suggest exchange from residual mantle to proto-crust, and the open arrows indicate flow with the opposite sense.

In the interests of self-consistency, the primary growth curve is taken to be that derived earlier in this paper. In fact, the only points on the primary growth curve that are relevant to the numerical results are the isotopic ratios of lead at the start of differentiation and at $x=18.322$, the point at which the lead-207 discrepancy is observed. As in the publications of RUSSELL (1972), and RUSSELL and BIRNIE (1973), the differences between the isotopic ratios in the proto-crust and in the total system are ignored.

Strictly speaking, according to the assumptions used to construct the model, the conformable ores should be used to define the proto-crustal system and not the total system. Because of the small value used for α, 0.05, the total system is very heavily weighted toward the proto-crustal system. (The weighting is approximately 95% towards the proto-crustal system and only 5% toward the residual mantle (Fig. 5).) Therefore, the numerical calculations give substantially the same results if the conformable ores are used to determine the total system and this procedure has been used here to simplify computation.

All calculations are made with a common set of physical parameters, as given above. The origin of time is taken to be the beginning of differentiation time t_a ago, so that the present is indicated by $t=t_d$.

4. Enunciation of the Models

All the models considered here are of the sort that evolve two quasi-isolated reservoirs, the proto-crust and the residual mantle, from a single system supposed to represent the primordial mantle of the earth (or at least, the part available for interaction with the surface). They assume that as the proto-crust is formed, the elements uranium, thorium and lead are transferred to it from the residual mantle. Prior to the beginning of the evolution all of these elements are presumed to be contained within the primordial mantle. To result in a lead-207 deficiency in the residual mantle, as is observed, uranium (and presumably thorium) must be transferred more rapidly than lead, and the conclusion of the evolution must result in non-zero concentrations of these elements in the residual mantle. In this way the lead remaining in the residual mantle will have experienced in turn, residence in a higher-μ environment, in a very low-μ environment and in a higher-μ environment.

Evolution is supposed to take place through a mechanism that may be described by first order rate constants, as described above. In the present paper the rate constants are assumed to be time independent. Taken together the proto-crust and residual mantle are assumed to form a closed system with respect to the three elements concerned.

Two classes of model are considered. In the first of these the concentration of materials into the proto-crust is strictly unidirectional. No return flow is permitted and the isotopic composition of the lead in the residual mantle is not influenced by the isotopic changes in the lead in the proto-crust. According to this class of model, transport of the uranium, thorium and lead-204 to the proto-crust is limited by the fact that some fraction of the total number of atoms of these elements is not available to the transport process. The transfer of the other isotopes of lead is fixed by the isotopic composition of the lead in the residual mantle and the rate of transfer of lead-204. This is the model developed by Russell (1972) and referred to by him as an "evolutionary model".

If the rate constant for lead is denoted by a, and the fraction of the lead-204 not available for transport is α, then the number of atoms S_M and S_C in the two systems are described by the pair of equations

$$-\dot{S}_M=a\{S_M-\alpha(S_M+S_C)\}, \tag{1}$$

$$S_M+S_C=S_0. \tag{2}$$

Similarly, using for the parent uranium the symbols b, β, and λ to represent the rate constant, the fraction unavailable for transport and the decay constant, the number of atoms P_M and P_C are given by the pair of equations

$$-\dot{P}_M=\lambda P_M+b\{P_M-\beta(P_M+P_C)\}, \tag{3}$$

$$P_M+P_C=P_0e^{-\lambda t}. \tag{4}$$

For the evolutionary model, the transport of the daughter atoms, D, is controlled by the isotope ratio in the mantle, defined here by the relationship

$$x_M = D_M / S_M. \tag{5}$$

The result can be expressed by the pair of equations

$$\dot{D}_M = \lambda P_M + x_M \dot{S}_M, \tag{6}$$

$$D_M + D_C = D_0 + P_0(1 - e^{-\lambda t}). \tag{7}$$

These equations, though, are very awkward because the quantity x_M is also a variable, and it is much simpler to use the pair of equations

$$\dot{x}_M = \lambda P_M / S_M, \tag{8}$$

$$S_M x_M + S_C x_C = D_0 + P_0(1 - e^{-\lambda t}). \tag{9}$$

In this paper the rate constants a and b are assumed to be independent of time. This assumption is not required by the differential equations above, and some solutions for time-dependent constants were reported by RUSSELL and BIRNIE (1973). The initial concentrations, at the beginning of differentiation are taken to be (P_0, S_0, D_0) for the number of atoms in the residual mantle, and zero for the numbers of the corresponding atoms in the proto-crust. Then it is possible to integrate Eqs. (1) to (4) to give

$$S_M = S_0\{e^{-at} + \alpha(1 - e^{-at})\}, \tag{10}$$

$$S_C = (1 - \alpha) S_0 (1 - e^{-at}), \tag{11}$$

$$P_M = P_0 e^{-bt} + \beta(1 - e^{-bt}) e^{-\lambda t}, \tag{12}$$

$$P_C = (1 - \beta) P_0 (1 - e^{-bt}) e^{-\lambda t}. \tag{13}$$

Equations (10) to (12) can be substituted into Eqs. (8) and (9) which then, upon numerical integration, will give the isotopic ratios x_M and x_C. Because P_M is linear in β and S_M is independent of β, it follows that the isotopic ratios in the residual mantle (and consequently in the proto-crust) will vary linearly with β. To permit β to take a range of values is the most convenient way of providing for a variable μ in the residual mantle.

Table 3 shows representative solutions for this model. The youngest time for the start of differentiation occurs at approximately 2,410 Ma and corresponds to very rapid equilibration of uranium and thorium. Exact solutions could not be found corresponding to a beginning time of differentiation, t_d, greater than about 2,900 Ma. Care must be taken in interpreting these limiting times, for there is significant uncertainty in the slope and intercept of the assumed residual mantle isochron.

For the bi-directional model, equation (8) does not apply, but it can be shown as follows that the solutions for P and for S remain valid. Suppose that the rate constant describing the transport of S for residual mantle to proto-crust is taken to be $(1-\alpha)a$ and the constant for transport in the return direction, from proto-crust to residual mantle, is taken to be αa. Then we may write

$$\dot{S}_M = \alpha a S_C - (1 - \alpha) a S_M, \tag{14}$$

and

$$\dot{S}_C = (1 - \alpha) a S_M - \alpha a S_C. \tag{15}$$

Table 3. Solutions for the "evolutionary" model giving a Residual Mantle isochron of slope 0.1158 and passing through the abundances (18.322, 15.453).

Start of differentiation			End points of secondary isochron					
	a	b	$\beta=0$			$\beta=1$		
t_D (10^9 yr)	10^{-9} yr^{-1}	10^{-9} yr^{-1}	x	y	z	x	y	z
2.41	4.72	∞	14.301	14.989	33.913	72.430	21.721	101.335
2.45	4.33	31.22	14.295	14.987	33.907	71.762	21.642	100.581
2.50	3.90	14.61	14.302	14.987	33.915	70.842	21.535	99.537
2.55	3.54	9.92	14.317	14.989	33.931	69.778	21.412	98.329
2.60	3.21	7.61	14.338	14.992	33.954	68.519	21.266	96.896
2.65	2.92	6.19	14.366	14.995	33.985	67.046	21.095	95.216
2.70	2.66	5.20	14.401	14.999	34.023	65.306	20.894	93.224
2.75	2.41	4.45	14.442	15.004	34.068	63.155	20.645	90.758
2.80	2.17	3.85	14.492	15.009	34.124	60.457	20.332	87.655
2.85	1.92	3.33	14.556	15.017	34.195	56.729	19.900	83.353
2.90	1.55	2.77	14.645	15.027	34.295	48.924	18.997	74.307
$\geqq$2.95		No exact solutions could be found						

Table 4. Solutions for the "bi-directional" model giving a Residual Mantle isochron of slope 0.1158 and passing through the abundances (18.322, 15.453).

Start of differentiation			End points of secondary isochron					
	a	b	$\beta=0$			$\beta=1$		
t_D (10^9 yr)	10^{-9} yr^{-1}	10^{-9} yr^{-1}	x	y	z	x	y	z
2.60	0.91	112.6	14.959	15.064	34.660	26.485	16.395	48.142
2.80	0.95	6.70	15.141	15.085	34.864	28.526	16.635	50.539
3.00	1.01	4.09	15.410	15.116	35.166	30.989	16.920	53.436
3.20	1.09	3.44	15.719	15.152	35.513	33.506	17.211	56.406
3.40	1.16	3.40	16.024	15.187	35.855	35.625	17.457	58.918
3.60	1.23	3.79	16.301	15.219	36.164	37.104	17.628	60.682
3.80	1.29	4.85	16.541	15.247	36.428	37.972	17.729	61.731
4.00	1.34	8.30	16.739	15.270	36.646	38.393	17.777	62.251
$\geqq$4.10		No exact solutions could be found						

Equations (14) and (15) are equivalent to Eqs. (1) and (2), even if the quantity a is not independent of time. In a similar manner it can be shown that for rate constants $(1-\beta)b$ and βb the equations (3) and (4) remain valid.

Replacing Eq. (6), we may write

$$\dot{D}_M = \lambda P_M + \alpha a D_C - (1-\alpha) a D_M. \tag{16}$$

Equation (7), of course, remains valid, but not Eq. (8).

RUSSELL and BIRNIE (1973) showed that the equations lend themselves well to representation and manipulation in a matrix form. The results are summarized as follows, for the bi-directional model:

$$\frac{\mathrm{d}}{\mathrm{d}t}\begin{pmatrix} P_M \\ P_C \end{pmatrix} = -\lambda \begin{pmatrix} P_M \\ P_C \end{pmatrix} + b \begin{pmatrix} -(1-\beta) & \beta \\ (1-\beta) & -\beta \end{pmatrix} \begin{pmatrix} P_M \\ P_C \end{pmatrix}, \tag{17}$$

$$\frac{\mathrm{d}}{\mathrm{d}t}\begin{pmatrix} S_M \\ S_C \end{pmatrix} = a \begin{pmatrix} -(1-\alpha) & \alpha \\ (1-\alpha) & -\alpha \end{pmatrix} \begin{pmatrix} S_M \\ S_C \end{pmatrix}, \tag{18}$$

$$\frac{\mathrm{d}}{\mathrm{d}t}\begin{pmatrix} D_M \\ D_C \end{pmatrix} = \lambda \begin{pmatrix} P_M \\ P_C \end{pmatrix} + a \begin{pmatrix} -(1-\alpha) & \alpha \\ (1-\alpha) & -\alpha \end{pmatrix} \begin{pmatrix} D_M \\ D_C \end{pmatrix}. \tag{19}$$

Define hypothetical Sigma (Σ) and Delta (Δ) systems, by the defining equations of the form

$$\begin{pmatrix} \Delta \\ \Sigma \end{pmatrix}_I = \begin{pmatrix} 1 & -1 \\ 1 & 1 \end{pmatrix} \begin{pmatrix} I_M \\ I_C \end{pmatrix}. \tag{20}$$

$$(I = P, S, D)$$

Upon substitution of equations of the form of (20) into Eqs. (17), (18), and (19), there result the equations

$$\frac{\mathrm{d}}{\mathrm{d}t}\begin{pmatrix} \Delta \\ \Sigma \end{pmatrix}_P = -\lambda \begin{pmatrix} \Delta \\ \Sigma \end{pmatrix}_P - a \begin{pmatrix} 1 & 1-2\alpha \\ 0 & 0 \end{pmatrix} \begin{pmatrix} \Delta \\ \Sigma \end{pmatrix}_P, \tag{21}$$

$$\frac{\mathrm{d}}{\mathrm{d}t}\begin{pmatrix} \Delta \\ \Sigma \end{pmatrix}_S = -a \begin{pmatrix} 1 & 1-2\alpha \\ 0 & 0 \end{pmatrix} \begin{pmatrix} \Delta \\ \Sigma \end{pmatrix}_S. \tag{22}$$

For the evolutionary model

$$\frac{\mathrm{d}}{\mathrm{d}t}\begin{pmatrix} \Delta \\ \Sigma \end{pmatrix}_D = \lambda \begin{pmatrix} \Delta \\ \Sigma \end{pmatrix}_P - a x_M \begin{pmatrix} 1 & 1-2\alpha \\ 0 & 0 \end{pmatrix} \begin{pmatrix} \Delta \\ \Sigma \end{pmatrix}_S, \tag{23}$$

and for the bi-directional model

$$\frac{\mathrm{d}}{\mathrm{d}t}\begin{pmatrix} \Delta \\ \Sigma \end{pmatrix} = \lambda \begin{pmatrix} \Delta \\ \Sigma \end{pmatrix}_P - a \begin{pmatrix} 1 & 1-2\alpha \\ 0 & 0 \end{pmatrix} \begin{pmatrix} \Delta \\ \Sigma \end{pmatrix}_D. \tag{24}$$

The latter forms, in terms of the Sigma and Delta systems, are easier to handle in the case where the constants are time dependent.

For the bi-directional solution, one needs to solve Eqs. (21), (22), and (24). Unlike the evolutionary model, the equations all lend themselves to simple analytical solutions and therefore the use of numerical integration is unnecessary. For various values of the constant t_d, solutions were sought for a and b that would give a residual mantle isochron consistent with the slope and intercept given by RUSSELL and BIRNIE (1973). The values found are summarized in Table 4. Exact solutions could only be found for t_d-values in the range 2,600 Ma to 4,000 Ma. There are some significant differences between these values and the optimum solutions suggested by Russell and Birnie that exceed those that might be expected from the use of newer decay constants. The present investigation has used much more sophisticated and powerful computer programs for optimizing sets of nonlinear equations and it appears that they were able to find minima that were missed in the trial-and-error approach that was used in the earlier publication.

The times for the start of differentiation found for the bi-directional model seem more consistent with the ages of the oldest crustal rocks as reported above. Therefore this model will be explored more rigorously.

One matter of concern is that there are significant uncertainties attributable to the assumed slope and intercepts of the residual-mantle isochron. In attempting to take this uncertainty into account a measure of misfit to the lines has been defined as the quantity

$$R^2=\left(\frac{\text{slope}-0.1158}{0.01}\right)^2+\left(\frac{\text{intercept}-15.543}{0.01}\right)^2,$$

where the intercept is the y_M-intercept at $x_M=18.322$. The quantity r has been plotted in Figs. 6, 7, and 8. In each of these figures the darkly shaded areas correspond to values for R which are less than unity ($R<1$). The lightly shaded areas correspond to values for R greater than about three. These graphs, like Table 4, demonstrate that acceptable solutions can be obtained with differentiation periods beginning as early as 4,000 Ma.

5. Conclusions

The principal conclusion the author would like to draw from the calculations reported in this paper is the fact that it is quite possible to develop a simple two-reservoir model that will reproduce exactly the secondary isochron inferred by RUSSELL and BIRNIE (1973) for the oceanic basalts. In particular, it can provide a quantitative explanation for the observed lead-207 deficiency that is apparently a fundamentally important characteristic.

It will be apparent from what has been written, that the calculations do not lead to a unique solution. Still, some broad generalizations can be made. The evolutionary (unidirectional) models lead to younger estimates for the period of proto-crustal formation than do the bi-directional forms. The two models together provide exact solutions for starting times of evolution ranging from 2,400 Ma to 4,000 Ma. The reciprocals of the rate constants, $1/a$ and $1/b$, may be taken as representative durations for the differentiation processes. From the figures reproduced in Tables 3 and 4, one can see that the differentiation periods for uranium and thorium range from 0 Ma to 360 Ma, and for lead range from 210 Ma to 1,100 Ma. Subtracting the quantities $1/a$ from the starting times, t_a, one can show that the differentiation process for lead extended well beyond the first quarter of the earth's history.

If the existence of ancient rocks with ages greater than about 3,600 Ma is interpreted to show that the start of the differentiation began prior to that time, one is forced to

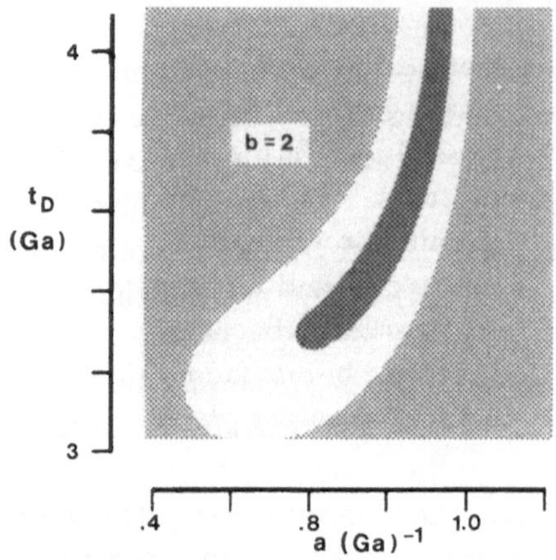

Fig. 6. Regions of acceptable (heavily shaded) and unacceptable (lightly shaded) values for the parameters t and a. The parameter b is held at 2.0 (Ga)$^{-1}$.

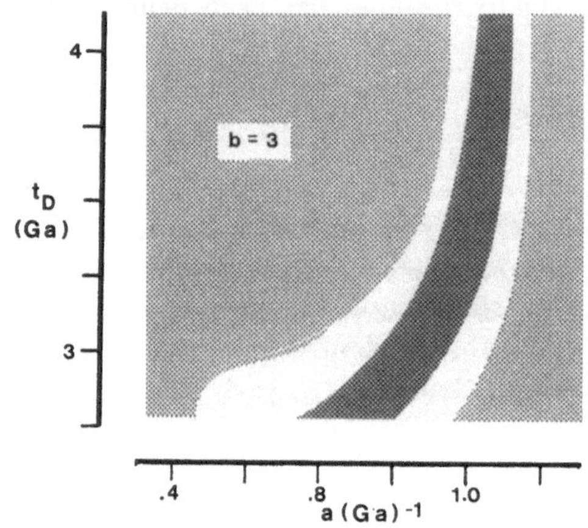

Fig. 7. A similar plot to Fig. 6, but with b held at 3.0 (Ga)$^{-1}$.

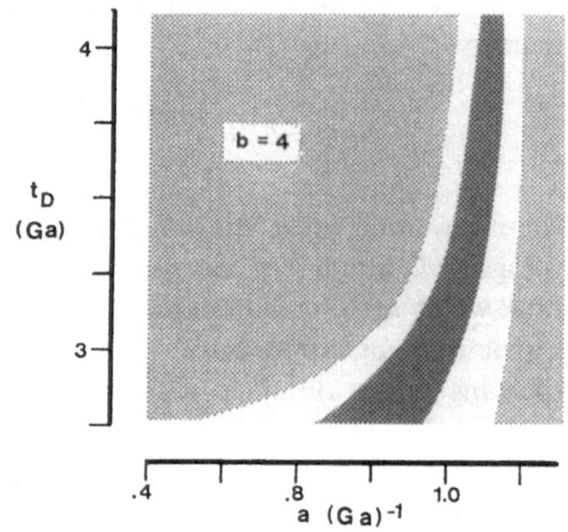

Fig. 8. A similar plot to Fig. 6, but with b held at 4.0 (Ga)$^{-1}$.

choose the bi-directional model over the evolutionary model. It is interesting to work out some numerical values for the particular bidirectional model having the earliest starting date t_d=4,000 Ma. The essential numerical parameters are shown in Table 5.

Table 5. Values inferred from the parameters corresponding to the earliest evolving bi-directional model.

Quantity	Time-related parameters	Geochemical parameters	
		Residual mantle	Proto-crust
Start of differentiation	4,000 Ma		
Rate constant for lead	1.34×10^{-9} yr^{-1}		
Mean transport interval for lead	4,000–3,250 Ma		
Rate constant for uranium and thorium	8.3×10^{-9} yr^{-1}		
Mean transport interval for uranium	4,000–3,880 Ma		
Range of μ-values		7.41–21.1	7.83–9.20
Geometric means of μ-values		14.4	8.7
Range of κ-values		1.81–3.72	3.69–4.14
Geometric means of μ-values		2.70	4.0

A point of interest is the comparison of the apparent ages of the conformable galenas, as inferred from their isotopic compositions, and the accepted ages of the environments in which they occur. In the present calculations, no ages were assigned to any of the ore leads and no parameters used in the model calculations that rest on such assumed ages. Therefore an objective comparison seems possible. Table 1 gave the apparent ages for the ores used to obtain the primary growth curve, as well as some values for assumed geological ages. Even accepting the criticism of CUMMING and RICHARDS (1975), that such 'independent' ages are rather uncertain and that they seem in some cases to rest on interpretations of their isotopic ratios, it still seems that young ages appear significantly too young and older ages, too old. A complete proto-crustal growth curve has not been evaluated for any of the evolutionary or bi-directional models elaborated here, but it would be expected that this age discrepancy would be even greater for such curves, particularly in the case of the older ores. If one wished to extend the present models to minimize or eliminate this apparent age discrepancy, this could be done by replacing the assumption that the total of the two reservoirs comprise a closed system of the form used here. For example, one could use the parameters for an open-system model as given by CUMMING and RICHARDS (1975) or else the parameters for the multistage model of STACEY and KRAMERS (1975) to define the 'Sigma' system. Such calculations are planned.

The research described in this paper has been supported through grants from the National Research Council of Canada, especially Grant A-720. It depends heavily on the facilities of the Computing Center of the University of British Columbia and on its library of programs for the manipulation of non-linear equations. The author is indebted to many colleagues for their comments on the material presented in this paper, particularly to Professor Minoru Ozima and Dr. Igor Tolstikhin. The constructive criticisms of Dr. Y. Hamano have lead to a consideration of the manner in which the models might be extended to give better agreement with the apparent ages of the conformable ores. The author is particularly indebted to the sponsors of the U.S.-Japan Seminar on Rare Gas Abundances and Isotopic Constraints on the Origin and Evolution of the Earth's Atmosphere for making it possible for him to participate as a third country representative.

REFERENCES

ALLAN, D. W., R. M. FARQUHAR, and R. D. RUSSELL, A note on the lead isotope method of agedetermination, *Science*, **118**, 486–488, 1953.

ALLSOPP, H. L., L. O. NICOLAYSEN, and T. J. ULRYCH, Dating some significant events in the Swaziland sequence by the Rb-Sr method, *Can. J. Earth Sci.*, **5**, 605–619, 1968.

BAADSGAARD, H., U, Th, Pb dates on zircons from the early Precambrian Amitsoq Gneiss, Godhaab District, West Greenland, *Earth Planet. Sci. Lett.*, **19**, 22–28, 1973.

BAADSGAARD, H., Further U-Pb dates on zircons from the early Precambrian rocks of Godthaabsfjord area, West Greenland, *Earth Planet. Sci. Lett.*, **33**, 261–267, 1976.

BLACK, L. P., N. H. GALE, S. MOORBATH, and R. J. PANKHURST, Isotopic dating of very early Precambrian amphibolite facies from the Godthaab District, Western Greenland, *Earth Planet. Sci. Lett.*, **12**, 245–259, 1971.

COOPER, J. A., P. H. REYNOLDS, and J. R. RICHARDS, Double spike calibration of the Broken Hill standard lead, *Earth Planet. Sci. Lett.*, **6**, 467–478, 1969.

CUMMING, G. L. and J. R. RICHARDS, Ore lead isotope ratios in a changing earth, *Earth Planet. Sci. Lett.*, **28**, 155–171, 1975.

DICKINSON, W. R. and W. C. LUTH, A model for plate tectonic evolution of mantle layers, *Science*, **174**, 400–404, 1971.

DOE, B. R. and J. S. STACEY, The application of lead isotopes to the problems of ore genesis and ore prospect evaluation: A review, *Econ. Geol.*, **69**, 757–776, 1974.

DZIEWONSKI, A. M., Seismological earth models: A critical assessment, *EOS* (Abstract), **57**, 282, 1976.

GOLDICH, S. S. and C. E. HEDGE, 3,800 m. yr. granite gneiss in S. W. Minnesota, *Nature*, **252**, 467–468, 1974.

GOLDICH, S. S., C. E. HEDGE, and T. W. STERN, Age of the Morton and Montevideo gneisses and related rocks, Southwestern Minnesota, *Bull. Geol. Soc. Am.*, **81**, 3671–3695, 1970.

HSUI, A. T., M. N. TOXSOZ, and D. H. JOHNSON, Thermal evolution of the Moon, Mercury and Mars, *EOS* (Abstract), **57**, 271, 1976.

JAFFEY, A. H., K. F. FLYNN, W. C. GLENDENIN, W. C. BENTLEY, and A. M. ESSLING, Precision measurement of half-lives and specific activities of ^{235}U and ^{238}U,*Phys. Rev.*, **C4**, 1889–1906, 1971.

JAHN, B. and C. Y. SHIH, On the age of the Onverwacht Group, Swaziland sequence, S. Africa, *Geochim. Cosmochim. Acta*, **38**, 873–885, 1974.

KANASEWICH, E. R., The interpretation of lead isotopes and their geological significance, in *Radiometric Dating for Geologists*, edited by E. I. Hamilton and R. M. Farquhar, pp. 147–223, Interscience Publishers, New York, 1968.

LE ROUX, L. J. and L. E. GLENDENIN, Half-life of thorium-232, *Proc. Natl. Conf. on Nuclear Energy*, pp. 83–94, Pretoria, April 1963.

MIZUTANI, H., T. MATSUI, and H. TAKEUCHI, Accretion process of the moon, *The Moon*, **4**, 476–489, 1972.

MOORBATH, S., Geological interpretation of whole rock isochron dates from high grade gneiss terrains, *Nature*, **255**, 391, 1975.

MOORBATH, S., R. K. O'NIONS, and R. J. PANKHURST, Early Archaean age of the Isua Iron Formation, West Greenland, *Nature*, **245**, 138–139, 1973.

MOORBATH, S., R. K. O'NIONS, N. H. GALE, and V. R. MCGREGOR, Further rubidium-strontium age determinations on the very early Precambrian rocks of the Godthaab District, West Greenland, *Nature Phys. Sci.*, **240**, 78–82, 1972.

OSTIC, R. G., R. D. RUSSELL, and P. H. REYNOLDS, The age of the earth, in *Problems in Geochemistry*, edited by N. I. Khitarov, pp. 37–49, Nauka, Moscow, 1965; English Translation by Israel Program for Scientific Translations, 35–48, Jerusalem, 1969.

OVERSBY, V. M., New look at the lead isotope growth curve, *Nature*, **248**, 132–133, 1974.

PATTERSON, C. C., Age of meteorites and the earth, *Geochim. Cosmochim. Acta*, **10**, 230–237, 1956.

ROBERTSON, D. K., A model discussing the early history of the Earth based on a study of lead isotope ratios from veins in some Archean cratons of Africa, *Geochim. Cosmochim. Acta*, **37**, 2099–2124, 1973.

RUSSELL, R. D., Abundances of meteoric lead isotopes, *Nature*, **179**, 92, 1957.

RUSSELL, R. D., Evolutionary model for lead isotopes in conformable ores and in ocean volcanics, *Rev. Geophys. Planet. Phys.*, **10**, 529–549, 1972.

RUSSELL, R. D. and D. J. BIRNIE, A bi-directional model for lead isotope evolution, *Phys. Earth Planet. Inter.*, **8**, 158–166, 1973.

RUSSELL, R. D. and R. M. FARQUHAR, *Lead Isotopes in Geology*, pp. 243, Interscience Publishers, London, 1960.

STANTON, R. L. and R. D. RUSSELL, Anomalous leads and the emplacement of lead sulphide ores, *Econ. Geol.*, **54**, 588–607, 1959.

SHARPE, H. N., D. W. STRANGWAY, and W. R. PELTIER, The nature of planetary magnetism, *EOS* (Abstract), **57**, 271, 1976.

SIPKIN, S. A. and T. H. JORDON, Lateral heterogeneity of the upper mantle from multiple ScS travel times, *EOS* (Abstract), **57**, 283, 1976.

SLAWSON, W. F., E. R. KANASEWICH, R. G. OSTIC, and R. M. FARQUHAR, *Nature*, **200**, 413–414, 1963.

STACEY, J. S. and J. D. KRAMERS, Approximaton of terrestrial lead isotope evolution by a two stage model, *Earth Planet. Sci. Lett.*, **26**, 207–221, 1975.

STACEY, J. S., M. H. DELEVAUX, and T. J. ULRYCH, Some triple filament lead isotope ratio measurements and an absolute growth curve for single-stage leads, *Earth Planet. Sci. Lett.*, **6**, 15–25, 1969.

SUN, S. S. and G. N. HANSON, Evolution of the mantle—geochemical evidence from alkali basalt, *Geology*, **3**, 297–302, 1975.

SUTTON, J., Strontium isotopes and crustal evolution, *Nature Phys. Sci.*, **254**, 382, 1975.

TATSUMOTO, M., R. J. KNIGHT, and C. J. ALLEGRE, Time differences in the formation of meteorites as determined from the ratio of lead-207 to lead-206, *Science*, **180**, 1279–1283, 1973.

ULRYCH, T. J., G. BURGHER, and L. O. NICOLAYSEN, Least radiogenic terrestrial leads, *Earth Planet. Sci. Lett.*, **2**, 179–184, 1967.

WETHERILL, G. W., Beginning of continental evolution, *Techtonophysics*, **13**, 31–45, 1972.

Matter Accretion into the Solar System

Satio HAYAKAWA

Department of Physics, Nagoya University, Nagoya 464, Japan

The mass accreted into the solar system while the sun moved in a dark cloud is estimated to be comparable to the mass of the total planetary system. Continual accretion of intracloud matter, whose elemental abundances vary from one part to another due to different degrees of pollution by synthesized and/or spallogenic elements, may account for elemental heterogeneities observed in the present solar system. Accreted rare gases may be subject to fractionation due to the photoionization of rare gas atoms by sunlight.

1. Introduction

It has been suggested on the basis of the isotopic composition of meteoritic matter that matter in the solar system is not of a single origin but is from several sources with different elemental abundances; see CLAYTON *et al.* (1976) and references therein. The present author (HAYAKAWA, 1978) has proposed continual matter accretion into the solar system to explain such material sources.

Stars are being efficiently formed in dark clouds of the density typically of 10^{-20} g·cm^{-3}. A few of the stars can evolve to supernovae within the lifetime of a dark cloud ($\sim 10^7$ yr) and can eject heavy elements. Hundreds of stars are formed in a typical dark cloud, so that some of them contain supernova ejecta. The sun must be one such star, because it contains nuclides which are synthesized by supernova explosion.

After the solar system was formed, it moved around in the dark cloud and accretes matter therein. For a spherical dark cloud of mass M and average density ρ, the rate of accretion is estimated to be equal to the mass flux times the area of a circle with the accretion radius as

$$\dot{M}_a = (12\pi\, G\rho)^{1/2} M_\odot^2/M = 6.5\, n_4^{1/2} M_\odot^2/M \text{ Myr}^{-1}, \tag{1}$$

where G is the gravitational constant, $M_\odot$ is the solar mass, n_4 is the density of hydrogen atoms in 10^4 cm^{-3}, and the relative velocity of the sun to the diffuse matter is assumed to be equal to the virial velocity $(4\pi\ \rho/3)^{1/6}\ G^{1/2}\ M^{1/3}$. For $M = 10^4\, M_\odot$ and $n_4 = 1$, the mass accreted during 10 Myr is $6.5 \times 10^{-3}\, M_\odot$, which exceeds the total mass of the present planetary system, $1.34 \times 10^{-3}\, M_\odot$.

2. Elemental Heterogeneities

Since the mass of accreted matter is comparable to the total mass of the planetary system, the accreted matter mixing with primordial matter may cause elemental heterogeneities, depending on the mixing ratio, if different relative abundances of the elements occur. Although the matter forming the solar system and that accreted later belong to the same dark cloud, which may be initially homogeneous, the solar system is strongly polluted by supernova ejecta, so that its chemical composition may be considerably different from the average composition of the dark cloud.

The accretion rate is high when the solar system encounters the ejecta of other supernovae, of giant stars with large mass losses, and of close binary systems exchanging matter. In all these cases the composition of accreted matter differs from the primordial composition of the solar system because of the differences in chemical evolution.

There may be several such occasions during the lifetime of the dark cloud, and the interval between great accretion events may well be longer than the growth time of dust grains which is estimated by KUSAKA *et al.* (1970) to be as long as 1 Myr. Hence a solid body may consist of dust grains of different abundances.

3. Fractionation of Rare Gases

In addition to the elemental heterogeneities caused by continual accretion as discussed above, there exists another mechanism which gives rise to heterogeneities. This is a fractionation process associated with photoionization of accreting matter. If atoms are ionized on a path towards the sun the ions thus formed do not follow ballistic orbits but are trapped by interplanetary magnetic fields. Atoms with smaller probabilities of ionization can propagate closer to the sun.

Application of this processes to rare gases results in heavier rare gases mixing with interplanetary matter farther from the sun, since the ionization potential decreases as the rare gas atom becomes heavier. The probability of survival against ionization is approximately expressed as exp $(-\lambda)$, and λ is the photoionization rate integrated over the free fall time,

$$\lambda(r)=2C(r)(r/2GM_{\odot})^{1/2}r. \tag{2}$$

Here r is the distance from the sun, and C is the photoionization rate by sunlight. Since $C(r)\propto r^{-2}$, λ is proportional to $r^{-1/2}$. For Ar at the earth's orbit $\lambda\simeq3$, whereas for He $\lambda\simeq0.3$. Thus the survival probability of heavier rare gases is smaller than that of lighter ones. A difference in the relative abundances of rare gas elements would be observed between the inner and outer planets.

REFERENCES

CLAYTON, R. N., N. ONUMA, and T. K. MAYEDA, A classification of meteorites based on oxygen isotopes, *Earth Planet. Sci. Lett.*, **30**, 10–18, 1976.

HAYAKAWA, S., Continual matter accretion into the solar system, *Astrophys. Space Sci.*, 1978 (in press).

KUSAKA, T., T. NAKANO, and C. HAYASHI, Growth of solid particles in the primordial solar nebula, *Prog. Theor. Phys.*, **44**, 1580–1595, 1970.

Subject Index

* page number for an article.